CISM COURSES AND LECTURES

The series presents lecture notes, monographs, edited works and proceedings in the field of Mechanics, Engineering, Computer Science and Applied Mathematics.
Purpose of the series is to make known in the international scientific and technical community results obtained in some of the activities organized by CISM, the International Centre for Mechanical Sciences.

INTERNATIONAL CENTRE FOR MECHANICAL SCIENCES

COURSES AND LECTURES - No. 435

REFURBISHMENT OF BUILDINGS AND BRIDGES

EDITED BY

FEDERICO M. MAZZOLANI
UNIVERSITÀ DI NAPOLI "FEDERICO II"

MIKLÓS IVÁNYI
BUDAPEST UNIVERSITY OF TECHNOLOGY AND ECONOMICS

Springer-Verlag Wien GmbH

This volume contains 388 illustrations

SPIN 10883947

In order to make this volume available as economically and as
rapidly as possible the authors' typescripts have been
reproduced in their original forms. This method unfortunately
has its typographical limitations but it is hoped that they in no
way distract the reader.

ISBN 978-3-211-83690-3   ISBN 978-3-7091-2570-0 (eBook)
DOI 10.1007/978-3-7091-2570-0

PREFACE

This book reports the main contents of the lectures held at CISM – Udine from 2 to 6 October 2000 within the Advanced Professional Training Course entitled "Refurbishment of Buildings and Bridges". As customary of CISM policy, the main task of the course was to provide an highly developed educational programme for young engineers and architects in the field of structural rehabilitation. The preservation of cultural heritage and, hence, the refurbishment of existing works, in fact, represent today emerging activities of the modern building practice, deserving a particular attention by people involved in both design and constructional process.

"Improving both knowledge and awareness of people involved in refurbishment of existing constructions is a fundamental step for a deeper exploitation of precious and invaluable resources spread all over European countries". This concept, thoroughly stressed throughout the course by all lecturers and widely shared within participants, can be considered as the main educational purpose of this activity and is constantly highlighted throughout the book. It summarises in the best way the objective of such initiative with regard to its fall out in the public domain. This aspect is to be evaluated not only at the light of the great importance of restoration in practical applications, which is clear by itself, but also in the view of the relatively scarce availability of similar advanced educational programmes in regular courses held at university. This is the main reason why the course was organised and, in this framework, the role played by CISM as educational and scientific institution has been fundamental in giving this initiative the necessary international and interdisciplinary character, making it fully liable to a wide-range spreading not only among technicians but also within general public.

During the course many innovative aspects were tackled, going from the seismic protection of existing buildings to the improvement of global urban environment and life quality, so as to promote and encourage a wider conscience about refurbishment problems in some strategic application fields, namely buildings, bridges, but also industrial and monumental constructions. Correspondingly, a great emphasis was given to the application of new materials and techniques in the structural rehabilitation of buildings. In particular, recent developments in restoration activities have pointed out the use of metal materials, and in particular steel, as an appropriate solution from both structural and architectural point of view. This results from typical refurbishment requirements for constructional systems able to guarantee flexibility of execution, ease of erection and reversibility. Practical examples collected from all over the world and discussed in detail by lecturers have shown that metal materials are widely and successfully used at all levels of strengthening, namely safeguard, repairing, reinforcing and restructuring, and are also very suited to seismic upgrading.

Concerning the approach followed in the book, all peculiar aspects of interventions considered in the course are carefully examined, extrapolating basic design and execution rules, useful for engineering purposes. Special attention is also paid to the seismic retrofit of buildings and bridges by means of innovative techniques based on the use of special devices fitted with energy dissipation capabilities. To this purpose, many study cases are presented and thoroughly discussed. Furthermore, a special emphasis has been given to the problem of strengthening of buildings damaged by earthquake. The participation of trainers coming from different cultural areas and, in particular, that of an expert coming from Japan allowed to compare different intervention criteria according to relevant current intervention practices and existing design codes. This caused outstanding results and experiences to be evaluated together with similar data achieved in Europe, allowing a direct comparison in terms of materials, techniques and technologies. In this context, the results of recently developed European research projects dealing with restoration topics (TEMPUS SJEP 2747 "Civil Engineering", TEMPUS SJEP 09524 "The Use of Steel in Refurbishment as an Environmentally Friendly Activity", CNR - Progetto Strategico Beni Culturali, CNR - Progetto Finalizzato Beni Culturali, etc.) represented a valuable basis for the topics dealt with in the course. In the same way, the valuable contribution of the direct experience of trainers, both professional and scientific, made for a clear and effective presentation of results, appealing to the general public, too.

The editors are confident that the whole of contributions provided within the course and summarised in this book can help to a significant extent the development of the European educational activity in the field of refurbishment. Also, the presentation and discussion of different restoration activities, ranging across several application fields, can contribute to the improvement of both knowledge and awareness of readers with respect to a number of problems deeply felt all around Europe. To this purpose, the editors wish to thank all people who contributed to make this course a success and, most of all, the lecturers, for the excellence of the work they carried out before and during the course, as well as for the precious synopsis they provided in the chapters of this book. The editors are also very grateful to the CISM General Secretary Prof. G. Bianchi, to the CISM Rector Prof. S. Kaliszky, to the Editor of the Series Prof. C. Tasso, as well as to all the CISM staff in Udine, for their professionalism and factual contribution to the execution of the course.

Federico M. Mazzolani
Miklós Iványi

CONTENTS

Page

Preface

Chapter 1
Principles and Design Criteria for Consolidation and Rehabilitation
by F.M. Mazzolani 1

Chapter 2
Refurbishment of Steel Bridges
by M. Iványi...61

Chapter 3
Refurbishment of Single-Storey Buildings
by H. Pasternak 151

Chapter 4
Strengthening Techniques for Buildings
by A. Mandara..197

Chapter 5
Use of Special Techniques in Refurbishment
by J-P. Muzeau...265

Chapter 6
Strengthening of Steel Frames for Seismic Resistance
by H. Akiyama 325

CHAPTER 1

PRINCIPLES AND DESIGN CRITERIA FOR CONSOLIDATION AND REHABILITATION

F.M. Mazzolani
Università di Napoli "Federico II", Napoli, Italy

1 General Overview

1.1 Introductory remarks

The refurbishment of existing constructions (buildings and bridges) is today an emerging activity. It can be observed that the building industry is getting more and more devoted to these activities of consolidation, rehabilitation and modernisation of old buildings. (Mazzolani, 1986, 1990 a, b, 1992, 1994, 1996, 2000, 2001 b).

The old masonry buildings are very often damaged by the age and by the ravages of time and, therefore, they require structural consolidation and functional rehabilitation.

Under the definition of *architectural constructions* we want to basically consider the historical monuments built in the past centuries, without neglecting more recent buildings, characterised by an architectural value. (Fig. 1)

Figure 1. Monumental building under restoration

It is well known that Italy is the Country having the largest amount of cultural heritage in the World. Therefore, many consolidation and restoration systems have been experienced in the last decades for monumental constructions in this Country.

In addition, also more recent buildings made of reinforced concrete very often need refurbishment operation due to their bad state of preservation. Some of them can be also considered as architectural constructions.

1.2 Operational aspects

The restoration and consolidation operations, particularly in the more delicate case of structural restoration of monumental constructions, require a very careful selection of new constructional materials which must be chosen according to the perequisites raising from the materials of the past to be consolidated.

Distinction can be made between the new materials which represent the "medicine" and the old ones which are the "sick". As a "medicine" we can use both the traditional materials, like cement, mortar, reinforced concrete and steel, and the innovative materials, like special mortars, polymeric and composite material, special metals (high strength steels, stainless steel, aluminium alloys, titanium alloys, etc.) as well as some special devices belonging to advanced systems of seismic protection by means of passive control technologies.

The problem, therefore, can be faced by considering a complex matrix, where all the "medicine" materials and the "sick" ones are located. The selection of the suitable combination, among all the possible one given by this matrix, represent the mail goal of the structural consolidation.

A statical overview of the operational activities developed in South of Italy during the year 70s and 80s for the consolidation of historical constructions has been done in order to emphasise the different types of consolidation intervention. They have been subdivided in buildings and churches, because of the different structural scheme.

The results are shown in Figs. 2 and 3 for buildings and churches, respectively. Each figure gives the percentage of the different types of work used in the consolidation (left), and also which kind of structural element and in which percentage has been consolidated by means of steelwork (right).

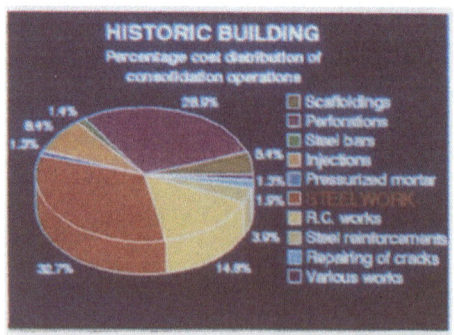
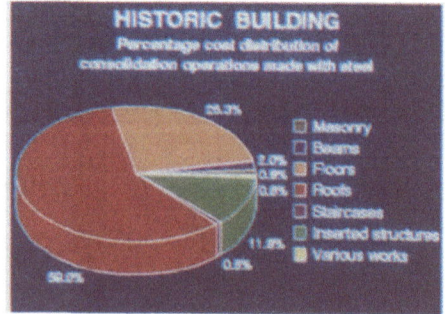

Figure 2. Results of a statistical analysis on consolidation technologies for historical buildings in South of Italy

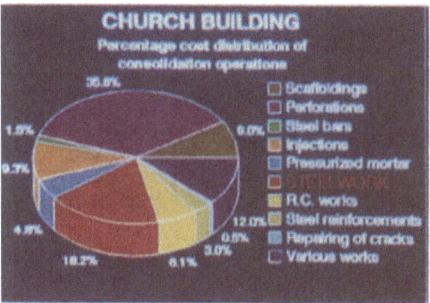 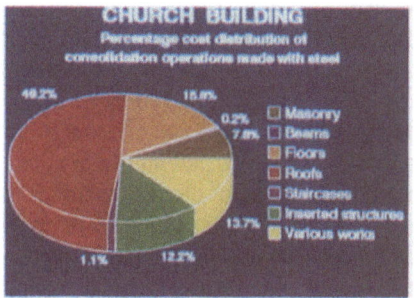

Figure 3. Results of a statistical analysis on consolidation technologies for church building in South of Italy.

From a general overview we can recognise that
- the consolidation systems based on cement and concrete materials are widely used, especially in seismic upgrading operations, under form of injections and/or r.c. elements, but their compatibility with the masonry of the old architectonical constructions is questionable;
- the consolidation systems based on polymeric and composite materials are very new and now-a-day there is not enough experience to validate their durability;
- the consolidation systems based on steelwork are widely and successfully used both in case of monumental constructions and for usual buildings made of masonry and r.c. (Mazzolani &Mandara, 1989, 1992);
- the experience on the use of special devices is now at a pilot level, but it is promising for a more and more wide application in the future.

From the structural point of view, the analysis of several practical examples collected from all over the world shows that the above techniques are selected in order to fulfil the requirements rising at the different levels of consolidation of the already existing structures which need consolidation. These levels are differentiated in order of importance according to the main constructural phases, which are commonly defined in order of importance as safeguard, repairing, reinforcing, restructuring (see Section 2).

1.3   Restauration Prerequisites

When the building to be consolidated is of historical interest, its restoration is a very delicate operation. The criteria, upon which restoration operations must be based, tend basically towards the conservation of pre-existing buildings and their integration with new works necessary to ensure their return to functionality. Such new works must be clearly modern in features; they must be surely distinguishable and they must be reversible, by using technologies and materials which can be removed without damaging the existing structure.

The various international restoration "Charts" in fact state the incongruity of reconstruction by using the methods of the past, which can no longer be reproduced for many reasons, above all technological. Other reasons are related to sentimental feelings of construction tradition, to new functional requirements and lack of availability of old materials. At the same time these "Charts", especially in cases where the restoration operation involves restructuring with partial reconstruction, indicate the need to use well adapted technologies and materials in a clearly modern way.

A logical applications of these principles undoubtedly shows that steel, as a material and its technology, have the necessary advantages of being a modern material with "reversible" characteristics, particularly suited to reconcile with the materials of the past and to form integrated structural systems. In addition, its choice is substantially based on its high mechanical performance and on the flexibility of the constructional systems.

Constructional systems based on other materials (cement, mortar, concrete, polymeric, composite) do not fulfil the above important prerequisite of reversibility.

In conclusion, the use of structural steel in rehabilitation of old monumental buildings is perfectly in line with the recommended criteria of the modern theory of restoration.

Steel is therefore widely used in restoration works in all kind of ancient monuments and historical buildings, also under form of special devices for seismic protection (Mazzolani, 1997, 2001 a).

1.4   Ranges of application

Several examples of refurbishment, rehabilitation and extension can be found in all over the world (Mazzolani, 2000).

Old industrial constructions have been transformed in apartments or in offices. Monumental buildings have been entirely degutted by keeping the original facades and the interior has been completely substituted with a new skeleton. Self supporting structures have been inserted into historical monuments by providing a suitable integration with modern stylistic values. This kind of application is becoming more and more common for museums and exhibitions hall.

Many ancient churches have been covered by steel roofing systems, composed by trusses and trapezoidal sheets. Other important buildings have been restructured by vertical and horizontal extensions, which are harmonised both from structural and aesthetic point of view.

Entire districts of old towns in Italy have been completely rehabilitated after being seriously damaged by recent earthquakes; steel components have been used in order to improve the seismic resistance of old masonry buildings. Concrete structures have been repaired by means of steel elements after being damaged or when heavier serviceability conditions are required.

Reinforced concrete structures have been also transformed by changing the structural scheme from the original one, when a reduction or an increase of the storey height is carried out or steel bracings are introduced for seismic upgrading purposes.

2   Classification of consolidation levels

Facing the problem of the structural consolidation of a building, distinction can be done among the different levels, which correspond to the quality of the interventions and, sometimes, also to the chronological order in which the consolidation operations have to proceed.

The proposed classification considers four levels; they are (Mazzolani, 1986, 1990 a): safeguard, repairing. reinforcing, restructuring.

Safeguard represents the first level, also in chronological order, when action of a temporary nature is required – usually urgently – prior of starting any other intervention of a final nature; its aim is to achieve adequate safety at a temporary stage for both the public and the site.

What needed are constructional systems which offer flexibility of execution and ease of erection. For this purpose, steel provides "ad hoc" solutions under the classical form of scaffoldings or in more sophisticated and personalised versions of steelwork (Fig. 4).
Scaffoldings can be used to create a provisional protection of the site during restoration operations (Fig. 5).

Figure 4. Provisional reinforcement by means of steelworks (Berlin)

Figure 5. Protection during restoration operations by means of scaffolding structure

In Munich a reticular girder has been provisionally set-up to provide the contrast between two buildings during the reconstruction of the intermediate one (Fig. 6).
Scaffolding structures (Fig. 7) has been largely used in the Historical center of Naples after the earthquake of November 1980, in order to provisionally reinforce the façades of the damaged building (Mazzolani & Mandara, 1993).
In the same period other damaged buildings were provisionally consolidated by means of steelworks and stone barbicans (Fig. 8).
During the earthquake of 1997 in Umbria the famous church of St. Francesco in Assisi has been damaged; a provisional intervention of safeguard by means of steel scaffoldings has been immediately done (Fig. 9).
A steel structure made of hot-rolled steel sections was designed for the temporary shoring

Figure 6. Provisional reticular girder as a contrast element between two existing buildings (Munich)

Figure 7. Scaffolding structures as provisional reinforce after an earthquake in Italy

Figure 8. Provisional reinforcing system
for a damaged building (Campania, Italy)

Figure 9. The safeguard operation in the
St. Francesco church of Assisi

up of the stone columns at the entrance hall of Palazzo Carigliano in Turin during restoration
work involving the substitution of the old floors with new ones (Fig. 10).
The rebuilding of the Waring and Gillow's Store in London has been done by means of a
steelwork to preserve the old façade (Fig. 11).
Similar systems have been very often used in many cases of degutting, during the initial phase
when the facades stand alone without the internal part of the structure (Fig. 12).

Figure 10. Provisional structure during the consolidation of Carigliano Palace in Turin

Figure 11. Rebuilding of the Warmg and Gillow's Store in London

Figure 12. Degutting operation of a building in Bari (Italy)

The case of a large building unit in the center of Lisbon, which was completely empted by a fire, represented a significant example of safeguard by means of steelwork (Fig. 13) .

Where the old façades represent a valuable heritage, they are carefully conserved, as it can be observed in Montreal-Canada (Fig. 14).

After the safeguard aspect has been satisfied, the next stages in the logical and chronological chain are repairing and reinforcing.

Repairing involves carrying out a series of necessary operations on the structural elements of a building to restore its former structural efficiency, before the damage occurred.

Reinforcing, on the other hand, involves improving structural performance in order to enable the building to fulfil new functional requirements (for example, heavier load conditions along with a new type of use) or environmental conditions (such as the location in an area recently declared to be subject to seismic conditions).

Figure 13. Provisional support of a facade in Lisbon after a fire

Figure 14. Old facade isolated and protected in Montreal, Canada

Reinforcing operations can in turn be subdivided into:
- improvement operations involving a variety of work on individual structural elements of a building in order to achieve a higher degree of safety, but without significantly modifying overall performance;
- fitting operations, including the series of works required to make the structure capable of withstanding the new design actions. This is compulsory:
  - in the case of vertical additions or horizontal extensions;
  - when changes in use entail increases in the original loads;
  - when transformations lead to a different structural system from the original one;
  - in all cases generally involving a change in overall performance.

Within the repairing and reinforcing stages, there are numerous technological consolidation systems, which can be used to restore both masonry and reinforced concrete constructions(Section 3).

A level of consolidation of a more general nature takes place in *restructuring*, which consists of the partial or total modification of the functional distribution and volumetric dimensions, together with all the other characteristics of the building, including the change of the original structural system. It is carried out when a new intended use requiring a different arrangement with new volumes and new areas is planned. Degutting, insertion, extension and lightening operations are all part of restructuring(Section 4.2).

3   Main features of consolidation systems

3.1   Consolidation of masonry elements

The load bearing capacity of masonry elements must be improved whether when they are cracked due to the damaging of external unexpected actions (i.e. earthquake) or also when the structure as a whole must be upgraded in order to resist new loading conditions which are imposed in case of re-use of the building.

A classical system for improving the load carrying capacity of masonry elements consists on injections of pressurised mortar or cement, which in some cases can be integrated by means of anchor steel bars. In the last case the use of stainless steel is advisable in order to avid future damage due to corrosion.

Masonry columns, when damaged, are usually repaired by means of steel hoopings. The lateral restraining of the material produces a sensible increasing of the vertical load bearing capacity.

In case of circular columns the hooping can be made by means of vertical bar with rectangular cross-section, which are forced by horizontal steel rings (Fig. 15a). In the past, this forcing operation was made by heating these rings at high temperature and using the shortening due to cooling for introducing the lateral prestressing of the column. Nowaday, two half rings can be forced by means of bolts.

In case of square or rectangular cross-sections, angle shapes can be used as vertical elements in the corners (Fig. 15b). They can be connected in different ways: by means of internal ties integrated by batten plates, by means of channels connected by external ties or by means of horizontal rings.

When it is necessary to transfer an important part of the total vertical load acting on the masonry panel to a new steel structure, the new steel columns can be inserted in proper grooves or simply placed in adherence to the masonry (Fig. 15c). Both hot rolled shapes and cold-formed sections can be used to this purpose, according to statical and esthetical requirements (Mazzolani &Mandara, 1991).

As an alternative to steelworks, the masonry vertical elements can be repaired and reinforced by creating on the surface a reinforced concrete thin slab.

Recently, the composite materials under form of rigid sheets or textile glued to the surface can be used to confine the masonry elements.

3.2 Consolidation of wooden floors

The masonry buildings are usually integrated by floor structures made of wood. It is very often necessary to strengthen the wooden parts (beams and deck) because they are usually in a bad state of conservation.

Many systems have been proposed to improve the bending capacity of beams (Fig. 16). Two main ways can be followed, according to whether it is convenient to introduce from the bottom or from the top of the beams the additional steel elements.

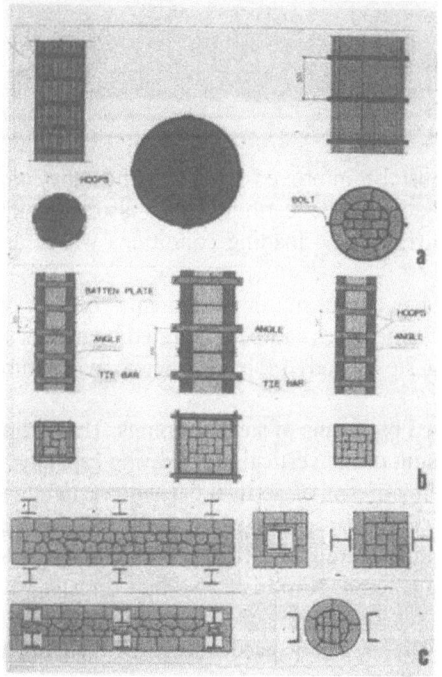

Figure 15. Consolidation techniques for masonry elements

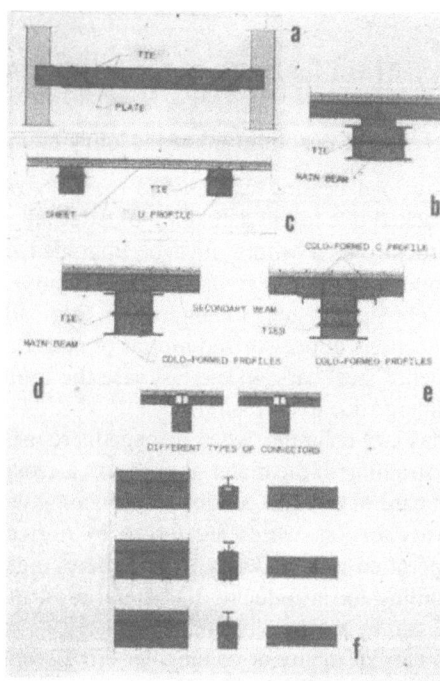

Figure 16. Consolidation techniques for wooden floors

In the first case, the steel reinforcements can be added in different forms, from the simple plate at the bottom (Fig. 16 a) to the hot-rolled double T sections (Fig. 16 b) or to the more advanced use of cold-formed profiles, which can be designed case by case according to the feature of the structure to be consolidated (Fig. 16 c, d, e).

When the original shape of the beam must be conserved because it has particular interest from the historical point of view, it is necessary to follow the second way, namely by operating on the top of the beam (Fig. 16 f).

The final result corresponds to a composite wood-steel system, which considerably increases strength and rigidity of the original structure. In all cases, such co-operation between the new and the old material must be guaranteed by using appropriate connecting systems from simple ties to different types of studs (Mazzolani & Mandara, 1991).

Recently, the consolidation of the wooden beams can be done by means of the insertion of composite poltruded bars in drilled holes.

3.3 Consolidation of steel floors

At the beginning of this century, the use of wooden beams for floor structures has been gradually substituted by the ancient I sections (Fig. 17). The steel beams were integrated firstly by a wooden deck (Fig. 17 a, b) and, later on, by clay blocks (Fig. 17c), concrete (Fig. 17 d) or stones (Fig. 17 e).
In all these cases the necessity to increase the section modulus can be very easily fulfilled by adding appropriate steel shapes to the lower flange, under form of square bars, T sections, cold-formed profiles with Ω shape or box sections.
When it is not allowed to operate from the bottom, the additional steel element can be connected to the top flange (Fig. 17 f).
The connection between the old and the new steel requires to pay particular attention to its state of conservation. In many cases welding is not allowed due to the umpure composition of the old material and the use of bolting is therefore advisable (Mazzolani &Mandara, 1991).

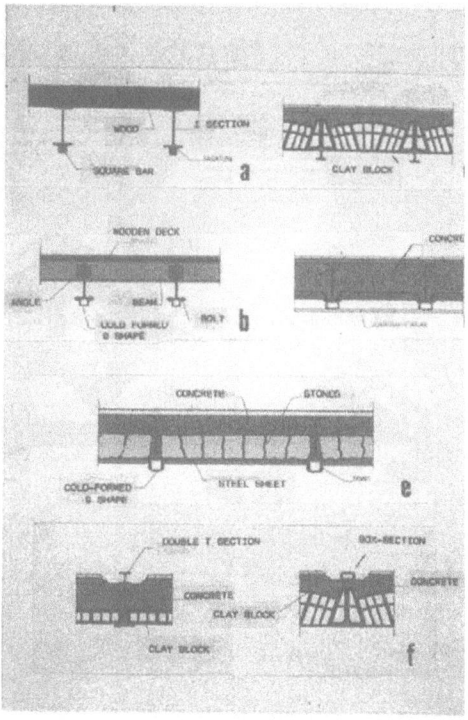

Figure 17. Consolidation techniques for old steel floors

3.4 Consolidation of r.c. structures

The increase of the load carrying of r.c. columns can be obtained by adding, in one or in two directions, a couple of hot-rolled steel sections (Fig. 18), which are connected together by means of appropriate ties. The use of cold-formed shapes (channels, angles and plates) allows to obtain a continuos resisting perimeter (Fig. 19), where the forcing effect is given by means of bolts. To improve the adherence between the concrete surface and the external steel sheet, the injection under pressure of gluing materials is recommended.

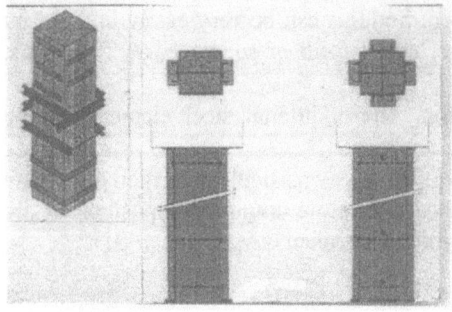

Figure 18. Consolidation techniques for r.c. columns by means of steel shapes

Figure 19. Consolidation techniques for r.c. columns by means of cold formed profiles

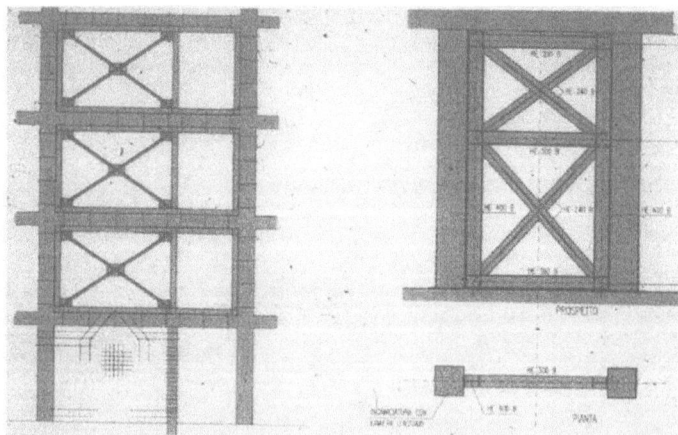

Figure 20. Seismic up-grading of r.c. frames by means of steel braces

When the seismic upgrading of a r.c. skeleton is requested, we can use steelworks in order to provide shear resistant elements, which increase the capacity of the structure to resist new horizontal loading conditions. A reticular shear-wall is obtained as a composite structure, where the r.c. frame is integrated by cross-bracings made of steel profiles (Fig. 20).

Each steel bracing is inserted in the mesh of the r.c. frame and the connection between both materials must be guaranteed by means of bolts or ties alongside the perimetral frame of the steel diagonals.

Together with the advantage of an easy erection, this system gives the possibility to have openings for doors or windows, by using – if necessary – appropriate shapes for the diagonals or introducing only one diagonal per mesh.

The strengthening as well as the reparation of r.c. beam-to-column joints is usually fulfilled by means of angles and batten plates, which are located around the r.c. members (Fig. 21 a). The steelwork is welded and, possibly, glued to the concrete surface. The size of the additional elements depends upon the requested amount of increase in shear and bending capacity.

The increase of the inertia of r.c. beams can be obtained by integrating the r.c. section with steel plates or profiles, which are connected to the concrete by means of bolts or ties and glue (Fig. 21 b). The same system can be used for strengthening the floor structures composed by r.c. and clay blocks (Fig. 21 c).

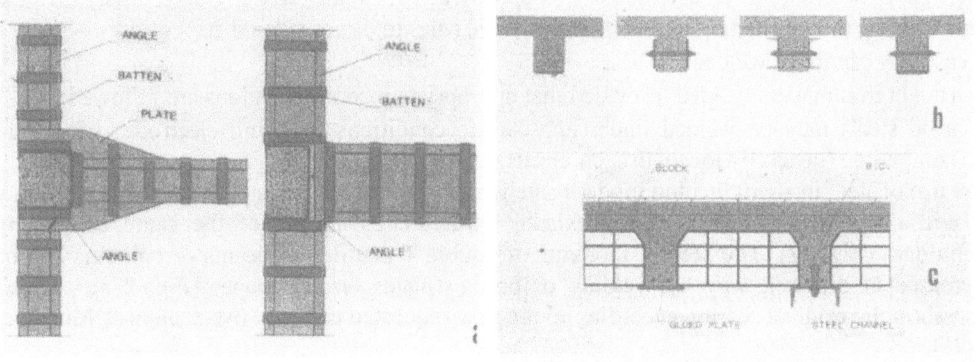

Figure 21. Consolidation techniques for r.c. joints (a) and sections (b, c)

Also in case of r.c. elements the use of composite materials is going to be more and more popular (Mazzolani & Mandara, 1991).

Mixed concrete and brick floors can be strengthened by the following methods:
-   plating the bottom of the individual concrete beams by means of steel plates, without breaking the tiles;
-   reinforcing the individual concrete beams by means of cold-formed steel sections;
-   inserting double T profiles between concrete beams in suitable housings;
-   strengthening with double T beams placed and forced below each concrete beam.

3.5. Consolidation of iron and steel structures

Resistance of iron and steel in building have gradually increased as improvements in manufacture and production have taken place. In the later 1800s allowable stresses for cast iron were around 20 Nmm^{-2} and for wrought iron around 100 Nmm^{-2}. Current allowable stresses for steel, which are given in the latest standards for design of steelwork, are very much higher. The strength of existing iron and steel structures obviously needs to be considered in relation to the standards in force at the time of original construction, although with extensive testing it may be possible to justify a slight increase in the allowable stresses specified at that time (ESDEP, 1990).

When considering the strengthening of existing steel beams various techniques can be employed:

- plates or angles can be welded to top and bottom flanges;
- channels or double T-sections may be welded onto flanges;
- plates may be welded between top and bottom flanges to form a box section;
- working from above, a reinforced concrete slab can be cast, joined to the beams below by suitable connectors (angles, T-sections, bars, studs, etc.) welded to the top flange to develop composite action.

In all cases the connection between the new and existing must be carefully considered. If bolting is to be used the initial loss of strength of the existing member whilst the bolt holes are drilled will need consideration, as this temporary condition may prove to be critical. If the alternative of welding is used then the specification of the welding technique must be compatible with the existing material. A few basic rules to be considered are:

- cast iron cannot be welded;
- wrought iron may be welded, provided that appropriate recommendations are followed;
- mild steels may be welded under appropriate conditions by using electrodes which are compatible (generally low hydrogen electrodes).

The use of steel in strengthening modern steelworks is the simplest case. In fact, it is very easy to add integrative elements to the existing structure by means of the same connecting techniques (Fig. 22). The section modulus of double T profiles of beams or columns can be increased in different way by welding or bolting plates or/and shapes (Fig. 22 a), which transform the original section according to the new requested capacity (Mazzolani & Mandara, 1991).

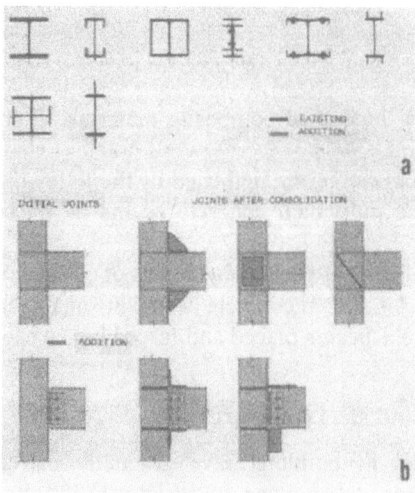

Figure 22. Consolidation techniques for steel structures

It happens, for instance, when a structure must be upgraded to resist seismic actions, because of the recent inclusion of the building in a new seismic area. In this case, not only strength but also ductility must be improved, particularly in the joints.

Appropriate systems for strengthening the two classical types of joints (rigid joint and pin-ended joint), by means of the introduction of stiffening elements, can be used.

In the first case, the moment capacity is improved. In the second joint the integration is designed to introduce a given capacity to resist bending actions, which is practically unexisting in the original joint (Fig. 22 b).

The improvement in resisting horizontal actions can be easily obtained by increasing the cross-section of diagonal bracings in case of braced structures or introducing new bracings in case of moment resistant structures.

4   Structural restoration of historical constructions

4.1 Consolidation Activities

Monumental buildings are usually made of masonry. They can be consolidated by using several techniques at the different stages of intervention, going from simple safeguard, through repairing and reinforcing, up to final restructuring operations (see Section 2).

A summarised overview is given in the following.

The load bearing capacity of masonry walls can be improved by several methods (see Section 3.1):

- encircling damaged masonry with vertical steel profiles and cross stiffening brackets (Fig. 15 a, b);
- insertion of new steel columns in suitable cavities or simply placed alongside the wall to be consolidated (Fig. 15 c);
- restoring the strength of the wall around openings by means of steel girders above the hole or frames inserted into the opening (Fig. 23).

The capacity to withstand horizontal actions, due to subsidence of foundations, geometric asymmetry, unevenness of load, or – a far more heavy circumstance – seismic quakes, can be conferred in different ways, such as:

- enchaining facade walls with steel profiles arranged to form horizontal hoops at each level joined together by tie-beams;
- enchaining corners by means of girders or tie-beams;
- insertion of steel bracings in the main walls.

As far as the consolidation of the horizontal structures is concerned, the types of floors usually found in old buildings belong to two main categories, which are characterised by the structural materials used, such as:

a) wooden beams;
b) steel beams.

As an alternative to complete substitution, the first type a) can be reinforced by two constructional methods (Fig. 16):

- working from the bottom upwards, each wooden beam is strengthened with a pair of steel profiles integrated by metal sheets or plastered nets to support the secondary elements;

working from the top downwards, if the wooden beams are in good condition and are worth exposing, the steel beams are connected to the wooden beams by means of a special system of connectors.

In the last case, in addition to the steel-wood composite system, the secondary beams can be substituted by a trapezoidal sheeting filled by casted r.c. (Fig. 24).

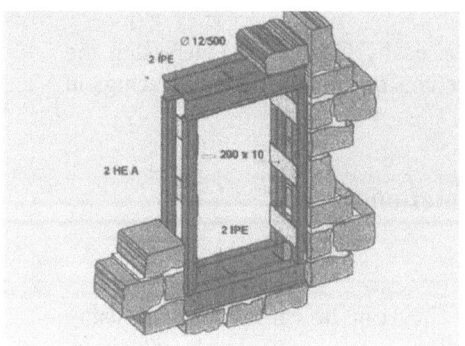

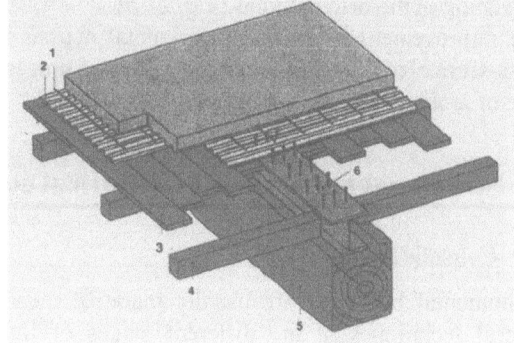

Figure 23. Opening reinforcement

Figure 24. Composite system for the consolidation of a wooden floor

Floors of type b) are frequently found in buildings dating back to the beginning of this century and are made of I-beams in conjunction with brick or tile vaults or hollow blocks (Fig. 17).

Due to gradual degradation of the restraint conditions, increasing of rigidity is usually required, what can be obtained by two constructional methods:

- working from the bottom upwards, the modulus of resistance of the beams can be increased by welding to the bottom flange of the I-beams a suitable steel section;
- working from the top downwards, a reinforced concrete slabs can be created, joined to the beams below by suitable connectors and cast on corrugated steel sheets.

Going to roof structures, masonry buildings are generally covered by wooden trusses, which deterioration is always made worse by direct contact with atmospheric agents.

The reparation of the deteriorated parts of the truss can be done by means of steel plates (Fig. 25).

An optimum solution can be attained by substituting the old structure with new steel structures (Fig. 26), completed by corrugated steel sheets. This method is very frequently used for the roof of church buildings.

The Cathedral of Naples (Fig. 27) was one of the first examples of complete sostitution of the old wooden roof structure with steel trusses.

If the church is located in an earthquake prone area, it is also advisable to create a steel grid below the trusses in order to obtain an horizontal diaphragm (Fig. 28), which provides a rigid connection between the top of the walls.

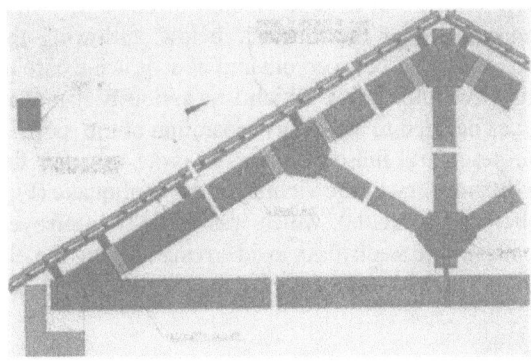

Figure 25. Reparation of a wooden truss

Figure 26. A new steel structure for roofing a church building

Figure 27. The steel roof of the Cathedral of Naples (Italy)

Figure 28. Horizontal diaphragm for the seismic upgrading of a church building(Solofra)

When the church is regarded as irrecoverable because of a large amount of damage, a new roof structure can be made completely independent of the masonry below, following the typology of insertion and giving rise to a strong contrast between old and new. It is the case of the church of St. Rocco in Morra de Sanctis (Avellino, Italy), which was seriously damaged by the earthquake of 1980 and, therefore, it was decided to avoid any reparation of the roof. The new steel roof is inserted inside of the perimetral masonry ruins, in order to cover the internal space as a sacrarium consecrated to the memory of the victims of the earthquake (Fig. 29). Another example is the church in Quebec city (Canada); which was partially destroyed by a fire and now is restored by means of some simple steelworks inside, remembering the old features (Fig. 30).

Figure 29. The roof structure of the Sacrarium of Morra de Sanctis (Avellino – Italy)

In the field of wooden structures, an interesting example is given by the Academia bridge in Venice (Fig. 31), which was erected in 1933 as a provisional arch bridge after the demolition of the iron bridge built during the austrian domination. It was decided to demolish this symbol of the detested period and to built a new masonry bridge after an international competition. But nothing happened and, fortunately, the provisional bridge became definitive.
A very common activity is now-a-day the rehabilitation of the old steel buildings of the 19th century, like in particular the markets (Fig. 32).

Figure 30. Church in Quebec City

Particular attention must be paid to the connection between the old and the new steel in relation with the state of conservation (see Section 3.5). In many cases welding is not advisable due to the unpure composition of the old material and the use of bolting is therefore suggested. From the structural point of view it must be observed that the old steel schemes are usually well conceived for carrying vertical loads, but they are weak for horizontal actions and therefore require the integration with new bracing systems. These problems have been faced during the restoration of the St. Lorenzo market in Florence (Fig. 33).

Figure 31. The Academia bridge in Venice (Italy)

Figure 32. An old market in Oporto (Portugal)

Figure 33. The St. Lorenzo market in Florence (Italy)

The same activity is devoted to old steel bridges. A challenging example is given by the old very famous bridge (Paderno bridge) built in 1886 on the Adda river near Milan (Fig. 34). Its structural scheme is very similar to one used by Gustav Eiffel in the bridges of Garabit (France) and Maria Pia (Portugal) (Fig. 35) .

Figure 34. The Paderno bridge on the Adda river(Italy)

Figure 35. Two important arch bridges of 19th century in Europe: the "Don Luis" and the "Maria Pia" bridges in Oporto(Portugal)

The retrofitting of suspension bridges can take profit of the specific aluminium alloy characteristics for obtaining the maximum structural effectiveness, particularly if compared to classical solutions based on the use of steel. During the seventies a rehabilitation program of ancient suspension bridges, constructed between the end of the 19th century and the beginning of the 20th century, has been developed in France (Fig. 36).

The old structures were made of wooden deck, masonry piers, steel girders and steel suspension chains. The adoption of aluminium alloy for girder and deck in the retrofit project of three bridges (the Montmerle and the Trevoux bridges on the Sôane river; the Groslée bridge on the Rône river), allowed for conserving as much as possible some of the old structural elements, and brought to very effective solutions, both from the cost and the structural performance points of view.

In the Montmerle bridge (two 80 m bays), the use of aluminium both for the two truss beams with bolted connections, and for the deck slab, led to the possibility of increasing the weight of the road vehicles, while preserving both the existing cables and piers without significant strengthening (Fig. 37) .

Figure 36. The Trevaux aluminium bridge on the Sôane river (France)

Figure 37. The Montmerle aluminium bridge on the Sôane river (France)

In the retrofit of the Groslée bridge (a single 174 m long bay) the floor structure is made of three longitudinal aluminium truss girders, connected to a light reinforced concrete slab (Fig. 38). In this scheme a remarkable example of co-operation among different structural materials, each of them utilised in an optimum working condition, can be observed: the old masonry piers, the harmonic steel suspension cables, the stainless steel suspension ties, the aluminium alloy reticular girders with high strength steel bolted joints and the light reinforced concrete slab floor.

Starting from the experience of the retrofitting of old suspension bridges in France by means of aluminium structures, a similar solution has been proposed and used in the structural restoration of the "Real Ferdinando" bridge on the Garigliano river near Naples, which was the first suspension bridge built in Italy in 1832 (Fig. 39). The structural restoration was completed in 1998 and the new bridge is characterised by an aluminium deck, first in Italy of this kind (Mazzolani, 1998).

Figure 38. The Groslée aluminium bridge on the Rône river (France)

Figure 39. The Real Ferdinando aluminium bridge on the Garigliano river (Italy)

4.2   Restructuring operations

According to the assumed classification(Section 2), restructuring is the highest level of consolidation, which involves a complete change of the structural scheme. In case of historical buildings, many examples has been collected and classified following the proposed classification, which considers restructuring subdivided into four categories: degutting, insertion, extension and lightening (Mazzolani & Mandara, 1992; Mazzolani, 2001 b).

Degutting consists of the total or partial substitution of the internal structures of a building by others of different type, by keeping the existing perimetral walls. It is resorted to when architectural and/or town-planning reasons require the complete conservation of the facades of a building, whilst the layout of the interior is changed.

Steel frames have been successfully used in many examples of degutting of buildings with facades of notable architectural merit.

Such type of operation is typical for buildings located in the hystorical centers of the towns and, also due to their strategical position, are mainly selected for commercial or public activities.

Many masonry buildings have been retrofitted by using this system in the City of London and some also in many italian towns (Milano, Modena, Bologna, Bari, Ancona).

The Law Court Palace of Ancona is a monumental building of renaissance style built at the beginning of the 20th century (Fig. 40). It has been completely empted. The new structure inside is composed by four r.c. cores containing stairs and elevators, on the top of which four main reticular steel girders are located (Fig. 41).

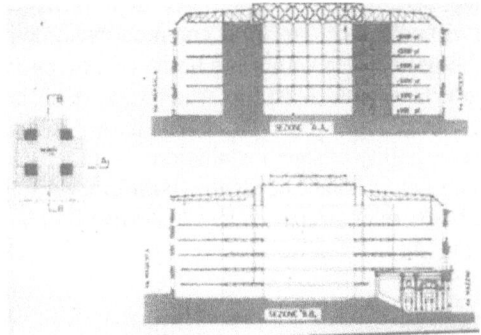

Figure 40. The facade of the Law Court Palace in Ancona (Italy)

Figure 41. The structural scheme of the Law Court Palace in Ancona (Italy)

They support the horizontal grid which covers the internal courtyard with small translucent domes for lightening (Fig. 42).

The floor structures on the perimeter are supported by the existing walls of the facade and in the central part they are suspended to the top girders by means of ties.

Figure 42. The internal courtyard of the Law Court Palace in Ancona (Italy)

Insertion comprises all those interventions providing for integration of the existing structure with new structures or structural elements inserted inside the same overall volumetric dimensions. The internal areas thus acquire new features derived both from their more rational layout and from the presence of new structural elements endowing the building with new stylistic values. The most common example is the additional floors created in order to increase the usable area within the limits of a given volume. In these cases, due to the necessity not to interfere with existing structures, steel is the most suitable and efficient material for constructing inserted structures, thanks to its special characteristics: high strength, low weight, reversibility of steel installations.

The Ducal Palace of Genoa, built in the 13th century, is a comprehensive example of insertion of new steel structures in a historical building, characterised in the past by many vicissitudes (Fig. 43).

Figure 43. The facade of the Ducal Palace of Genoa (Italy)

Figure 44. Insertion of mezzanine in the Ducal Palace of Genoa (Italy)

After the fire of 1591, it was reconstructed as the Dodge Palace during the Maritime Republic of Genoa. Because of a new fire in 1777 it was partially destroyed. In the second half of 19th century the roof was substituted with iron trusses having a Polanceau scheme. During the World War two it was seriously damaged by bombing and after it was in very bad condition until 1973, when the new restoration operation started.

Many steel structures have been inserted under form of mezzanine made of steel supported by the existing masonry and partially suspended to upper new steel beams (Fig. 44). They are connected by many elicoidal staircases, whose banisters have a structural role. There is an amphitheatre, whose gradine is made of heavy curved double T sections (Fig. 45).

The most interesting insertion is a steel ramp, which is suspended to a steel umbrella created on the top of a r.c. core (Fig. 46). The curious reason, why a ramp not a stair has been designed, is due to the fact that the Duke was used to go to bed riding on horseback and the bedroom was at the second floor.

Figure 45. An amphitheatre with steel gradine in the Ducal Palace of Genoa (Italy)

Figure 46. Insertion of a suspended ramp in the Ducal Palace of Genoa (Italy)

The structural restoration concept of the Ducal Palace of Genoa is characterised by a massive use of steelwork, whose dimensions are very impressive. In the suspended ramp the huge dimensions of the steel structure probably want to simulate the same sizes of the previous old wooden structure.

A nice suspended staircase has been inserted in the old castle of Rivoli near Turin. Contrary to the one of the Ducal Palace in Genoa, it is made of very small steel sections, which give a particular feature of lightness (Fig. 47).

A very famous building in Turin is the so-called "Mole Antonelliana", from the name of its designer Alessandro Antonelli. It is 167 m high, made of stone and brick with iron ties (Fig. 48). At that time (1863) it was the tallest masonry building in the World. Its first destination was as Jewish temple (the Synagogue of Turin), but during construction it was bought by the Municipality for a museum. This destination has been respected after about 140 years, being now after restoration the Museum of the Cinema.

Figure 47. Insertion of a stair-case in the Rivoli Palace near Turin (Italy) (courtesy V. Nascè)

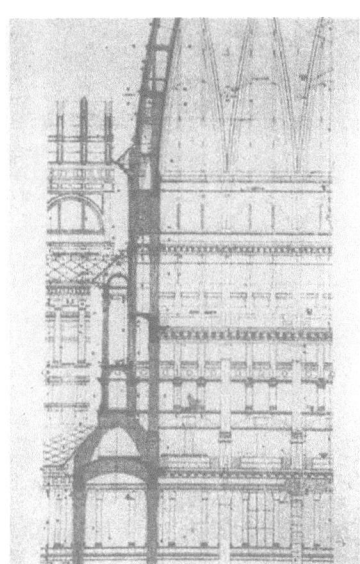

Figure 48. The "Mole Antonelliana" building in Turin (Italy) (courtesy V.Nascè)

From the beginning its life was very unhappy because of many troubles. Since 1870, many commissions of experts were charged to investigate its stability which was never convincing. In 1930 the first consolidation operations started in order to reduce the excessive stress state in the material, but this intervention was not satisfactory. Recently, more sophisticated analyses by means of FEM have been done and a new project of structural restoration has been executed and now completed.

During the construction, the organisation of the yard has been particularly studied in order to create a provisional insertion of a working plan at the level of 35 m, by means a steel deck made of reticular trusses, allowing to the consolidation of the masonry structure of the dome (Fig. 49). The permanent insertion consist on a large suspended elicoidal ramp made of steel which runs along the perimeter of the big hall (Fig. 50) .

Figure 49. Steel deck during the consolidation operations of the "Mole Antonelliana" building in Turin (Italy) (courtesy V. Nascè)

 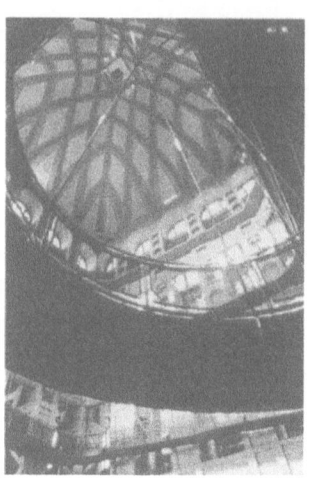

Figure 50. The elicoidal ramp in the Hall of the "Mole Antonelliana" building in Turin (Italy)
(courtesy V. Nascè)

Another example of insertion are the steel mezzanines and staircases of the new Faculty of Economy of the University of Turin, an old masonry building erected in 1887 by a charitable institution as the "house for the old" (Fig. 51).

Going to the South of Italy, the insertion of new steel structures has been used inside of the masonry ruins of the buildings of the Royal Bourbon ironware of Mongiana in Calabria, under form of portal frames (Fig. 52) and new steel and r.c. structures inserted into the old masonry ruins (Fig. 53) (Matacena & Mazzolani, 1981, 1986).

Some mezzanines and staircases have been inserted in the Museum of "Sinopie" in Pisa (Italy), in order to allow for a better view of the frescos on the walls (Fig. 54).

Extension can be done by means of vertical and lateral additions.

Vertical addition consists of adding one or more stories above the existing structure, resulting in an increase of the overall volume of the building.

Depending on the size and height of the new additional masses, it is necessary to re-check the load-bearing capacity of the original structure in order to decide whether to take consolidation

Figure 51. Steel staircase in the new Faculty of Economy in Turin (Italy) (courtesy V. Nascè)

Figure 52. Portal frames used in the restoration of a building in the Royal Bourbon ironware of Mongiana (Calabria, Italy)

Figure 53. A new steel roof on a r.c. wall inserted into the masonry ruins of a building of the Royal Bourbon ironware of Mongiana (Calabria, Italy)

measures or not. The necessity to minimise the weight of the new structure added above makes steel the most suitable material due to its excellent mechanical performance (high "strength to specific weight" ratio).

In general vertical additions are not allowed in monumental buildings, but there are some exceptions.

Figure 54. The mezzanine of the "Sinopie" Museum in Pisa (Italy)

An old industrial building in Briatico (Calabria, Italy), which was active in different productions (sugar, wool and soap) from fifteen to seventeen centuries, has been retrofitted and re-used as a sport club (Fig. 55). After consolidation of masonry, a new level has been created by means of a steel framed structure (Matacena & Mazzolani, 1981, 1986).

Lateral addition does not require specific strength features, but it is sometime used in order to increase the functionality of the lay-out. In case of monumental buildings, steel structures provide an aesthetic combination between old and new materials (Fig. 56). There are many examples of lateral additions.

Figure 55. Vertical addition in a restored building in Briatico (Calabria, Italy)

Figure 56. Lateral addition in the "La Villette" area in Paris (France)

In the above mentioned new Faculty of Economy of Turin, beside the masonry building a new steel building has been erected for the auditoriums. The style of the facade of this additional volume is clearly new, but it is harmonised with the existing one (Fig. 57).

Contrary, the Episcopal building of San Andrea of Conza in Campania, original of the eighteen century, has been extended with a new building made of steel and glass, which strongly contrasts with the old architecture, but at the some time creates a mutual valorisation of styles (Fig. 58).

Ligthening is the opposite to vertical addition. It can include the demolition of the one or more levels at the top, when this is required because of the necessity to limit the loads involved, in order to reduce the stress state in the existing structure. In this sense, but less

Figure 57. Lateral addition in the New Faculty of Economy in Turin (Italy) (courtesy V. Nascè)

Figure 58. The Episcopal building of St.Andrea of Conza (Campania, Italy

drastically, lightening interventions are considered all operations substituting floors, roofs or other structural elements with lighter materials. In fact, the substitution of heavy wooden and masonry floors with light steel I-sections and corrugated steel sheets as well as the complete remaking of roofs with steel trusses are very common (Fig. 59).

This operation has been done in some important churches like the Cathedral of Naples and outstanding palaces, like the Royal Palace in Naples (Fig. 60).

Figure 59. Corrugated sheeting for a new steel roof

Figure 60. The Royal Palace and the Cathedral of Naples (Italy)

After the earthquake of 1980, many churches, which were damaged, have been retrofitted by means of steel trusses completed by trapezoidal sheeting with casted r.c. We remember the churches of San Vincenzo Ferrari in Dragonea (Salerno) (Fig. 26) and of San Giuliano in Solofra (Avellino) (Fig. 28), both in Campania.

An interesting example of lightening has been done in Succivo (Caserta) near Naples. It is a masonry building belonging to the municipality, whose roof structure made of heavy masonry tympanums and wooden beams has been completely demolished. The new structure consisted of light steel frames (Fig. 61) with four columns, completely prefabricated in workshop, transported, erected and completed with trapezoidal sheeting, giving rise to a new volume, like a penthouse, where the cultural center of the village has been located (Fig. 62).

Figure 61. The new light structure for the
roofing of a building in Succivo
(Caserta, Italy)

Figure 62. The new volume created on the top of
a masonry building in Succivo (Caserta,
Italy)

A very light steel roof has been built to cover the gallery of the Museum of Contemporary Art in the Rivoli Palace near Turin (Italy) (Fig, 63).

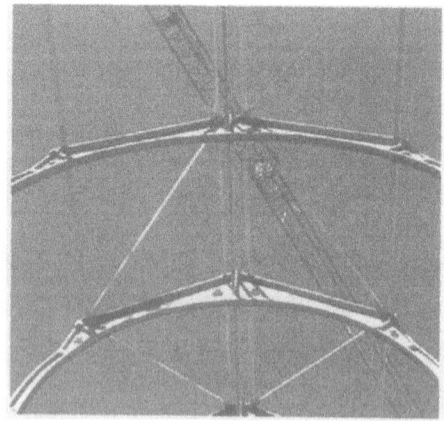

Figure 63. A light steel roof (courtesy V. Nascè)

5. Seismic up-grading

5.1 Bracing systems

The use of steel bracings is very effective in strengthening both masonry and reinforced concrete structures earthquakes (Fig. 64). It allows the introduction of shear walls with lattice scheme, which has the dual purpose of considerably increasing the resistance of the structure to horizontal forces and at the same time balancing the distribution of internal rigidity with respect to the shear centre, so as to minimise dangerous torsional vibrations.

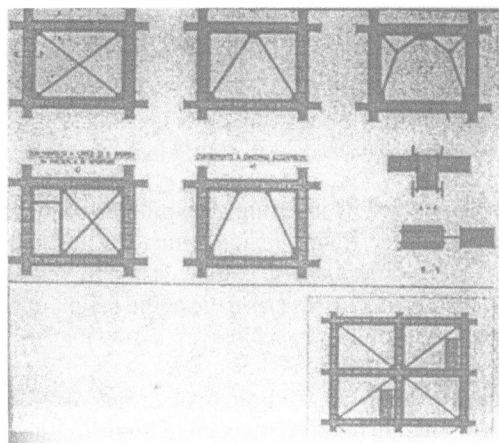

Figure 64. Steel braces for strengthening masonry and r.c. buildings

In case of masonry structures two significant examples are: the electrical power station near the Balaton lake in Hungary (Fig. 65) and the church of Montemarano near Avellino, Italy (Fig. 66) .

Figure 65. Seismic upgrading of an electrical power station near the Balaton lake (Hungary)

Figure 66. Steel braces on the masonry wall of the Montemarano church (Avellino, Italy)

Such steel bracings can be obtained by inserting steel profiles connected to the perimeter of the meshes of the reinforced concrete frame, inside which diagonals and counter-diagonals are arranged in the classical "St. Andrew's cross" pattern or in other patterns more suited to the use of the building. If the St. Andrew's crosses are made at the height of two levels, the presence of a single diagonal for each rectangular panel makes it possible to provide door or window openings (Fig. 67).

Several seismic up-grading operations have been done around over the World by means of the introduction of steel braces into the reinforced concrete frame meshes (Mazzolani, 1994).

The St. Andrew cross system on two levels has been used for upgrading many high rise reinforced concrete buildings in Mexico City after the earthquake of 1986 (Fig. 68).

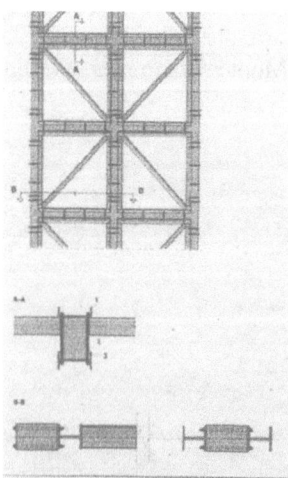

Figure 67. Steel braces for upgrading r.c. buildings

Figure 68. A r.c. building upgraded by mean of steel braces in Mexico City

Many r.c. buildings belonging to the University of California in Berkeley have been upgraded by means of steel bracings with different shapes, according to functional and aesthetic requirements.

Very often these reinforcing steel elements give an additional value to the façade, whose aspect is improved, like in the case of many buildings in Berkeley (California): the University Hall (Fig. 69), the Student dormitory (Fig. 70) and a multi-storey parking building (Fig. 71).

Other examples can be found in different places: St. Monica – Los Angeles (Fig. 72); Tessaloniki – Greece (Fig. 73).

Figure 69. The University Hall in Berkeley (California, USA)

Figure 70. A building of the campus of the University of Berkeley (California, USA)

Figure 71. A parking building in Berckeley (California, USA)

Figure 72. Steel cross braces in a building of St.Monica, Los Angeles (California USA)

An interesting example of seismic upgrading is given by the rehabilitation operation of the entire Capodimonte district of the historic centre of Ancona, the oldest quarter of the town where the fisherman people lived in the past centuries (Mazzolani & Mandara, 1992).

The masonry buildings were in an advanced state of decay caused by the vast damage suffered during the earthquake of 1972 in addition of the damaged caused by earlier earthquakes, such as in particular that of 1936, or by the bombardments suffered during the

Figure 73. Steel cross braces in a building of Tessaloniki (Greece)

second world war. This situation had led to the precautionary evacuation of practically all the inhabitants of the district.

In all the buildings with two or three stories above the ground floor, the solid brick and dressed stone walls showed many widespread cracks and the mortar had completely lost all consistency.

The need for a reliable method of restructuring these buildings led to rejection of the traditional methods of consolidation based on local strengthening of the individual constructional components and preference was given to a solution whereby the task of transferring the loads to the foundations was completely entrusted to a new structural system (Fig. 74) .

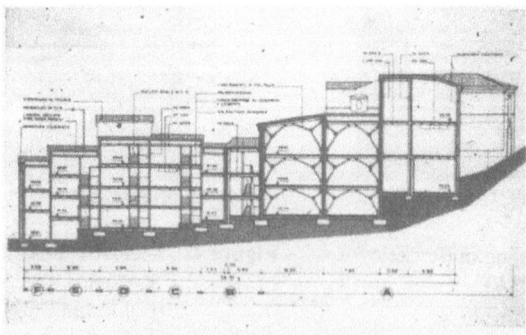

Figure 74. The entire Capodimonte district in Ancona (Italy) has been seismically upgraded by means of steelworks

The intervention carried out in fact provided a steel structure inserted into the perimeter and internal walls, integrated with horizontal structures of steel sections and corrugated steel sheets (Fig. 75). The new steel skeleton, suitably connected to the bracing walls, forms an independent structural system with regard to both vertical and horizontal loads, particularly designed to withstand the seismic actions.

The steel skeleton is completely autonomous and independent of the existing walls, which declassed to simple partition walls, without any load bearing capacity request.

The refurbishment works were carried out in the following stages:

- breaking openings in the lower part of the walls to contain the new reinforced concrete foundations (Fig. 76);
- positioning of the anchoring bolts and base plates;
- after breaking suitable vertical channels in the perimeter walls, erecting steel columns over the full height and temporary bracing of these at floor levels;
- constructing roof structures with trusses and purlins and finishing off with the existing covering of roof tiles (Fig. 77);
- from the top floor, demolition of internal walls and the corresponding floor and reconstructing the new floor with main and secondary beams, corrugated steel sheets and casted concrete (Fig. 78):
- building up the reinforced concrete walls of stairway cores with steps and landings cast on site;
- final connection of the steel framework to the existing walls and reinforced concrete staircases and fixing with sealing concrete;
- completion with partitions, plastering, floor coverings and finishes.

The external walls, suitably restored, still retain their architectural and sheltering function, but are relieved of main load-bearing functions (Fig. 79).

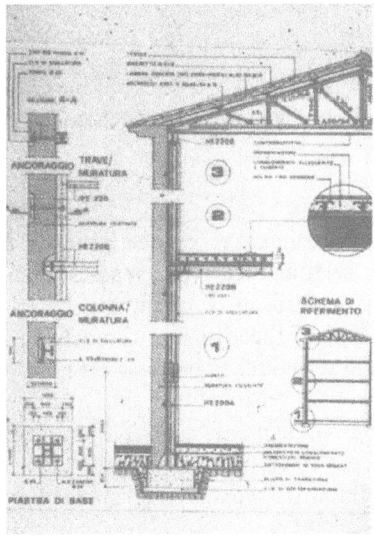

Figure 75. The new steel structure inserted into the old masonry walls of the Capodimonte district (Ancona, Italy)

Figure 76. The new r.c. foundations and the steel column in the Capodimonte district (Ancona, Italy)

Figure 77. The new roof structure of the
            Capodimonte district (Ancona,
            Italy)

Figure 78. The new floor structure of the
            Capodimonte district (Ancona,
            Italy)

Figure 79. The facade of the buildings of the Capodimonte district (Ancona, Italy)

5.2   Passive control systems

The control of the structural response produced by earthquakes can be done by means of
various systems being based on different concepts, such as modifying, masses, damping and
producing passive or active counter forces. This control is substantially based on two
approaches, namely providing either the modification of the dynamic characteristics or the
modification of the energy absorption capacity of the structure. In the first case, the structural
period is shifted away from the predominant periods of the seismic input, thus avoiding the risk
of resonance occurrence and usually leading to less severe dynamic actions. In the second case,
the capacity of the structure to absorb energy is increased through appropriate devices, which
preserve the structure from damage. Both these approaches can be implemented in passive,
active or hybrid systems.

Following the passive systems, which do not require an external power source, the properties
of structure (period and/or damping capacity) do not vary depending on the seismic ground
motion. The base isolation or energy absorbing devices serve as the first line against seismic

forces, as a filter which considerably reduces the seismic impact on the protected structure. The use of passive control techniques in the rehabilitation of existing buildings having monumental features represents a quite new issue (Serino et al, 1994; Serino & Mazzolani, 1994; Mazzolani, 2001a).

The first historical application of the base isolation technique has been experienced in 1989 on the City and County Hall in Salt Lake City (USA): hundred years old masonry building which has been located on about 200 cylindrical rubber paths at the foundation level (Fig. 80).

The second example is the Parliament House and Library in Wellington (New Zealand) (Fig. 81).

Figure 80. The first application of "base
isolation" in a historical building:
The City and Country Hall of Salt
Lake City (USA)

Figure 81. The Parliament House and Library
in Wellington (New Zealand)

After the Northridge – Los Angeles earthquake (1994), many important buildings, like City Halls, Court of Appeals, Hospitals have been seismically upgraded by means of special devices and new projects are now in progress.

Few applications have been done in Italy until now and their main data are listed in Table 1 (Mazzolani & Mandara, 2001; Mazzolani, 2001 a). The more significant examples are illustrated in the following.

Table 1: Monumental Buildings retrofitted with passive control in Italy

YEAR	STRUCTURE	LOCATION	TYPE OF DEVICES	N°
1990	St. Giovanni Battista Church	Carife (Avellino – Campania)	- Oleodynamic restraints	18
1996	University of Naples Federico II New Library	Napoli (Campania)	- Oleodynamic restraints - Neoprene bearings	24 34
1999	"La Vista & Domiziano" Schools	Potenza (Basilicata)	- Dissipative brace	224
1999	Bell tower of the St. George Church	Trignano (Emilia Romagna)	- Shape Memory Alloy devices (SMAD)	4
1999	Basilica of St. Francisco	Assisi (Perugia - Umbria)	- - Shape Memory Alloy devices (SMAD) - Oleodynamic restraints	47 34
2000	"Gentile Fermi" School	Fabriano (Ancona – Marche)	- Viscoelastic dampers	N.A.

Church of St. Giovanni Battista (Carife)

The rehabilitation project of the St. Giovanni Battista Church in Carife near Avellino (Campania) represents the first example of application of oleodynamic damping devices (shock transmitter units) in the field of monumental buildings (1990). Goal of the application was to improve both seismic resistance of the church and its behaviour under thermal loads (Candela et al, 1989; Mazzolani & Mandara, 1994).

As it is well known, the basic condition for ensuring a good seismic performance in a masonry building is to have one or more floors able to behave as rigid diaphragms: only if this condition occurs, an efficient transmission of the horizontal forces to the vertical walls is assured. On the other hand, if rigid links are used between masonry and roof structures, this may cause some problems to masonry due to thermal variations. In the St. Giovanni Battista Church a new steel roof structure, consisting of a plane gridwork and triangular trusses, was built to provide a box-like behaviour of the masonry structure under seismic loads (Fig. 82). At the same time, oleodynamic restraints were placed on one side of the gridwork , so to obtain a fixed or a free restraint situation at the base of the trusses according to the loading condition (Fig. 83).

Figure 82. The new roof system of the St. Giovanni Battista church in Carife (Avellino, Italy)

Figure 83. The oleodynamic restraint (shock block transmitter) of the St. Giovanni Battista church in Carife (Avellino, Italy)

Under thermal loads, whose speed of application is very slow, the oleodynamic devices behave as sliding bearings: the structural scheme of the roof is statically determined and no additional stress arises as a consequence of thermal variation. Under an earthquake, the devices behave as fixed restraints owing to the high speed of load application: in this condition the structural scheme becomes redundant, with a significant improvement of the overall seismic behaviour. The devices have a plastic threshold: when this is exceeded a significant energy dissipation occurs, which is able to reduce the effects of the seismic action on the masonry structure. The devices adopted for the church have been calibrated so to behave as fixed bearings under the action of a design earthquake corresponding to the Italian code, causing the dissipative behaviour to occur in case of a more severe earthquake.

The test results on the devices confirmed the design assumptions (Fig. 84).

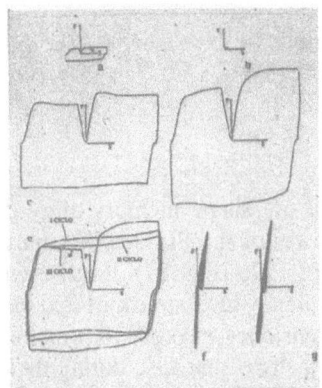

Figure 84. Test results on the oleodynamic devices

New Library (Ex-Mathematics Department) of the University Federico II (Naples)

The same concept of the previous Church has been applied later on in the re-use and structural rehabilitation of the building called ex-Mathematics Department of the University of Naples "Federico II" for creating a new Library. This intervention has been done in a wider operation of restoration of all the monumental building, more than one hundred years old, belonging to the original part of the old central University of Naples (Mazzolani & Mandara, 2001).

The upper floor structure (covering an area of 16 x 32 meters) was re-made during the Fifties by means of r.c. beams (16 m clear span) with mixed clay blocks and r.c. casted elements. This structure was in very bad condition due to the steel rebars corrosion and the superficial degradation of concrete. It was decided to demolish it and to build a new steel structure, made of castelleted beams and trapezoidal sheeting (Fig. 85). A system of 24 oleodynamic cylinders and neoprene bearing devices have been used for the support of the new steel beams on the top of the perimetral masonry walls (Fig. 86), providing an expected double behaviour under serviceability conditions and in case of earthquake. This intervention has been completed in 1996.

Figure 85. The steel structure of the new
          Library of the University of
          Naples (Italy)

Figure 86. The shock block transmitter at the
          support of the steel girders

The tower of St.George Church (Trignano)

This has been a pilot application of shape memory alloy devices (SMAD) in a structural restoration operation. The bell tower of the St. George Church in Trignano (Reggio Emilia) was damaged by the earthquake of 1996 (Fig. 87). It has been consolidated by means of four vertical ties in the corners of the plan, which are connected in series to the SMA devices. The system has been previously pre-tensioned in order to pre-stress the masonry walls, in such a way that they can keep the tension stress state also during the cyclic loading conditions due to the earthquake.

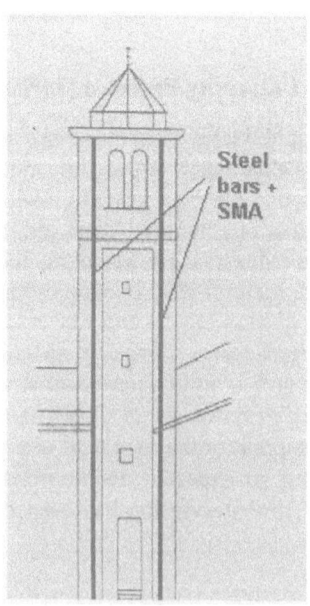

Figure 87. The bell tower of the St. George church in Trignano (Reggio Emilia, Italy)

Church of St. Francesco of Assisi

The Church of St. Francesco of Assisi suffered large damage during the earthquake of 1997 (Fig. 88). Its structural restoration has been done very quickly in two years and was based on two different innovative technologies for the seismic protection: the contemporary use of shape memory alloy devices (SMAD) and of oleodynamic cylinders (shock transmitter units), like in the previous applications of Carife and Naples. The SMADs are characterised by a variable stiffness in function of the imposed elongation, allowing to limit the maximum load transmitted to the structure which they are connected to.

Due to the pseudo-hysteretic feature of the stress-strain relation-ship of shape memory alloys, in case of seismic actions SMAD are able to dissipate part of the in-come energy. They provide the connection between the upper parts of the masonry facade and the roof structure (Fig. 89), allowing the masonry to undergo stronger seismic actions without damage.

Figure 88. Damaged facade of the St. Francesco church in Assisi

Figure 89. SMA devices in the roof of the St. Francesco church in Assisi

An additional connection has been created at an intermediate level by means of a steel truss, in order to provide the masonry wall with additional ductility. The steel truss connections have been obtained by the insertion of 34 oleodynamic cylinders.

6   Case Studies

6.1   Degutting: The Court of Justice in Ancona, Italy

The building was completely gutted and restructured to house the new court offices (Mazzolani & Mandara, 1992). The arrangement of the windows, cornices and all ornaments in the masonry facades characterising its neo-renaissance style was preserved (Fig. 40).

The main load-bearing structure consists of four reinforced concrete towers 9m x 9m, containing stairs, lifts and floor services and located at the corners of the inner covered courtyard (Fig. 41). These towers provide the vertical support to the roof and the five floors suspended from it, as well as horizontal stability to resist the effects of seismic activity.

The system of suspension in the roof consists of four pairs of truss girders supported on the inside edge of the four reinforced concrete towers, thus marking the perimeter of the covered courtyard (Fig. 90). Each pair of trusses forms a box girder 1.80 m wide, 4 m high with cross members at 3.0 m centres and X-shaped wall diagonals (Fig. 91).

Figure 90. The truss girder supported on the top   **Figure 91.** The cross-section of the
of the r.c. towers   truss girder

All the truss members (chords, vertical and diagonal bars) are made of steel I-sections, connected by means of bolted gusset plates (Fig. 92). The inner ring made up of four pairs of girders with a span of 21.40 m represents the key component of the steel skeleton which the other members of the structure are connected to:
- The beams supporting the dome skylights (Fig. 93) which illuminate the inner courtyard rest on the upper truss nodes of the inner box walls (Fig. 94)
- The cantilever beams which cover the zone outside the perimeter defined by the four towers are connected to the lower truss nodes of the outer box walls

The tension rods for the five suspended floors below start off in groups of four from the truss nodes of the inner bottom chords.

The five floors suspended from the roof girders are associated with the four zones of approximately 9 x 20 m between the four towers (Fig. 42). They consist of structural steel beams and joists supporting composite metal deck floor slabs. The main beams on the interior side are suspended by tie rods from the box girder ring (Fig. 95), whilst on the outer side they rest on the reinforced concrete structures forming the perimeter area between the four towers and the exterior facades of the building. They were connected by welding to suitable plates pre-set in the concrete. All the other structural components were assembled on site with bolted connections. The individual elements were fabricated in sizes convenient for transportation inside the historic city centre and erection within a densely built area.

Figure 92. Detail of bolted connection

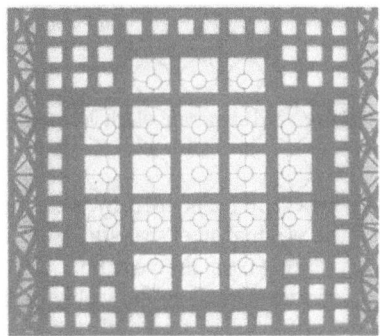

Figure 93. Dome skylight pattern

Figure 94. Support structure of the dome skylights

Figure 95. Suspension tie connection

6.2 Consolidation by prestressing: the Deutsche Bank Building in Naples, Italy

The building has been erected during the 50's years and its facade is covered by marble (Fig. 96).

The steel skeleton is composed by transverse frames, with external columns supporting an upper truss, which two internal tie-beams are connected to (Fig. 97). The intermediate floor

 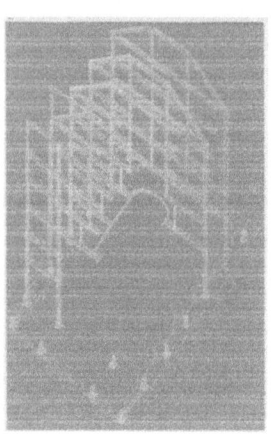

Figure 96. The facade of the Deutsche Bank   **Figure 97.** The structural skeleton of the Deutsche
             in Naples, Italy   Bank building in Naples, Italy

structures are suspended by these tie-beams, thus creating an empty space at the ground floor.
All the original steel structure was field welded in the 50s years (Fig. 98).
After forty years, the very poor conditions of the steel structure due to both corrosion (Fig. 99)
and serious defects in the welded connections required a drastic consolidation operation, which
was conceived in line with the basic principle of minimising the cost and inconvenience of
such operations, allowing the bank to continue operating also during the works (Mazzolani &
Mazzolani, 1996).

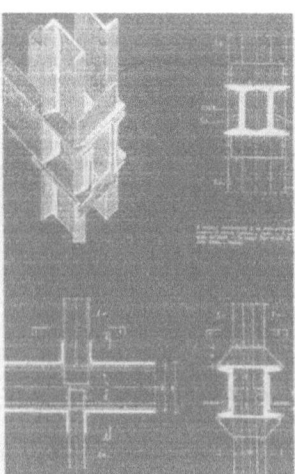 

Figure 98. Detail of the beam-to-tie   **Figure 99.** Corrosion effect in the roof structure
             connection

The consolidation system have been designed in order to increase the load carrying capacity of the structure as a whole and its single component parts, by assuming a minimum level of reliability in line with modern safety principles.

On the basis of the results of non destructive x-ray and ultrasonic tests on weldments, as well as of analysis and calculations, the interventions were identified in both tension and compression members where the stress value exceeded an assumed limit value. Three different types of operations were selected (Fig. 100).

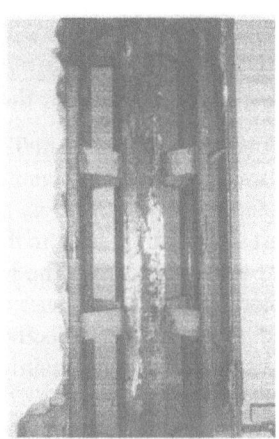

Figure 100. Different types of consolidation operations and detail of the intervention type B

System type A: This operation was used for the intermediate columns and the external columns of some frames; it consists in boxing the section by means of welded steel plates and filling inside with expanding cement, giving rise to a steel – concrete composite action.

System type B: This operation was used for the tie-beams of suspension frames; it consists in four pretension cables located into perimetral tubes which are connected to the corners of the existing tie-beams by means of batten plates.

System type C: This operation was used for some tie-beams at the second level which are off-set with respect to the upper ones; it consists of a single pretension bar of the Dywidag type in the centre of the member which is boxed and filled with expanding cement, as in system A.

The cables, 40 in total, used in system B have been pretensioned from the top by means of 40 jacks located at the roof level (Fig. 101).

Figure 101. Location of 40 jacks for pre-tensioning the cables

They were connected to the same oil pressure system, in order to guarantee a contemporary action on all the tie-beams, which were prestressed in such a way to eliminate the tension state and to introduce a small amount of compression.

During the prestressing operations the monitoring of the pressure in the jacks, of the strain in some key sections of the tie-beams and of the deflection of each floor has been carefully done.

We can observe that this structure was the first of its kind fortyfive years ago at least in Naples; today this consolidation system based on prestressing cables may also be considered completely new, as it is adapted to a special structural situation, which is very rarely found in the constructional practice.

6.3   Masonry consolidation: the Main Hall of "Mercati Traianei" in Rome

The building complex of "Mercati Traianei" (Emperor Traiano's Markets), placed close to the area of the ancient Imperial Forum in Rome, is presently undergoing a wide refurbishment programme aiming at improving its capability to serve as a museum (Fig. 102). It contains, in fact, many art works dating back to the Roman Age, which are not completely displayed owing to the lack of available rooms. The Main Hall of the Mercati has been interested in past times by several modifications of the structural layout, sometimes for consolidation purposes, which often proved to be neither effective, nor durable. As a result, the structure is presently interested by a spread damage, mainly consisting of cracks and surface degradation (Fig.103).

Figure 102. The Main Hall of the "Mercati Traianei"   **Figure 103.** Some details of the
                       in Rome   masonry structure

The intervention of consolidation proposed, consisting of an improved confining system made of stainless steel elements, has been assessed on the basis of a mechanical model of confined masonry developed for this purpose (Fig. 104a). A comprehensive series of non destructive tests on the bearing elements has been also planned. Together with a wide non-linear FEM analysis(Fig.104b), the tests are aimed at outlining a complete frame of the material conditions and to assess the degree of structural reliability (Mazzolani & Mandara, 1999).

The bearing structure consists of a double system of cylindrical vaults supported by regularly spaced males (Fig. 105). The main constitutive material for the vaults is an artificial conglomerate, a sort of *ante litteram* concrete, forming the upper part of the structural system. The intermediate part is made of the same conglomerate, clad by an external skin of brick masonry, in order to achieve a higher resistance at the vault impost. The lower part consists of travertine males, whose cross-section is smaller as compared to the above vault impost.

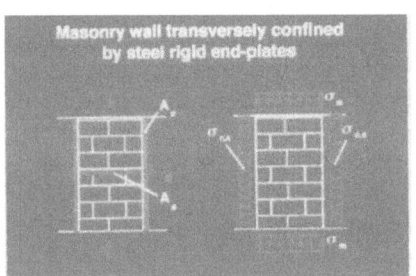
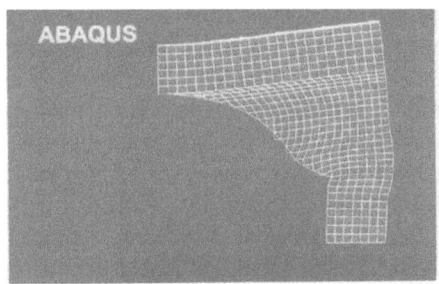

Figure 104. a) The mechanical model of confined masonry; b) the FEM model of the cylindrical vault

According to the original design, the male cross-section was connected to the impost by means of a projecting travertine element, loaded in bending and shear (Fig. 106). At the present state, this element is absent in all males except one on the side opposite to the hall main entrance. The consequent change in the vault profile at the impost involved a strong increase of the state of stress, resulting in a crush cracking at both the vault impost and male base. In order to correct this situation, a hooping system made of steel tied elements was preliminary arranged several decades ago, presumably after the inappropriate modification of the structural geometry which brought to the present damage.

Figure 105. The Main Hall of the "Mercati Traianei" museum in Rome

Figure 106. Detail of the support of the vault impost and of the previous reinforcement

The effectiveness of the reinforcing system is today very poor owing to both the absence of internal ties and the rupture of some elements.

Furthermore the strong stress concentration arising in the section angles has increased the damage of the masonry involving the spalling out of the confined core along the section side.

An approximate evaluation of the acting vertical loads has been done by assuming a specific weight for materials equal to 20 kN/m^3 and 25 kN/m^3 for conglomerate and travertine respectively. A live load of 4 kN/m^2 has been considered on the top floor above the vault, which is open to public use. Based on these values, the average state of stress in the most stressed zones has been determined as a preliminary investigation on the safety degree of the structures. The state of stress was relatively severe even before the reduction of the cross sectional areas, with average values of the compression stress in absence of live load not far from 60 and 90 N/cm² at the arch impost and male base, respectively. After the elimination of the projecting part, the stress at the impost has increased of about 50%, reaching values clearly incompatible with an artificial concrete dating back to almost two thousand years ago. By considering that the above values are only average values and that the actual peak stresses are far higher, it is easy to justify the onset of crush cracking in the most stressed parts. In order to get a more accurate evaluation of the state of stress in the vaults, a FEM analysis with the ABAQUS code has been carried out. Due to the geometrical symmetry, a quarter only of the actual structure has been analysed.

The numerical simulation has provided peak compression stresses equal to 157 and 227 N/cm² on the original and the modified structure, respectively.

The consolidation intervention has been conceived in such a way to follow as closely as possible the existing hoops. To this purpose, a new confining system has been designed, aiming at providing an improved structural performance as respect to the existing one (Fig. 107). As widely referred to in the technical literature, the confining technique, based on the creation of a tri-axial state of stress in the masonry, is highly effective in the strengthening of members in compression. In addition, it is characterised by low cost, very soft appearance and total reversibility. In order to guarantee a good protection against corrosion, the system adopted for this application is based on the use of stainless steel confining elements, connected at both masonry sides by passing bolted ties. In this way, a quite uniform confining density can be reached, avoiding the stress concentration in the section angles which characterised the previous intervention. The load bearing capacity of the confined element has been evaluated with a procedure similar to that commonly followed for hoped r.c. columns. (Mandara & Mazzolani, 1998).

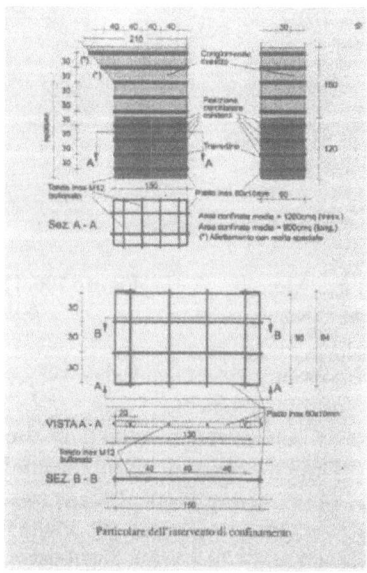

Figure 107. Consolidation system by means of stainless steel elements

The structural restoration design has been completed by space structural systems made of aluminium alloy (GEO systems), in order to provide a horizontal roofing to cover the roman ruins with a square on square double layer structure (Fig. 108) and to create relaxing areas for visitors, covered with two reticulated cylindrical vaults (Fig. 109) and one geodetic dome (Fig. 110) (Mazzolani et al, 2000).

 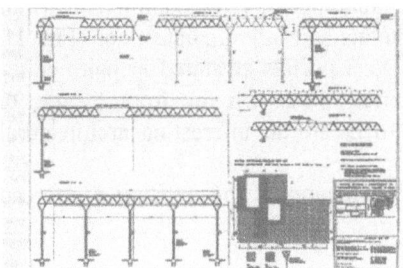 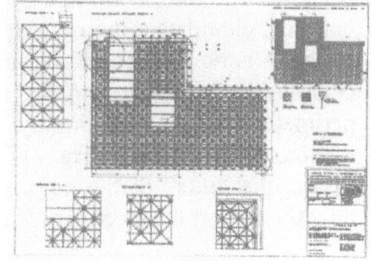

Figure 108. Reticular space structure made of the aluminium GEO system to cover the roman ruins

 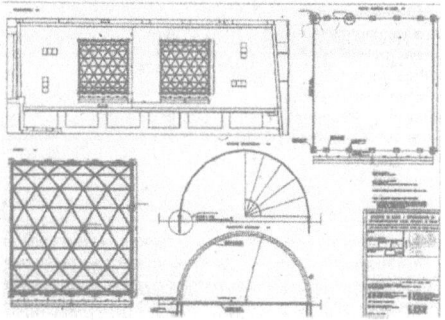

Figure 109. Cylindrical vault made of reticulated aluminium structure to be located on the roof of the Main Hall

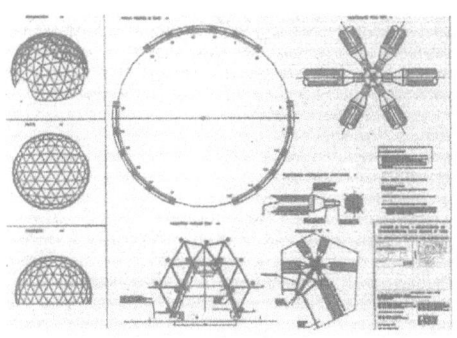 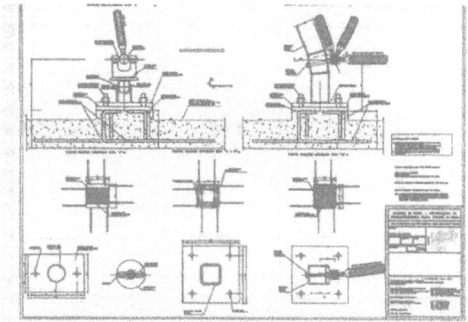

Figure 110. Geodetic dome made of reticulated aluminium structure

6.4 Change of structural scheme: Gymnasium in Cantu, Como, Italy

An old industrial building in Cantu in the province of Como has been converted into a gymnasium using structural steelwork to achieve a change of layout of the original reinforced concrete structure (Mazzolani, 1990a)

The existing layout consisted of a two-storey reinforced concrete frame with intermediate columns. The conversion to a gymnasium, necessitated completely stripping the interior of the building, eliminating the central columns and intermediate floor (Fig. 111). The existing roof structure is now supported by new steel portals arranged in pairs either side of the existing columns and inserted into the external walls. On the front façade, the portals pierce the perimeter walls in such a way as to create an interesting architectural motif relieving the monotony of the façade (Fig. 112).

Inside, the rafters of the new frame encompass the existing reinforced concrete roof truss directly support the secondary structures.

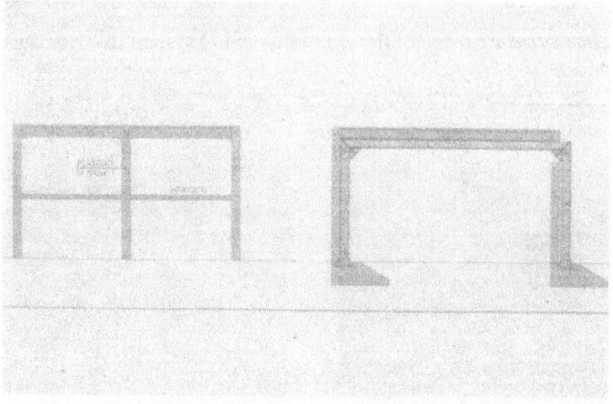

Figure 111. Change of structural scheme in a building in Como (Italy)

Figure 112. The new facade of the transformed building in Como (Italy)

6.5 Increase of area: Van Leer Office Building in Amstelveen, Netherland

The office building was built at the end of the fifties to accommodate about 500 employees, but due to the de-centralisation of the Van Leer organisation, only about 300 staff have been working here in the last few years (Mazzolani, 1990 a).

Furthermore, as with most buildings designed before the oil crisis, the energy costs were very high.

The building consists of a central hall, with a V-shaped two storey office wing at each end. Each office floor has a surface area about 1000 square metres. The service rooms are in the central office and in separate subsidiary buildings. The storey heights are very large: 5.6 m gross (4.3 m net) in the office wings, while the central hall is 7.2 m high.

The load-bearing structure is made of steel. There are 19 columns in each 1000 square meter wing. The frame is 8.0 m between centres. The distance between columns varies due to the form of the ground plan, from 8.15 to 9.0 m.

When the building was originally designed, the possibility of adding an extra floor to the end wings at a later stage was taken into account for the foundations and construction.

The main points in the programme of requirements were:

1. Reducing the storey height in the office wings from 5.6 m to 3.75 m, so that within the existing building volume the useful office floor space can be increased from 4000 square meters to 6000 square meters.
2. The design of a new, completely insulated façade, but retaining the original expression of the building.
3. The creation of new utility provisions in both wings, such as lift, stairs and toilets.

The following solution was adopted for realising the first phase (Fig. 113):

- Fit a steel construction for the new floor at 8.25 m;
- Assemble the temporary support construction under this;
- Shorten the bottom columns by 1.85 m and keep these column sections;
- Put the jacks in place;
- Release the column division and allow the floor to be lowered by 1.85 m;
- Replace the column sections and weld the whole structure together.

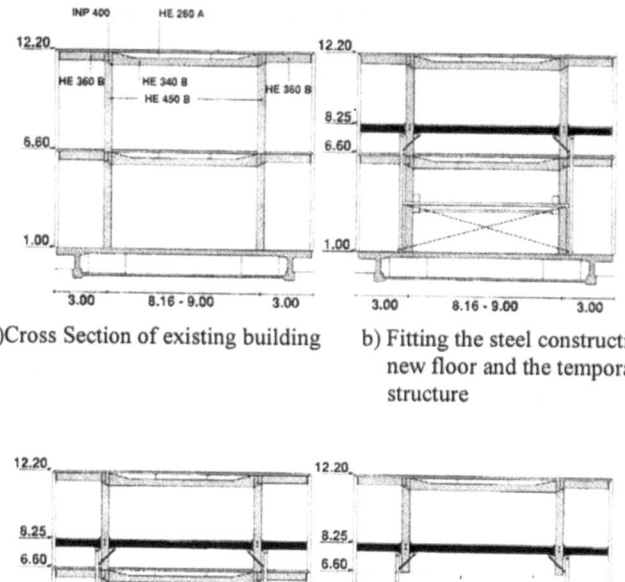

a)Cross Section of existing building   b) Fitting the steel construction for the
                                              new floor and the temporary support
                                              structure

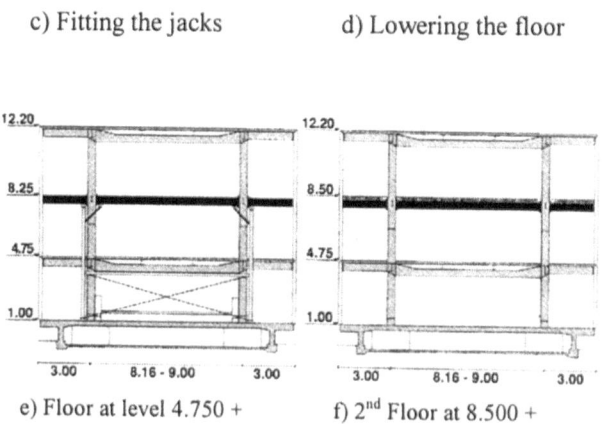

c) Fitting the jacks   d) Lowering the floor

e) Floor at level 4.750 +   f) 2nd Floor at 8.500 +

Figure 113. The operational phases in the transformation of a building in Amstelveen (Netherland)

6.6   Change of use: Rue de l'Ourcq Building, Paris, France

This building is situated at 135 to 145, rue de l'Ourcq and at 24 to 36, rue Labois-Rouillon in Paris. It was an industrial building originally used as a depot and baling plant for old paper and fabric, and later as a furniture warehouse (Mazzolani, 1990 a). The property had to be adapted for its new role as an apartment block, whilst keeping the feature of its late 19th Century industrial architecture (Fig. 114).

Figure 114. An industrial building in Paris before restoration

The depth of the building did not allow the whole of the floor area to be used for apartments. It, therefore, proved necessary to form a void in the central part.

The architects made use of this constraint to create a unique interior space, strongly defined yet highly differentiated. It formed a kind of backbone which services all of the apartments, allowing them to open onto a quiet garden area away from the noise of the street and providing them with natural day lighting. This arrangement gives the apartments an individual character with a private interior street (Fig. 115).

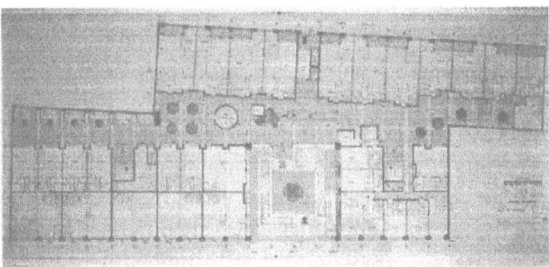

Figure 115. The new apartments lay-out

Small business premises have been built on the ground floor, along rue de l'Ourcq and on the little square. This position was chosen because of the easy access and the liveliness that it brings to the street.

All of the floors, beams and columns of the steel structure inside the building built at the beginning of the 20th Century were in an acceptable state with no major damage or excessive corrosion. The structure was very well suited to the change of building use since its components had been originally designed to support heavy industrial loads.

The internal columns supporting the floors are of cast iron construction on a structural grid measuring 4m by 4m. Where the new arrangement created small loads, the columns were left in their original condition (Fig. 116).

Figure 116. The original columns made of cast iron

For heavy loads the columns were encased in a square section of reinforced concrete. The columns are supported horizontally at mid height by the beams of the mezzanine floors or by the reinforced concrete façade.

The beams were too narrow and some of them were off-centre. In most cases they were arranged in pairs, a flange width apart. Occasionally, a main girder was made up of two beams of different depths. Sometimes the beams were jointed, sometimes single. The connections were as varied as the beams. All the joints have, therefore, been checked and strengthened where necessary and the beam supports at the columns reinforced.

The original floors were made of joists supporting brick and clinker vaults covered with reinforced cement mortar. In certain areas the floor was strengthened by covering with concrete over the whole depth of the joists. In other areas floors had to be demolished or reinforced.

The whole building is covered by a saw-tooth roof arranged parallel to the street (Fig. 117). The north slopes were glazed and the south slopes were tiled. The span of the saw-tooth trusses is double that of the floor beams at the lower level. The columns supporting the roof are generally IPN 260 sections.

Figure 117. The building vertical section

The orientation of the building and its saw-tooth roof form made it ideal for the installation of solar panels for heating water.

It was necessary to provide a fire resistance of half an hour for the floors and the supporting structure. The fire resistance was achieved in the apartments either by encasing in reinforced concrete approximately 70 mm thick where the columns fell within the party walls between apartments or by intumescent paint.

6.7   Vertical extension: Building in Victoria Street, Toronto, Canada

This example shows the potential of steel for improving the vertical additions (Mazzolani, 1990b, 2000). In Toronto an existing building structure of six stories made of reinforced concrete was designed to be super-elevated of more four stories in the same material (Fig. 118).

Contrary to the initial choice, it was later decided to use steel for the additional structure (Fig. 199). Thank to this choice, instead of four stories, it was possible to add new eight stories (Fig. 120). Therefore, at the end the super-elevated building is now composed by fourteen stories, instead of ten, with an important increase in volume (Fig. 121) with respect to the initial forecast.

Figure 118. The original r.c. building in Toronto (Canada)

Figure 119. The additional floors under construction

Figure 120. Eight additional floors made of steel frames

Figure 121. The building after superelevation

7   References

Candela, M., Mandara, A., Mazzolani, F. M. (1989). L'uso dell'acciaio nel restauro degli edifici di culto in Campania. In *Proceedings of the" Giornate Italiane sulla Costruzione in Acciaio "C.T.A., Capri, October 1989.*

ESDEP (1990). European Design Education Program.

Mandara, A. and Mazzolani, F. M. (1998). Confining of Masonry Walls with Steel Elements. *IABSE – Colloquium "Saving Buildings in Central and Eastern Europe",* Berlin (Germany), 4-5 June.

Matacena, G. and Mazzolani, F. M. (1981). Interventi di restauro con strutture di acciaio. *Proceedings of the "Giornate Italiane sulla Costruzione in Acciaio C.T.A.",* Palermo, October .

Matacena, G. and Mazzolani, F. M. (1986). Aspetti evolutivi della prevenzione sismica in Calabria: due esempi di restauro. *L'industria delle Costruzioni,* July/August .

Mazzolani F. M. (1985,1986). L'acciaio e il consolidamento degli edifici. *Acciaio, n.* 12/85 & 1/86.

Mazzolani, F. M. (1990a). Refurbishment. In *Arbet-Tecom.*

Mazzolani, F. M. (1990b). Refurbishment and Extensions: The case for steel. *Proceedings of the International Symposium of I.C.S.C.,* Luxembourg, May .

Mazzolani, F. M. (1992). The use of steel in refurbishment. *1st World Conference on Constructional Steel Design,* Acapulco, November .

Mazzolani, F. M. (1994). Il consolidamento strutturale. *In the volume "La progettazione in acciaio", Ed. Crea,* September .

Mazzolani, F. M. (1996). Strengthening options in rehabilitation by means of steel works. *5th Int. Colloquium on Structural Stability, SSRC Brazilian Session,* Rio de Janeiro, August 5-7.

Mazzolani, F. M. (1997). Tecniche di protezione passive per edifici monumentali (keynote lecture). *Proceedings of the V National Congress ASSIRCCO,* Orvieto, May .

Mazzolani, F. M. (1998). Il ponte "Real Ferdinando" sul Garigliano. *Restauro n.. 146.*

Mazzolani, F. M. (2000). Steel in structural rehabilitation (keynote lecture). *Proceedings of the ITEA Symposium,* La Coruña (Spain), April.

Mazzolani, F. M. (2001 a). Passive control technologies for seismic resistant buildings in Europe. *Progress in Structural Engineering and Materials,* Wiley (under press).

Mazzolani, F. M. (2001 b). Die Anwendung von Stahl bei der Restauriering von Gebäuden in Italien. *Bauingenieur* May.

Mazzolani, F. M. and Mandara, A. (1989). L'uso dell'acciaio negli interventi di restauro nell'edilizia monumentale nell'Italia meridionale. *Acciaio n. 10.*

Mazzolani, F. M. and Mandara, A. (1991). L'acciaio nel consolidamento. *ASSA.*

Mazzolani, F. M. and Mandara, A. (1992). L'acciaio nel restauro. *ASSA.*

Mazzolani, F. M. and Mandara, A. (1993). Structural preservation of the "Centro Storico" of Naples. *IABSE, Structural Engineering International,* January .

Mazzolani, F. M. and Mandara, A. (1994). Seismic upgrading of churches by means of dissipative devices. *Proceedings of the STESSA '94 Congress,* Timisoara,4-5 June.

Mazzolani, F. M. and Mandara, A. (1999). Methodology for the Structural Rehabilitation of the main hall of "Mercati Traianei" in Rome. *2nd Int. Congress on "Science and Technology for the Safeguard of Cultural Heritage in the Mediterranean Basin",* Paris (France), 5-9 July .

Mazzolani, F. M. and Mandara, A. (2001). Advanced Metal Systems in Structural Rehabilitation of Monumental Constructions (invited lecture). *Proceedings of the International Conference on Structural Engineering, Mechanics and Computation*, Cape Town (South Africa), 2-4 April .

Mazzolani, F. M. and Mazzolani, S.M. (1996). Structural retrofitting of the Deutsche Bank building in Naples. *Costruzioni Metalliche n. 4.*

Mazzolani, F. M., Mazzolani S. M. and Mandara, A. (2000). Aluminium Structures in the Restoration Project of Mercati Traianei in Rome. *5th Int. Congress Restoration of Architectural Heritage*, Firenze, 17-24 September.

Serino, G. and Mazzolani, F. M. (1994). Innovative techniques for seismic retrofit: design methodologies and recent applications. *Proceedings of the 2nd Franco-Italian Symposium of Earthquake Engineering – Strengthening and Repair of structures in Seismic Areas*, Nice (Francia).

Serino, G., Mazzolani, F. M. and Zampino, G. (1994). Seismic protection of italian monumental buildings with innovative techniques. *Proceedings of the International Workshop on Application and Development of Base Isolation*, Shantou (China).

CHAPTER 2

REFURBISHMENT OF STEEL BRIDGES

M. Iványi

Budapest University of Technology and Economics, Budapest, Hungary

Abstract. This chapter studies with the refurbishment of steel bridges.

In the first part the definition of notions in the field of the refurbishment are summarized and the developed different activities are classified.

In the second part the activity fields of the surveying and diagnostic of bridges are presented, the important considerable aspects and procedures are summarized.

In the third part the damages in steel bridges are presented, sketching that reasons, which basically influence the damage symptoms. The typical places of occurrence are presented: sway bracing, diaphragms, web stiffeners, components of orthotropic decks and crossing connections.

We studied the main typical cause of the damages, thus:

- the development of traffic,
- inadequate fatigue design of structural details and fabrications defects,
- inadmissible simplification of statical systems, mainly because the three-dimensional behaviour of bridges and secondary stresses usually neglected in design practice.

We studied in this part the connection between the fatigue problems and the post-buckling behaviour of plated structures. On the bases of American and Sweden case studies the consequences of effects arising from the localised distortions of bridge elements are summarized. On the base of Japanese experiences the fatigue situations in structural elements and the proposed strengthening methods are examined. On the base of Dutch studies we deal with the effect of the shape of orthotropic steel decks on the fatigue symptoms. One of the most important questions of the last decade is also examined, mainly based on German studies, which analyses the effect of the distortion of cross-sections of bridges on the statical model.

The fourth part summarizes the problems of strengthening during the relations of some typical shaping.

The fifth part presents the history, the design, the construction and the refurbishment of Danube bridges. During a short summary we deal with the river Danube and the types of Danube bridges, refer to the "Catalogue of Danube Bridges" and, as an example, we summarize the history, the design, the construction and the refurbishment of the "Szabadság" (Freedom) bridge. During the history of the bridge it could be refer to the notions defined in the first part, and to the particular appearances.

1 Definition of Notions

Refurbishment of bridge is a term covering all the actions that to be carried out to thoroughly repair and improve a bridge (Fig. 1.1) [Ghost, 2000]. Generally refurbishment has been considered as a beginning only when a bridge has been built and is brought into service. However, much can go wrong with a bridge as a result of actions and as decisions taken at the concept, design and construction stages.

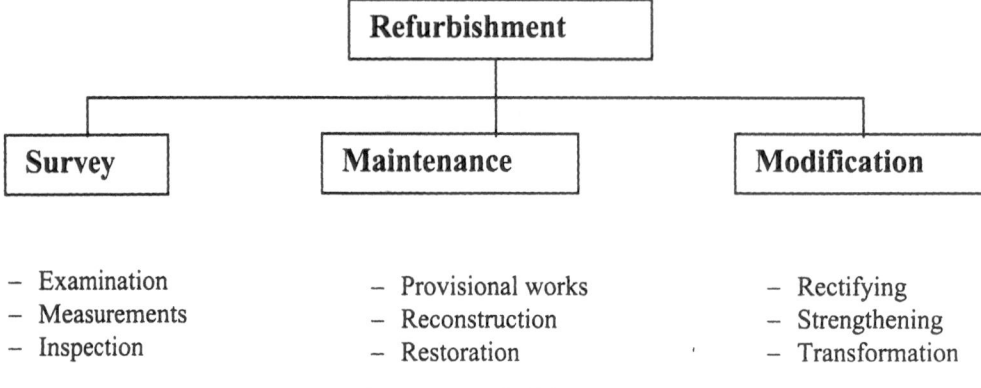

Figure 1.1. Refurbishment

(a) Conceptual stage: Fortunately thousands of bridges have been built successfully in the past and therefore errors in concept are rare if are follows the well-trodden paths.

Errors in concept are more likely when producing a new and innovative design. Such cases need to be properly reviewed and tests undertaken where necessary to verify the adequacy of the concept.

(b) Design stage: It is standard practice for the principle of design, including the concept, to be approved, at least from the safety and serviceability viewpoint, and for the design to be then certified and independently checked in appropriate cases. This tends to minimise errors in concept and the design of the structure.

One of the main cause of damages are inadmissible simplification of statical systems, mainly because the three-dimensional behaviour of bridges and secondary stresses usually neglected in design practice.

(c) Construction stage: It is necessary to ensure that materials and components manufactured on and off the site, and in fact the whole of the construction process, is carried out in accordance with the specification.

(d) In-service stage: Once a bridge is completed and brought into service, it not only starts to carry traffic but is also exposed to the environment. It is subject to wind, rain and temperature changes, and also chemical and (in some cases) biological attack. In time deterioration may occur and/or the bridge may have to carry heavier traffic loading than it was originally designed for.

The first part of the refurbishment of the bridge is the surveying (Fig. 1.2). The second part is the maintenance (Fig. 1.3) and the third part is the modification (Fig. 1.4). .

The efficiency of refurbishment can be shown in the Fig. 1.5 [Bancila, 1996].

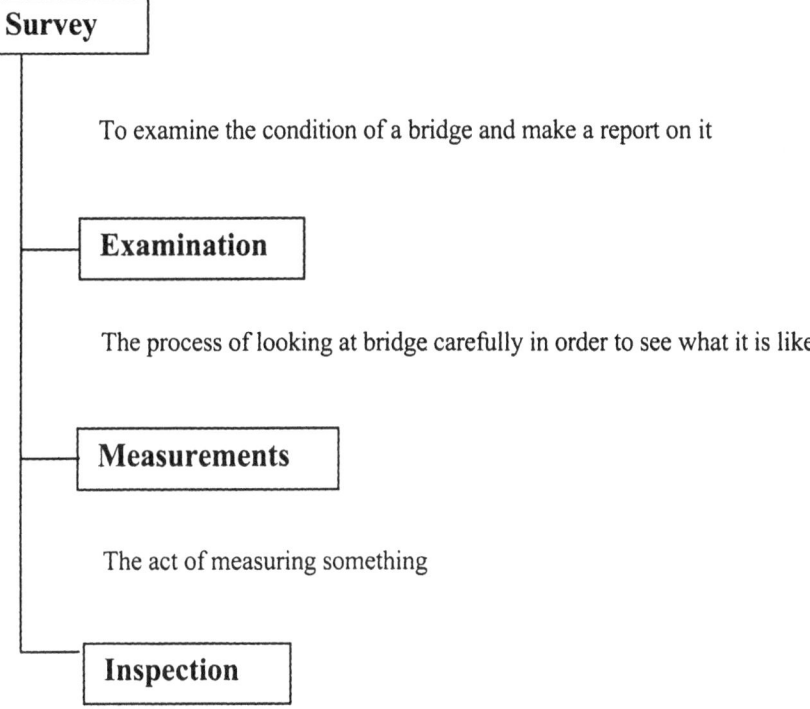

Figure 1.2. Surveying

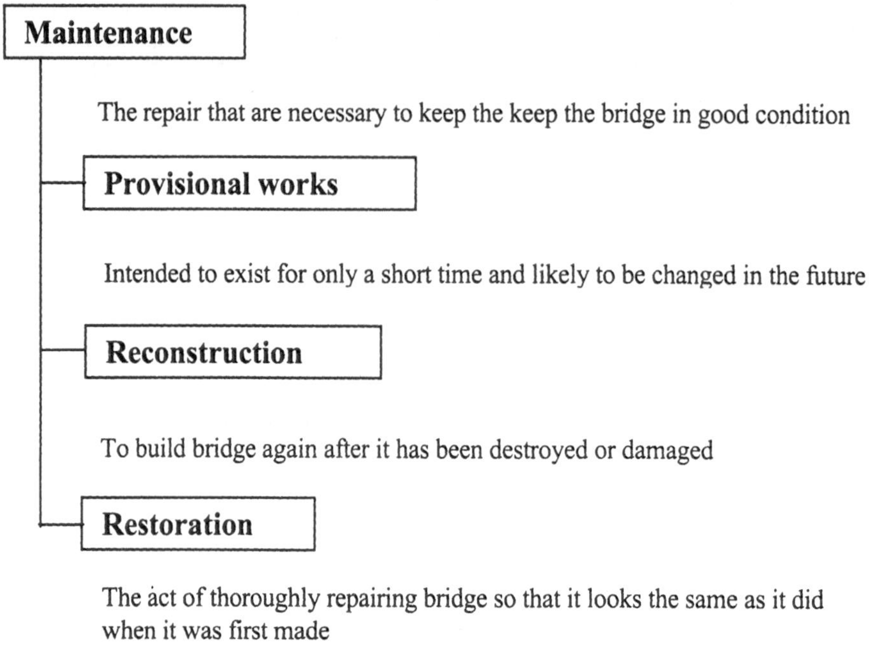

Maintenance

The repair that are necessary to keep the keep the bridge in good condition

Provisional works

Intended to exist for only a short time and likely to be changed in the future

Reconstruction

To build bridge again after it has been destroyed or damaged

Restoration

The act of thoroughly repairing bridge so that it looks the same as it did when it was first made

Figure 1.3. Maintenance

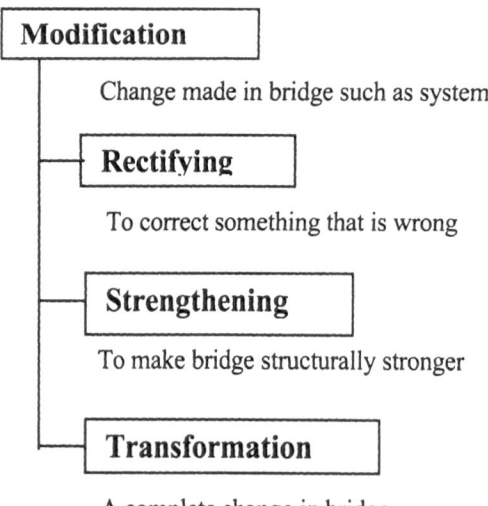

Modification

Change made in bridge such as system

Rectifying

To correct something that is wrong

Strengthening

To make bridge structurally stronger

Transformation

A complete change in bridge

Figure 1.4. Modification

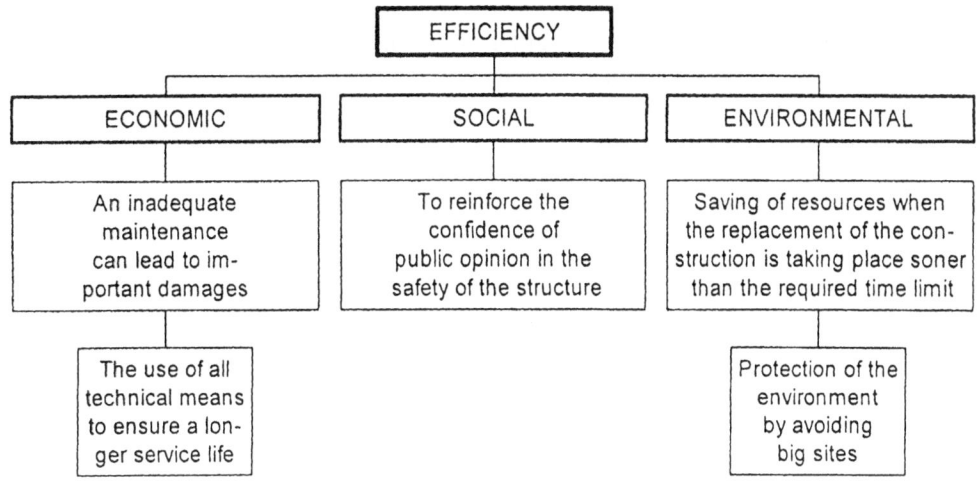

Figure 1.5. The efficiency of refurbishment

2 Surveying of Bridges

2.1 Summary

Load carrying constructions are to be designed, operated and maintained with regard to the corresponding standards and service rules.

The state of more important constructions should be checked regularly and the main important findings of this activity should be included into the complete documentation of the corresponding structure. Eventual checks are necessary in case of accidents or of any unforeseen changes in the structural and service conditions.

Surveying is a more or less new term for the definition of this process.

This subchapter is containing a short summary about the different levels of surveying activities, about their correlation and a list of the instrumentation generally used. [Agócs, 1996], [Hegeds, 1997].

2.2 Introduction

Overall general technical conditions of existing structures (bridges and buildings) are generally checked along with regularly and periodically repeated preventive and more or less detailed inspections.

This activity is mostly starting at the very beginning of the service life, when a controlled load test series is carried out, mainly to make a comparison between the designed and 'measured' behaviour.

Additionally, a non-regular, eventual or occasional inspection can be necessary in case of accidents, local or global defects, or in cases, when prolongation of service life of the construction is required.

As nowadays old constructions are in use in a very great number, a detailed check of the structural state before the intended refurbishment is necessary to ensure the background knowledge to help the decision for prolongation of the service life of the construction.

The chart in Figure 2.1 is illustrating the different kinds of activities, which involve surveying in a certain level.

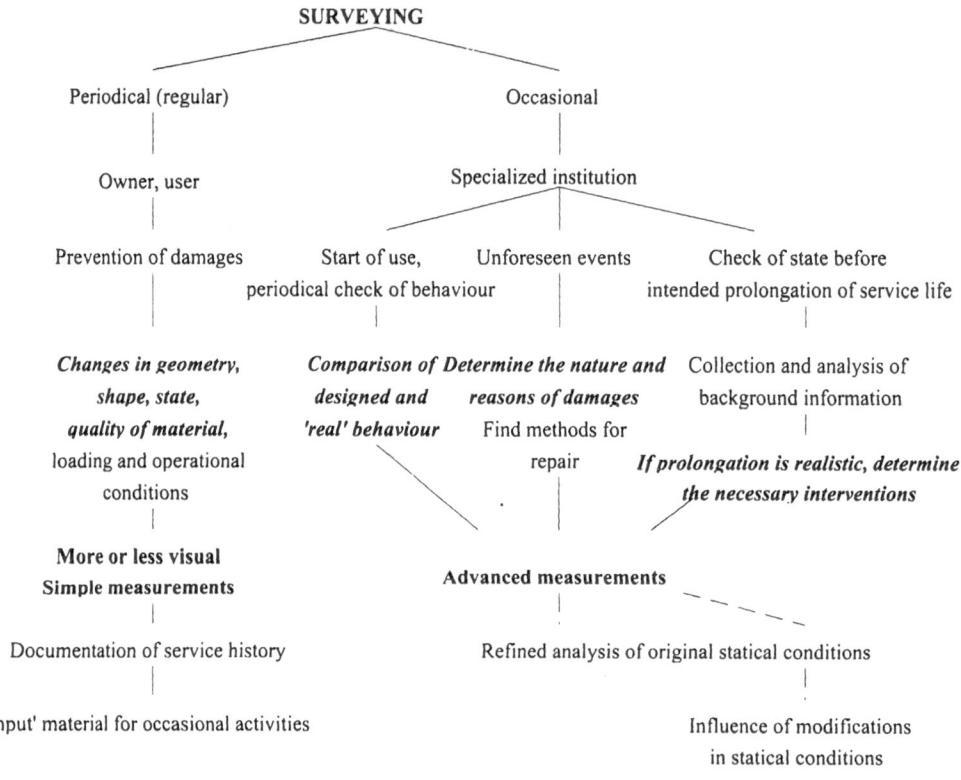

Figure 2.1. Surveying activities

Depending on the situation, the most important tasks of surveying can be the followings:

- increase or assure the reliability of operation,
- prolong the service life,
- find the reasons or effects resulting structural defects,

2.3 Backgrounds for Surveying

Surveying is generally carried out according to the technical standards and other regulations being in force at the time of the activity. Those ones from the design and construction period of the structure are of informative importance.

At the beginning the aims and purposes of the process should be clearly determined and agreed by the user or owner.

Although regulations prescribe the frequency of regular inspections and the basic maintenance requirements, it is a rather negative experience, that they are not realised. Similar rules state, that the original documentation and those of the intermediate interventions, modifications and reconstructions are to kept together. In most cases they all are not available together and this way, the history of the construction is not complete.

During the introductory period of surveying, efforts are to be done to collect all materials available in order to have a comprehensive knowledge about the structural history.

Most efficient background materials of surveying are:

– original design documentation (statical calculation, structural and technological plans, files of construction and erection period),
– standards being in force in the design period,
– reports of load tests, if any,
– protocol of taking into service,
– documentation of maintenance, functional changes, repairs, reconstruction, strengthening, etc.

2.4 Methodology of Surveying

As shown in Figure 1, surveying activities can be classified into two main groups:

– periodical (regular)
– occasional

The intervals of periodical (regular) surveying are given in the corresponding standards in one to three years, depending on the importance of the construction. More detailed checks should be carried out in every five - ten years. As these later ones generally require to use more skilled methods and techniques, they are mentioned in the second group.

An other respect, which indicates this separation, that

– in the first case the surveying activity is carried out by the owner or the user of the structure, it has first of all a preventive nature, it uses simple methods;
– while in the one a specialized institution is making the supervision, using advanced techniques and methods.

2.5 Periodocal (Regular) Surveying

Because of its preventive nature, periodical surveying contributes to preservation of operational reliability, decreases the maintenance costs and helps to extend the service life of the structure.

The most important checks, carried out in this process, are concerning to:

- changes of overall and local geometry, detection of distortion, vibrations, etc.
- general state of structural elements, connections, supporting structures (bearings, foundations)
- corrosion damages either in steel or in concrete,
- any kinds of other damages (loss of sections, rails, development or increase of cracks, etc.),
- changes in technological or loading conditions influencing the safety of the structure.

In this way the general physical state of the structure and its associated parts can be determined.

In case of any disadvantageous experience additional inspections are necessary, which are to be carried out by a specialised institution.

2.6 Occasional Surveying

Occasional surveying has three main tasks:

- In case of important constructions at the start of use and later on in a greater period a systematic comparison of the designed and the real (or the actual and anticipated) behaviour is compared, generally in form of load tests.
- In case of any unforeseen events (accidents, collisions, failures, etc.) a systematic check to measure the amount and influence of damages should be carried out to be able to determine the preventive actions.
- If the service life of a structure is to be prolonged, its state and suitability for the further use can be estimated.

2.7 Instrumentation of Surveying

The instrumentation is varying according to the tasks of the surveying. The following list is containing the most important groups of measuring devices used on the different levels of surveying:

- instruments to measure the overall geometry and geometrical data
- instruments to measure losses and damages of the structure
  - material and joint defects
  - changes of the structural behaviour of the material
  - corrosion losses
- instruments to measure static and dynamic response of the structure
  - temperature conditions
  - measurement of stresses, strains,
  - measurement of deflections, displacements
  - dynamical properties, eigen frequencies, amplitudes of vibrations
  - measurement of load intensities
- instruments to save and store the measured data
- instruments to analyse and evaluate the measured data

3 Damages in Steel Bridges

3.1 Fatigue of Thin-Walled Plated Structures

3.1.1 Critical and Post-Critical Behaviour. A new aspect is coming into being in the methods of stability analysis of steel and metal construction [Massonnet, 1977] and [Halász, 1977]. The classical stability analysis based its methods of calculation on the occurrence of bifurcation of equilibrium and the "critical" load parameter connected with the latter or on determination of stresses (P_{cr}, σ_{cr}). These quantities can be computed by exact or approximating formulas in closed form in easier instances and by the help of properly elaborated algorithms.

The content of information of critical load parameter is strongly limited. Essentially it only refers to the fact that unavoidable deviations (for example geometric inaccuracies) – which can be designated as "initial imperfections" – between real structure and model taken for calculation's basis, under the effect of critical load parameter, influence the result of calculation only to a limited degree; on the other hand, this statement is not valid in the vicinity of critical load parameter.

The non-linear examination of occurrence of bifurcation [Thompson and Hunt, 1973] [Gioncu and Ivan, 1978], for general case Figure 3.1 and for special cases Figures 3.2 – 3.4, which takes the occurrence to stable, unstable and asymmetrical cases, it enables the effect of deviations from ideal model of calculation to be better estimated - unfortunately only within the elastic region. The followings signify peculiar complications in case of local buckling of a plate.

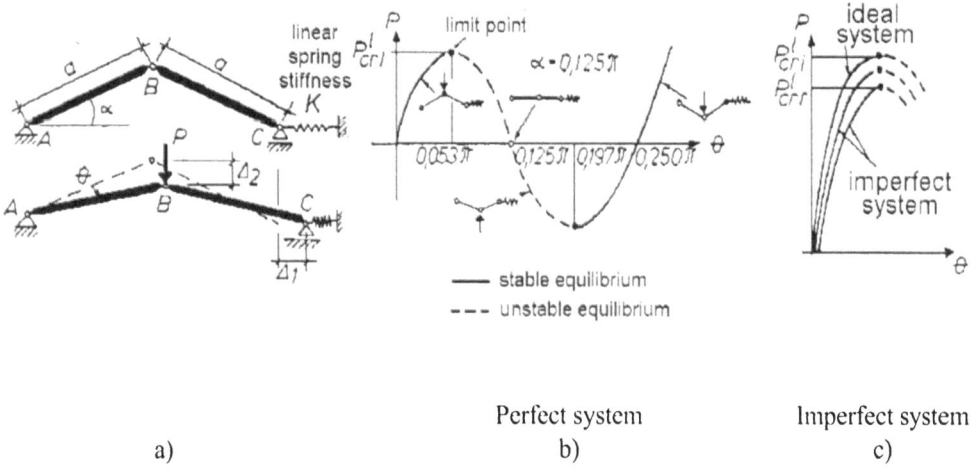

Figure 3.1. Limit point instability

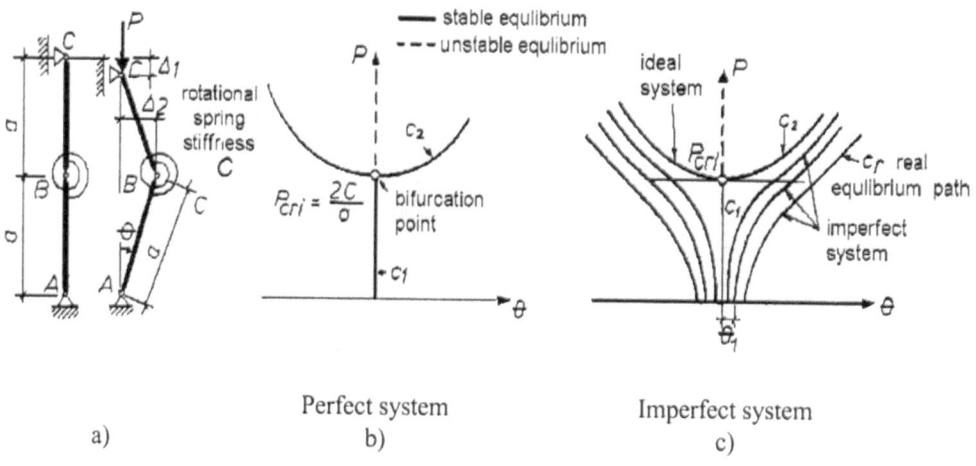

Figure 3.2. Symmetric stable post-buckling behaviour

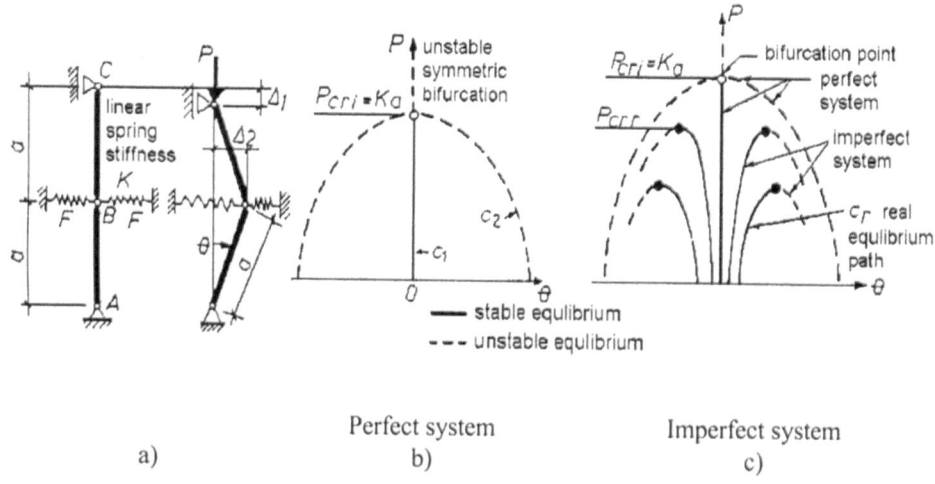

Figure 3.3. Symmetric unstable post-buckling behaviour

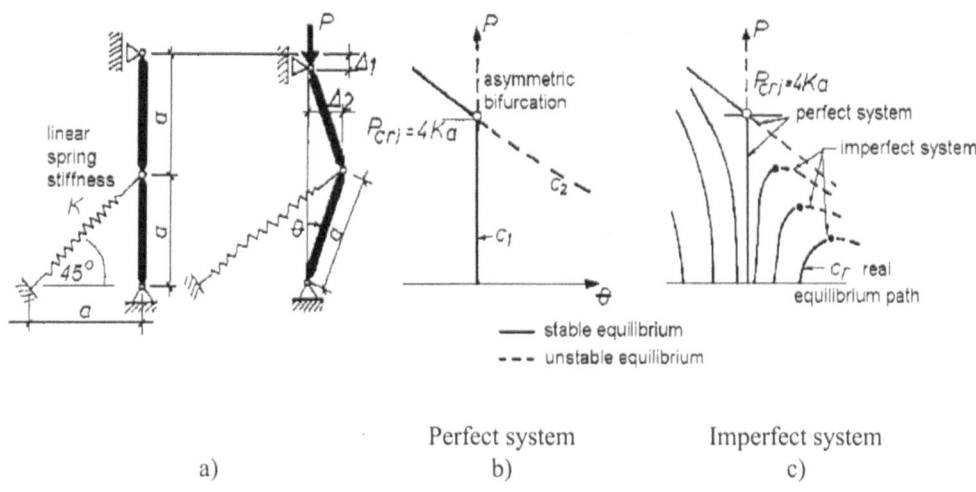

a)   Perfect system   Imperfect system
                                     b)   c)

Figure 3.4. Asymmetric post-buckling behaviour

This kind of bifurcation belongs to - in most cases - the kind, stable group: Because of it, there is an increasing post-critical way. Force-displacement diagrams of plates having initial disturbance (crookedness) are of the same type. Ultimate state of loading capacity at $\sigma_k = P_k/A$ average stress can be attributed in this way only to appearance of plastic regions and eventually to plastic instability. However, relation of σ_{cr} and σ_k is varying. In case of a thick plate (similarly to case of a rod): $\sigma_k < \sigma_{cr}$; reversely in case of a thin plate $\sigma_k > \sigma_{cr}$; moreover, $\sigma_k \gg \sigma_{cr}$ is possible (Figure 3.5). In this way the magnification factor is invalid, furthermore σ_{cr} critical stress as base of comparison also loses its significance a lot [Skaloud, 1978]; real loading capacity only can be characterized by the analysis of post critical behaviour.

In this manner the calculation of loading capacity of a plate loaded by "initial" disturbances and doing finite displacements requires a non-linear, elastic-plastic analysis.

Starting formulas of this scope of problem – for instance Kármán's non-linear buckling equations established in 1910 [Kármán, 1910]:

$$\Delta\Delta w = \frac{v}{D}\left(\phi_{yy} w_{xx} + \phi_{xx} w_{yy} - 2\phi_{xy} w_{xy}\right)$$

$$\Delta\Delta\phi = E\left(w_{xy}^2 - w_{xx} w_{yy}\right)$$

(where W is deflection, ϕ is Airy's function of stress), which were extended to imperfect plate in loaded condition by [Marguerre, 1937] and also to orthotropic by other authors

[Maquoi, 1971],. and which also can be constructed within the plastic region [Merisson et al., 1974] – is known and certainly they are solvable with adequate computer technical instrument.

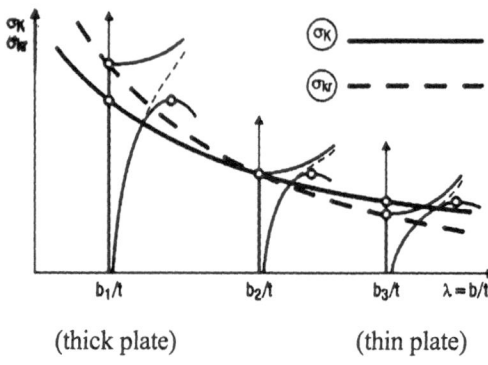

(thick plate)   (thin plate)

Figure 3.5. Critical and Post-Critical Behaviour of Plates

Nevertheless, design methods based on "exact" following in the wake of ultimate state of loading capacity, apart from special instances, do not seem to be practical because of the followings:

(i) Scattering of certain fundamental parameters (crookedness, residual stresses, etc.) is relatively high and their statistical characterizations are mainly based on estimations: thus their inaccuracies do not fit in with the method's mathematical accuracy. Moreover, magnitude of initial geometric disturbances, and besides the variety of their possible shapes and diverse effect of their shapes cause more complication.

(ii) An examined plate (plate strip, plate range) is usually one bridge element (generating plate, flange, web-section) of some complete construction. Thus in so far as the behaviour of a bridge element can be only defined by large mathematical apparatus and usually only through numerical way, analysis of a complete construction becomes difficult. Hence the complicated analysis of local instabilitical occurrence is hardly insertable in the compasses of global examination of the entire structure (Figure 3.6).

(iii) The real behaviour of structural elements supporting a plate considerably and dominantly influences local buckling of a plate and mainly the post critical condition: the phenomenon only can be analyzed through the simultaneous examination of the plate and its "flanging". The problem becomes unusually embarrassing if the equilibrium of bordering elements can show bifurcation as well - for example there is a possibility of simultaneous occurrence of local buckling of web and lateral-torsional buckling of a beam, of local buckling of a plate and buckling of a stiffener, of local buckling of a generating plate and column buckling of a rod. In this case not only the mathematical difficulties increase but also the coincide or approach of critical load parameters can significantly modify the essence of the phenomenon an separately pairings of "benevolence" cases produce "malevolence" ensembles [van der Neut, 1969].

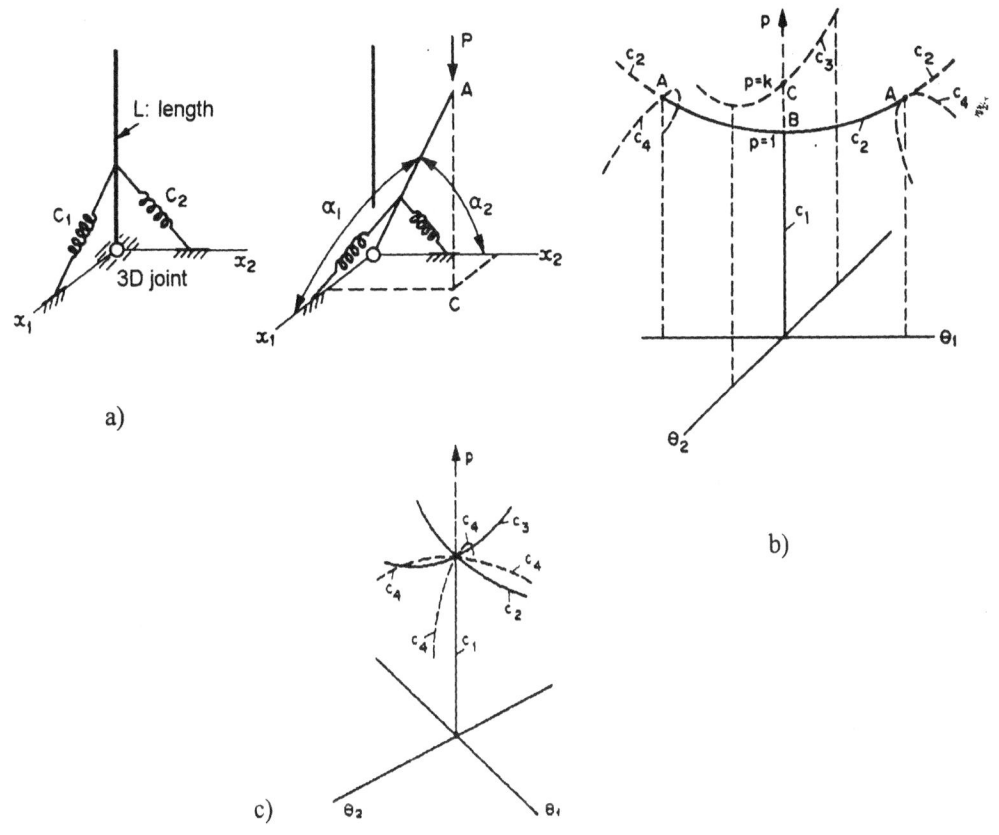

Figure 3.6. Interaction problems of bifurcation

Because of the difficulties above, instead of exact models, generally "target models" are put into practice that conform to practical demands, are valid in limited scope, an supply limited information. Because of it, function of experimental investigation is increasing, besides it serves as physical basis (and many times inspiration) necessary for creation of "target models", besides, it makes the termination of round of validity possible [Halász and Iványi, 1985]. The later is especially productive if - through international division of research work - an opportunity of relatively numerous investigations arises. It is interesting to mention that in the seventies, the validity of eight target models elaborated for the examination of one of the acute problems of steel bridge building, the local buckling of an orthotropic floor plate was controlled by 105 rather large-scale (in this way extremely expensive) tests. During these experiments, these experiment considerations of mathematical statistics and probability theory were also applicable [Dowling and Chatterjee, 1977]

The applied target models essentially can be divided into two groups. The first group is suitable for examination of plates that are rather this and in this way having essential post critical reverse; and common feature of them is that they only take the membrane stresses for their basis during the calculation process of loading capacity, assuming those act henceforward in the original medium plane of the crooked plate. The starting thought originates from [Kármán et al., 1932] in case of compression plates and from [Wagner, 1922] in case of shear plates. The rather productive idea of the former was the introduction of the notion of post critical "effective plate width" (Fig. 3.7.a). With use of

$$\frac{b_e}{b} = \sqrt{\frac{\sigma_{cr}}{\sigma_{max}}} \text{ and } \frac{b_e}{b} = \sqrt{\frac{\sigma_{cr}}{\sigma_{max}}}\left(1 - 0.25\sqrt{\frac{\sigma_{cr}}{\sigma_{max}}}\right)$$

- formed simple formulas which were created by the original one and by Winter on experimental basis (and thus reflecting the effect of "initial disturbances"), it affords a possibility of simple description of membrane stress rearrangement, the occurrence of yielding at plate edge and of "crease" (post critical loss of loading capacity) resulting from this kind of yielding. Interpretation of experimental examination of compressed structures built up from quite thin plates is based upon this idea; common examination of column buckling and local buckling of compression rods of box cross sections [Maquoi and Massonnet, 1967], [Skaloud, 1967], simultaneous analysis of local buckling of web and lateral-torsional buckling [Reis and Roorda, 1977].

Wagner's idea of post critical examination of a shear plate - accordingly in post critical state, a "tension strip" founded on inclined membrane stresses carries shear - has been extensively put into practice at structures on the basis of [Basler, 1961]'s suggestions and after a widespread experimental preparation it has developed into the new discipline of stiffened web plates' examination of local buckling. The summary of Dowling and Chatterjee (1977) has to be indicated in point of remarkably extensive examinations, emphasizing certain more important publications [Massonnet et al, 1977], [Rockey and Skaloud, 1972]. One of the accepted target models of post critical loss of loading capacity calculation is shown in Fig. 3.7.b.

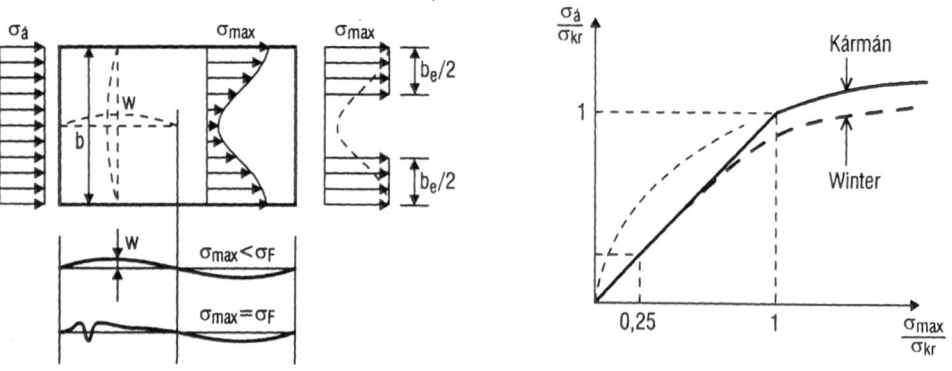

Figure 3.7.a Effective plate width (compression)

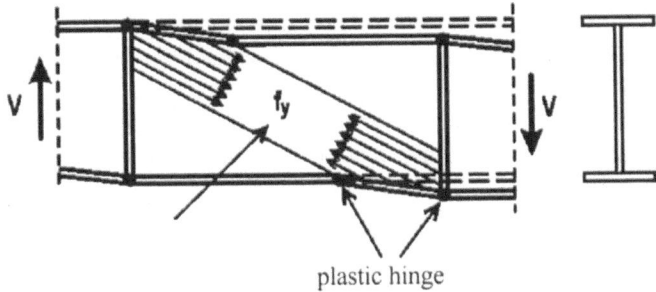

plastic hinge

Figure 3.7.b Effective plate width (shear)

The <u>second group</u> of target models aimed the description of behaviour of relatively thick plates. Since in these cases the reverse of post critical loading capacity is relatively small (possibly there is not at all), practically local buckling comes into being with yielding at the same time. The fundamental question is the characterization of strains after local buckling in order that the simplified strain regularities of a structural element with "buckling", increasing or decreasing loading capacity could be used for the analysis of the entire structure. The same is suitable for fixation of criteria for election of thickness of a plate that makes the examination of occurrence of local buckling omissible.

A feature of these target models is that they take also the function of bending moments occurring in a crooked plate into consideration besides membrane stresses. Simplification of examination can be done by the help of rigid-plastic "yielding mechanisms" - also applied at the plastic examination of loading capacity of rod assemblies (Fig. 3.8) [Iványi, 1979a and 1979b], [Ivány and Skaloud, 1995]. On the basis of these mechanisms, at least the region of decreasing loading capacity, the strain condition of a plate can be characterized by a simple relation. These models ca be extended to cases of elastic-strain hardening relations besides ideally elastic-plastic law of material.

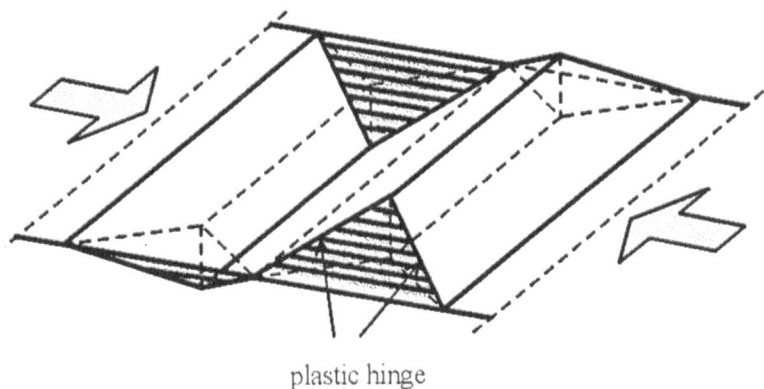

plastic hinge

Figure 3.8 Yielding mechanism

3.1.2 Serviceability – Fatigue. Although serviceability and fatigue represent two completely different limit cases, they are connected in terms of postcritical plate behaviour. Referring to Fig. 3.2 the postcritical domain is characterized by a marked increase of deflections w. Indeed only such deflections of the order of the plate thickness allow a redistribution of the membrane stresses and corresponding increase of ultimate load [Dubas and Gehri, 1986].

For very slender panels, the deformation occurring under service loads may therefore be relatively important and even unacceptable, although safety against collapse is adequately guaranteed by the postcritical strength reserve. For bridges, functional, constructional and architectural aspects may therefore prevent a full utilization of the ultimate strength. In some cases allowable deformations occurring during service will be determined by agreement between the owner and contractor. For structures subjected to repeated live loads, especially for highway bridges and particularly for railway bridges, the plate bending stresses accompanying these deflections may induce fatigue at the plate boundary. Panels are not hinged at their connection with the flanges or with the stiffeners. When calculating ultimate strength this classical assumption of simply supported edges is generally safe. For fatigue, however, the torsional stiffness of the flanges, or the continuity over the stiffeners, cause a restraint at the boundary which induces plate bending stresses in the connecting welds. It is now well known that the fatigue strength of transversally loaded fillet welds is very low. Even small stress differences may induce a fatigue failure if the number of stress cycles is high enough (Fig. 3.9).

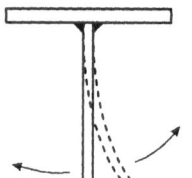

Figure 3.9 Web breathing

This problem of "breathing" or "oil canning" of slender panels subjected to repeated loading is not fully solved at the present time. One aspect is the calculation of the bending stresses in the deformed state, for which some solutions have been presented [Maeda and Okura, 1983] [Maeda and Okura, 1984]. However, the live loading to be introduced and its distribution during the service life of the structure are not well known. At the moment the simplest solution consists of limiting the slenderness of the panels but this results in discarding a proportion of the postcritical reserve. This approach considers only one, important factor. Depending on the load-histogram, the results may be too conservative and therefore uneconomical.

3.1.3 Fatigue of Plated Girders. As was summarised by Okura et al. (1993) when a thin walled plate girder is subjected to repeated loading, there are possibilities of initiation and propagation of fatigue cracks at the fillet welds connecting the girder web to the compression and tension flanges or the vertical stiffeners to the web. As shown in Fig. 3.10, the fatigue cracks are classified as depending the different loading conditions.

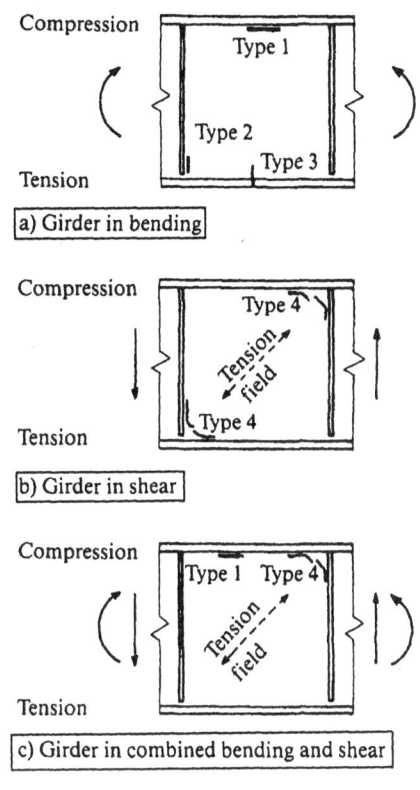

- **Type 1** crack occurs at a toe on the web side of the fillet weld to connect the web to the compression flange.

- **Type 2** crack is observed at a toe on the web side of the fillet weld connecting the vertical stiffener to the web.

- **Type 3** crack is initiated at a fillet weld to connect the web to the tension flange.

- **Type 4** cracks are created at toes on the web side of the fillet welds near the corners where a diagonal tension field is anchored.

a) Girder in bending

b) Girder in shear

c) Girder in combined bending and shear

Figure 3.10 Thin-Walled Plate Girders

(a) Bending
Figure 3.11 a and b show the relationship between α and β, and between w_{0max} / t_w and β, respectively, for the web panels of test girders in bending [Okura et al., 1993], where

$\alpha = a / b$ — rectangular plate aspect ratio,
$\beta = b / t_w$ — web slenderness ratio,
t_w — web thickness,
$w_{0,max}$ — maximum initial deflection.

Fig. 3.11.c shows the relationship between $\Delta\sigma_0$ and β for the web panels of the girders. The plate-bending stresses $\sigma_{b,max}$ and $\sigma_{b,min}$ are caused by the maximum and minimum in-plane bending stresses $\sigma_{0,max}$ $\sigma_{0,min}$, respectively. The range of inplane bending stresses $\Delta\sigma_0 = \sigma_{0,max} - \sigma_{0,min}$ / $\sigma_{0,max}$.
The allowable fatigue stress ranges for Types 2 and 3 cracks, which specified in the AASHTO Specifications (1989) are also drown in the figure for 2 million cycles.

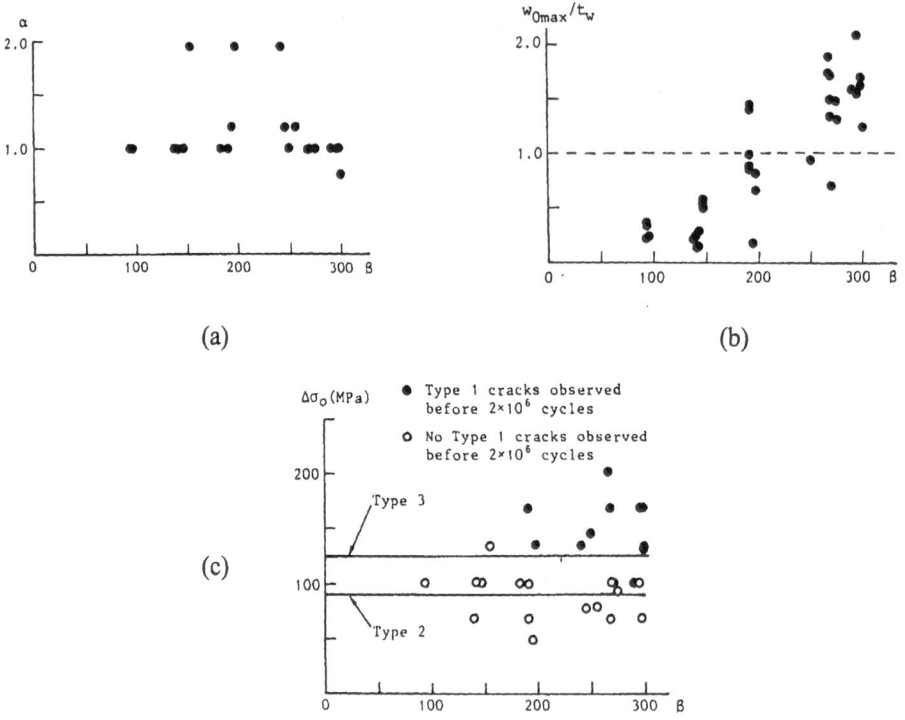

Figure 3.11 Fatigue in case of bending

As can be seen from Fig. 3.11.c Type 1 cracks do not occur below the allowable fatigue stress range for Type 2 cracks. This implies that the presentation of Type 2 cracks results in that of Type 1 cracks. The assessment for Type 2 cracks is always necessary in the fatigue design of vertical stiffened plate girders [Fisher, 1977].

(b) Shear

Figure 3.12 a and b show the relationship between α and β, and between w_{0max} / t_w and β, respectively, for the web panels of test girders in bending [Okura at al., 1993]

Fig. 3.12.c shows the relationship between $\tau_{0,max} / \tau_{cr}$ and β for the web panels of the test girders. In the Figure, Type 4 cracks were observed for the solid circles before 2 million cycles. No Type 4 cracks were found open circles. Only one solid cycle falls below the horizontal line of $\tau_{o,max} / \tau_{cr} = 1.0$. Moreover, there exist some open circles between the horizontal line and the upper solid circles. Thus, the shear buckling strength τ_{cr} for the simple-supported condition along four edges provides a conservative estimate for presenting Type 4 cracks.

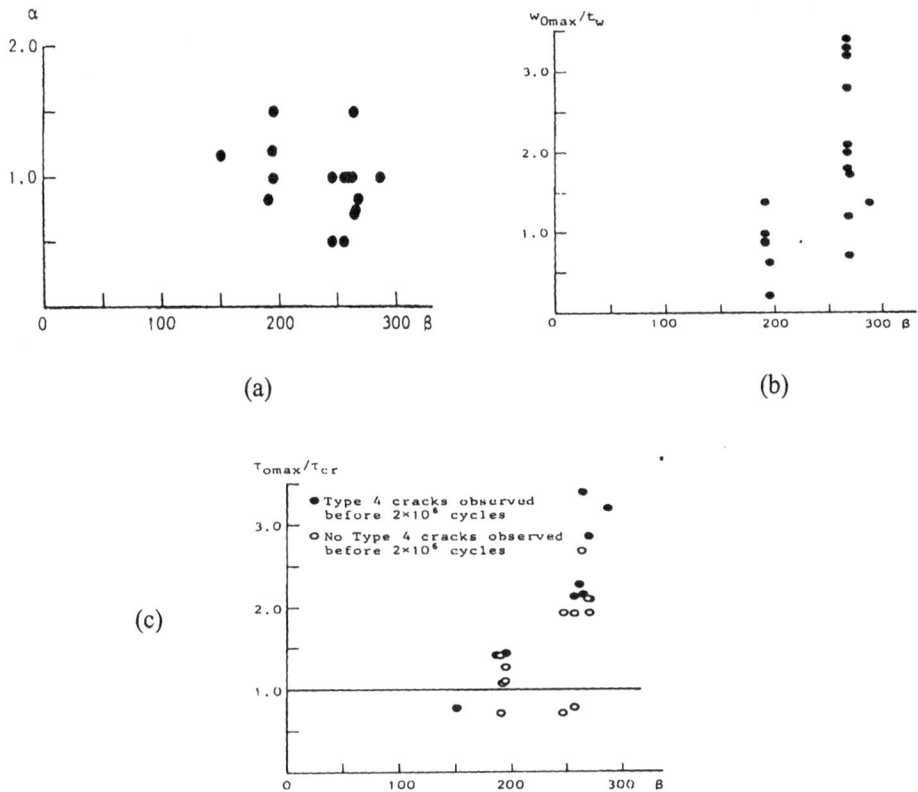

Figure 3.12 Fatigue in case of shear

(c) Combined Bending and Shear

Fig. 3.13 shows the relationship between $\tau_{0,max} / \tau_{cr}$ and $\Delta\sigma_0/89.63$ for the web panels of the test girders under combined bending and shear. Here, $\tau_{0,max}$ is the maximum applied in-plane shearing stress, τ_{cr} is the shear buckling strength, and $\Delta\sigma_0$ is the range of applied in-plane bending stress. $\Delta\sigma_0$ is divided by 89.63 MPa which is the allowable stress range at 2 million cycles specified by the AASHTO Specifications [AASTHO 1989] for Type 2 cracks. It can be seen that cracks occur in the web panels with larger values of $\tau_{0,max} / \tau_{cr}$ and $\Delta\sigma_0/89.63$. Since there are so few test results for web panels subjected to combined loading around ($\tau_{0,max} / \tau_{cr}$, $\Delta\sigma_0/89.63$) = (1.0, 1.0), it is difficult to draw a conclusion on the prevention of cracking of plate girders in combined bending and shear.

For vertically stiffened plate girders, the factors needed to prevent fatigue cracks, due to out-of-plane deformation of the web were introduced for bending and shear loadings.

For girders in bending, Type 1 cracks due to out-of-plane deformation of the web do not occur below the allowable fatigue stress range for Type 2 cracks which are classified in the AASHTO Specifications [AASHTO 1989]. Hence, fatigue assessment for Type 1 cracks is automatically accomplished by the fatigue assessment for Type 2 cracks which is inevitable for vertically stiffened plate girders. For girders in shear, the shear buckling strength for a rectangular plate with the simply-supported condition along the four edges gives a conservative guideline for the prevention of Type 4 cracks. Thus, it is recommended that to prevent Type 4 cracks, the maximum applied in plane shearing stress should be below the shear buckling strength for the simply-supported condition along four edges. These conclusions are applicable to thin-walled plate girders with web slenderness ratio not greater than 300.

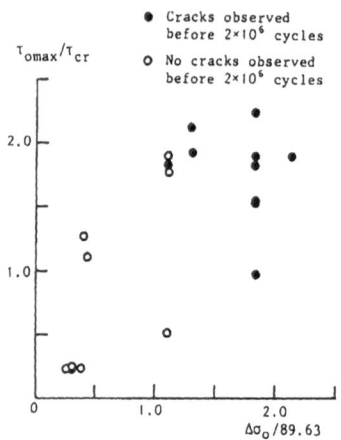

Figure 3.13 Fatigue in case of combined bending and shear

3.2 Localised Distortions of Bridge Elements

3.2.1 Steel Highway bridges fatigue cracks
Case studies from the USA.
In the last years many welded steel bridges, most of them relatively new, have cracked. Cause is lateral deflection of main-girder webs, caused by lateral loads imposed by connecting lateral beams and diaphragms or cross-bracing [Fisher and Mertz, 1985].

Fortunately, none of the cracks has led to collapse of a bridge, and in no case has a bridge had to be closed or load-limited. Relatively simple corrective steps appear to cure the problems, and new steel bridges are being designed to preclude repetition of the problems.

Bolted and riveted connections are less susceptible to such cracks than are welded ones. Many of the affected bridges underwent a relatively small number of stress cycles before significant crack development was observed. In a few special cases, cracking occurred before the bridges were put in service, indicating the cracks formed during shipping and handling.

Writing in 1940, W. M. Wilson divided steel railway bridge stresses into two categories, those due to load and those to deformation of the bridge or a member. Common terminology for the two classes is primary and secondary stresses.

Much research has been done on fatigue cracking of details in welded steel bridges due to load, but relatively little directly on displacement-induced or secondary fatigue of welded details.

Most such cracks have been found in the negative moment area of continuous-girder bridges (that is, over the piers) and only near the tension flange. That flange is constrained by the bridge deck above, and the same girder's web immediately below is freer to move laterally.

References on this subject, together with a collection of case studies are presented in J.W. Fisher's books [Fisher, 1984] [Fisher, 1977] and the paper of Castigliani et al. (1986). These books suggest that the most common kinds of fracture detected up to date in steel bridge members are the result of secondary stress states induced by scheme eccentricities and/or out of plane displacement components.

In a bridge structure, the interaction among longitudinal and transversal members does not heavily influence the global behaviour of the structure, and designing the single members for in plane actions and bending seems to be adequate. Some times, however, serious damage (growing in time) can be caused by localized distorsions induced by that interaction.

The transversal distribution of the loads causes, in fact, relative displacements (and/or rotations) between the longitudinal girders, opposed by the transversal members.

As a consequence, a stress state is induced at the principal to secondary members connections, the effects of which may be locally amplified by sudden variations of the stiffness of the connected members (Fig 3.14).

Web gaps between tension flange and cut short stiffeners (Fig. 3.14. a,b) sharp variations of the cross-sectional properties (Fig. 3.14.c), design and/or accidental eccentricities, web gaps between transversal stiffeners and gusset plates (Fig. 3.14.d) are typical details where fatigue fractures can develop. In fact, because of the sharp variations in the connected members stiffness, these locations represent weakness areas in the joint; as a consequence, localized strains take place (Fig. 3.15) that give raise to relevant stress concentration.

This kind of fracture was detected in bridges of different typologies [Fisher, 1984]; fractures were more frequent in welded structures, where the presence of the weldings in zones of relevant stress concentration, made crack initiation easier. The same problem was also detected in riveted and/or bolted structures, but in this case, because of the smaller mechanical imperfection and the greater adaptability of the connections, the cracks developed more slowly.

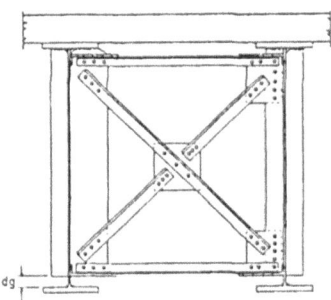
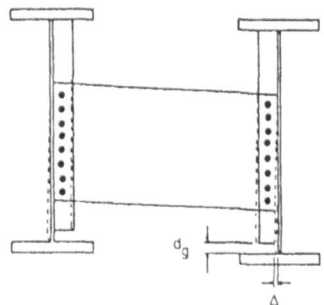

(a) Web gaps between tension flange   (b) Cut short stiffeners

Figure 3.14.a and b Localised distortion of longitudinal and transversal members

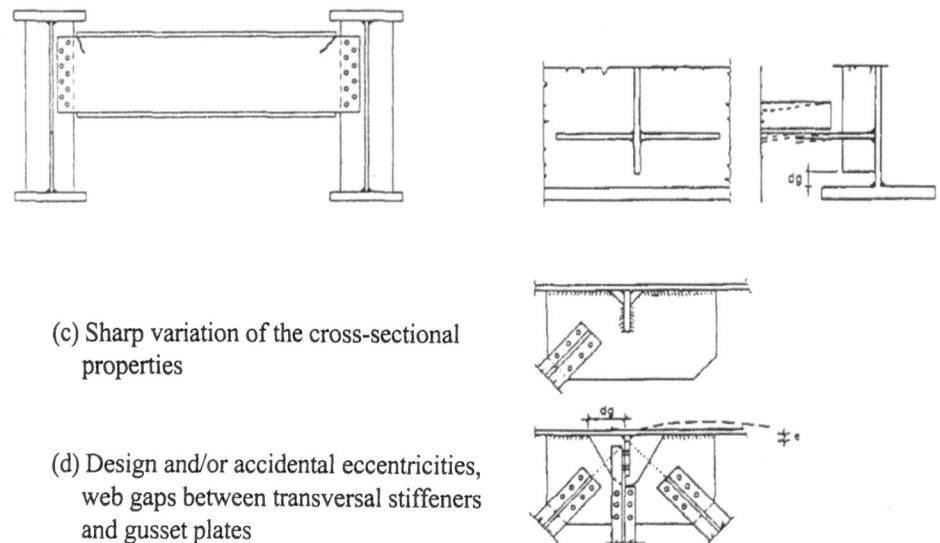

(c) Sharp variation of the cross-sectional
    properties

(d) Design and/or accidental eccentricities,
    web gaps between transversal stiffeners
    and gusset plates

Figure 3.14.c and d Localised distortion of longitudinal and transversal members

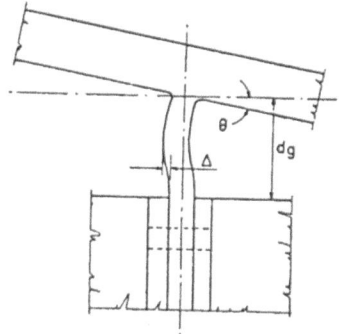

Figure 3.15 Localised strains give raise to relevant stress concentration

(a) Explanation of Cause. The secondary stresses are caused by bridge members moving in three dimensions. That is, not only does the girder move up and down due to live loads, but portions of it move laterally. This lateral movement or out-of-plane distortion is usually caused by lateral bracing or transverse beams. (Fig. 3.16.) [Fisher and Mertz, 1985].

This cracking has been observed in every kind of steel bridge, especially those with welded stiffeners, connection brackets or transverse beams.

Cracks form mainly in planes parallel to the longitudinal axis of the bridge. So they have not been detrimental to the structure's performance in cases where they were discovered and corrective action taken before the crack turned perpendicular to the original direction. In some bridges the cracks stopped growing when they had grown into low-stress areas and thus the crack now served to relieve the restraint condition.

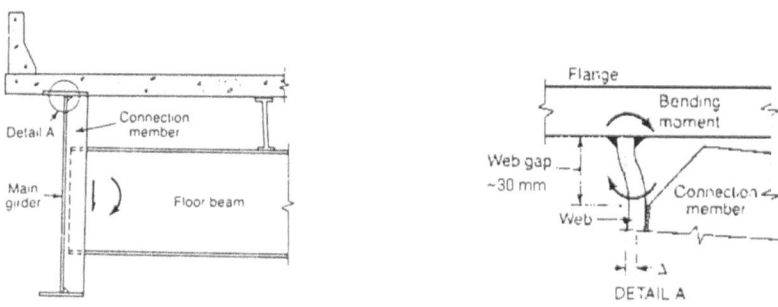

Figure 3.16 Web distortion

One of the first-discovered and most common sites of fatigue cracks from distortion is in the web gaps at the ends of floor-beam connection plates. These cracks occur in the small gap in the web between the end of the connection plate and the girder flange. That is, only in cases where the connection plate is not connected to the flange. (This situation is ironic, for the connection plate was purposely not connected to the flange to avoid another type of fracture-load-induced fatigue cracks of tension members at welded connections.)

The magnitude of out-of-plane movements depends on girder spacing, bridge skew, and type of diaphragm, or cross-frame. The cracking has occurred in skewed, curved and right bridges, but more severely and earlier in the first two types.

(b) Correcting the Problem. The problem can be attacked in two ways: by minimizing the movement or accommodating it [Fisher and Mertz, 1985].

Three techniques are used: Drill a hole at each end of the crack. Or remove a segment of the connection plate near the crack to lengthen the web gap. Or bolt the connection plate to the tension flange in the bridge's negative moment areas.

1. Drill holes at crack ends. When a fatigue crack forms in the short web gap, the gap's flexibility in accommodating out-of-plane displacement is increased, as the crack provides a horizontal slot.

Thus the displacement-induced stress is reduced. When it has been reduced sufficiently, drilling holes at the crack ends or tips will often stop further crack growth. Fig. 3.17 shows schematically the cracks, retrofit holes, and resulting displacement due to a through crack.

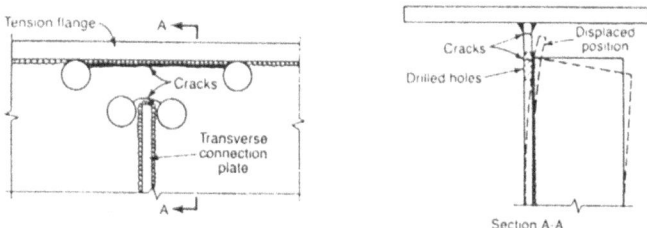

Figure 3.17 Retrofit holes

2. Lengthen the web gap. In some cases the cyclic stress at the retrofit holes may not be reduced enough, and cracking may re-commence from the holes. In such situations more involved corrective steps are needed. To increase the flexibility of the web gap, a portion of the connection plate (stiffener) adjacent to the web gap can be removed. See Fig. 3.18.

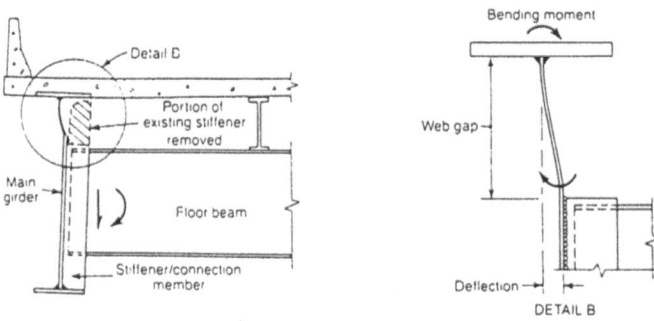

Figure 3.18 Lengthening the web gap

This lengthens the web gap, thus reducing the out of plane bending stress. In many cases this is enough to insure that cracks do not re-initiate from the holes drilled at the crack tips.

To remove a portion of the connection plate, 2 in. (50 mm) diameter holes are drilled or cut by hole saw in the connection plate, tangent to the girder web. The hole is placed at the depth of the stiffener to be cut back. The stiffener is then removed by cutting as shown. The rough surfaces on the two cut surfaces are ground smooth. Liquid penetrant inspection is used to identify any cracks in the web gap. Holes are drilled at the crack tips to prevent crack growth.

High-strength bolts were installed in the holes and tightened to further enhance the fatigue resistance of the drilled crack tips.

But this step does not work unless the amount of out-of-plane deformations grow no larger afterward. Unfortunately, in some cases these deformations increase as the gap length increases.

3. Attaching connection plate to flange. A more extensive retrofit may be required in cases where removing the transverse connection plates is unacceptable because lateral forces are not displacement-limited. This may be the case at piers where lateral forces are transmitted into the reactions. Here, movement of the web gap is prevented by attaching the connection plate to the girder's tension flange.

This can be done by welding or bolting. Welding generally is avoided for two reasons. First, there is concern regarding the quality of welds that can be provided under field conditions. Second, in older bridges the weldability and fracture toughness of the flange steel is unknown. Hence bolting is the choice.

If the retrofit is done with the concrete slab in place, it may be necessary to remove a piece of it to gain access to the embedded girder flange. Bolt holes are then drilled in the flange and connection plate, and splice angles are inserted, linking flange and connection plate. The resulting connection prevents distortion in the web gap.

Options two and three each have advantages and disadvantages. Bolting is usually more costly, so should be used only where removing a piece of the connection plate will not prevent further crack growth. And removing some of the connection plate is cheaper, but if you select this option it is necessary to go to the bother of predicting displacement-induced stresses in the web.

Even adding a bolted angle between flange and connection plate must be done carefully. In Fig. 3.19 the option with only one row of bolts in the bottom flange is unacceptable – it does not provide full fixity, so may permit some out–of–plane deformations of the web. Thus two rows of bolts should be used.

Further to optimize the connection's stiffness and resistance to distortion, friction-type connections and tightened bolts are essential.

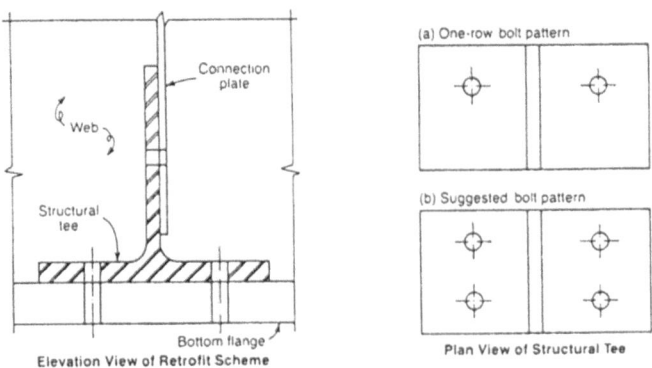

Figure 3.19 Splice angle

3.2.2 Steel Railway Bridge fatigue cracks
Case Study from Sweden

Fatigue cracks in the web at the ends of the vertical stiffeners were noted in the webs of some plate girders only a few years after completion of the welded railway bridges [Åkesson et al., 1997]. These cracks, located at the ends of the vertical web stiffeners, had propagated first horizontally and then gradually towards the neutral axis (Fig. 3.20).

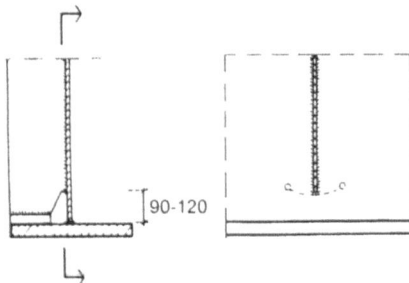

Figure 3.20 Typical fatigue cracks; propagation arrested by holes drilled at crack tips

The cracks were arrested by drilling holes at each crack tip, a method also used successfully for cracks that occurred later. No cracks so treated continued to propagate.

A total of five bridges of the same structural type exhibited 100-200 mm long fatigue cracks at the ends of the vertical stiffeners. These bridges were constructed of two parallel, simply supported welded steel I-beams, with the sleepers set directly on the top flange (Fig. 3.21). The bridge over the Maunojokk River, built in 1961, was selected for more detailed computer analysis. This bridge has a span length of 11.6 m and the track lies in a horizontal curve with a radius of 600 m. The two main girders, 1900 mm apart, are straight, 1000 mm in depth and connected by cross-framing approximately every 2.0 m.

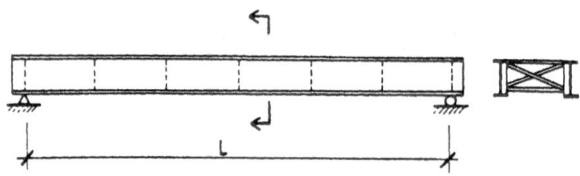

Figure 3.21 Bridge type where fatigue cracks occurred

The web plate of both girders is strengthened on the inside by a vertical stiffener at every cross frame. These stiffeners are not directly attached to the top and bottom flanges. Instead, they are in contact with the flanges through a steel plate fitted at their ends. This steel plate is welded to the stiffener only, and not to the flange, to ensure that the fatigue strength of the lower tension flange is not reduced. For horizontal forces like wind and centrifugal forces, the bridge is stabilized by a horizontal truss at the level of the top flange.

Due to the orientation of the fatigue cracks - almost horizontal and parallel to the main girder stresses caused by the vertical bending of the bridge - it is reasonable to assume that the cracks are the result of transverse bending of the web plate. The detailing of the vertical web stiffeners leads to increased local stress, especially in the region where the web plate is free to deform laterally, i.e., at the gap between the flange and the end of the weld of the vertical stiffener (Fig. 3.23). As Fig 3.22 shows, a small relative displacement p between the lower flange and the upper part of the unstiffened portion of the web at a vertical stiffener is enough to produce a substantial increase in transverse bending stresses in the web plate.

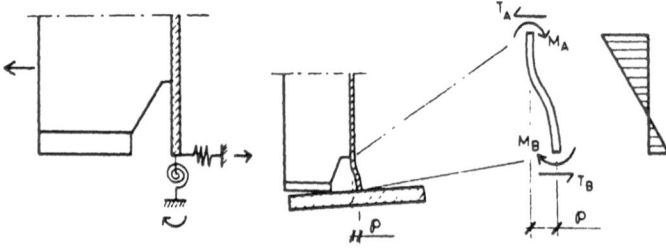

Figure 3.22 The effects of displacement p on the transverse bending stresses in the web plane

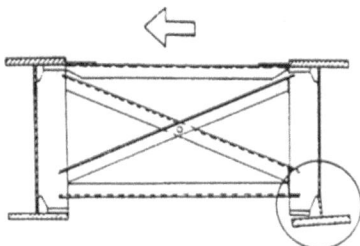

Figure 3.23 Lateral movement of a bridge cross section near mid-span induces an out-of-plane displacement of the web plate

With no direct connection between the lower part of the stiffener and the flange, the lower portion of the web plate must deform by lateral bending because of the difference p in out-of-plane displacement. When the bridge cross section is moved transversely (Fig. 3.23), this deformation is partially restrained by the lateral bending stiffness of the lower flanges. This behaviour is sufficient to produce fatigue cracking, even after only a few years of service, due to a large concentration of stresses in the web plate at the ends of the stiffener weld. Poor welding and inadequate ductility of the steel for today's requirements probably also contributed to the early initiation of fatigue cracks in these bridges. The primary factor, however, was the high localized stresses at the ends of the vertical stiffener welds.

If a horizontal force is applied to the upper flanges, a torque is introduced into the bridge due to the eccentricity between the applied force and the point of rotation. This point is located vertically quite close to the horizontal truss framing system at the upper flange position. This is due to the much higher stiffness of this stabilizing system against horizontal forces in relation to the lateral bending stiffness of the lower flanges.

The torsional moment is primarily carried by the girders due to their far higher bending stiffness and longer lever arm compared with those of the horizontal truss and the lower flanges. This behaviour invites comparison to Fig. 3.23, but under the action of a pure torsional moment the "right-hand" girder in Fig. 3.24 will deflect and rotate without web distortion, since such a distortion is prevented by the vertical stiffener, which stays in contact with the bottom flange. Instead, in this case it is the "left-hand" girder that suffers relative out-of-plane displacement (Fig. 3.24). Due to the rotation, the contact pressure between the vertical stiffener adjusting piece and the lower flange of the "right-hand" girder increases.

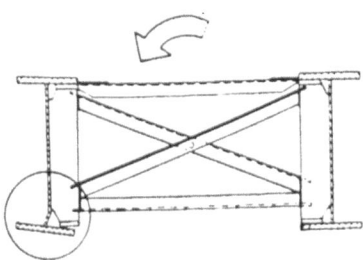

Figure 3.24 Applied torque produces overall rotation, but also relative out-of-plane displacement of the web plate near the bottom flange

A linear static analysis was carried out by modelling the bridge using a finite element analysis program. The horizontal truss and the vertical cross framing system were modelled by 36 bar elements. The bridge over the Maunojokk River has a span length of 11.6 m and in the FE model is loaded with two bogie axles of two adjacent 1000 kN wagons (Fig. 3.25).

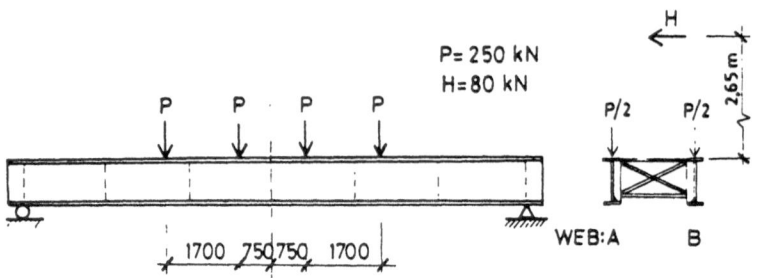

Figure 3.25 Loading positions

The transverse stresses σ_z due to bending around the longitudinal axis x of Webs A and B are summarized in Figs. 3.26 and 3.27. In Fig. 3.26, the transverse bending stresses for Web B in the unstiffened web (gap) region are much higher than the corresponding stresses for Web A. Fig. 3.27 shows that the stress gradient along a longitudinal line is more pronounced for Web B than for Web A.

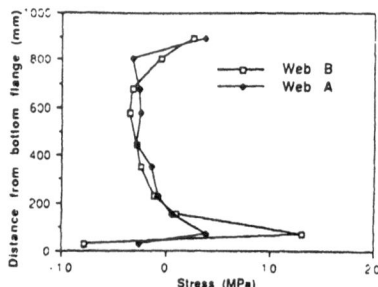

Figure 3.26 Transverse bending stresses σ_z in the web in the vertical direction along the mid-span vertical stiffener of Web A and Web B (original bridge design)

In order to reduce stress concentrations at the ends of the welds of the vertical stiffeners, thus improving structural behaviour, some alternative bracing systems were analysed as well.

Alternative 1: The horizontal bracing; of the as-built bridge consists of six horizontal diagonal truss bars. By adding six additional horizontal truss bars (reversed in comparison to the existing ones) the bracing, system will not only become stiffer (reducing the global displacement), but also prevent "local distortion" at the "insufficiently supported joints," those without diagonals in the original design.

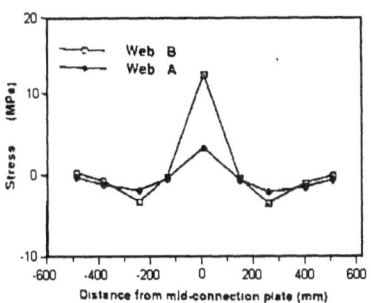

Figure 3.27 The variation of transverse bending stresses σ_z along the longitudinal x-direction at the level of the top of the stiffener gap (original bridge design)

Alternative 2: By removing all the vertical cross-framing diagonal bars between the supports, less horizontal forces are distributed down to the lower flange. Also, the stress concentration effect in the lower part of the web stiffener diminishes due to the loss of stiffness in that area.

Alternative 3: In the original bridge design the web stiffeners between the supports are not attached to the bottom flange in order to avoid welding to a tension flange. In Alternative 3, all vertical web stiffeners are attached to the bottom flange.

Alternative 4: The unstiffened web (gap) in the original design is 100 mm long. In Alternative 4 the effect of changing the gap length on the transverse stresses at the top of the web gap was investigated.

Alternatives 1-3 resulted in more uniform stress distribution, vertically and longitudinally, typified by the results for Alternative 2 in Figs. 3.28 and 3.29. Alternative 4 clearly showed that when the gap length increased, maximum transverse bending stress decreased (Fig. 3.30). However, the reduction was not as pronounced as expected. Probably the benefits from increased flexibility were counteracted by increased out-of-plane displacement due to decreased lateral stiffness.

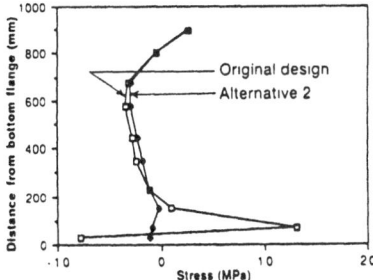

Figure 3.28 Transverse bending stresses σ_z in the vertical direction along the mid-span vertical stiffener of Web B (Alternative 2 compared to original design)

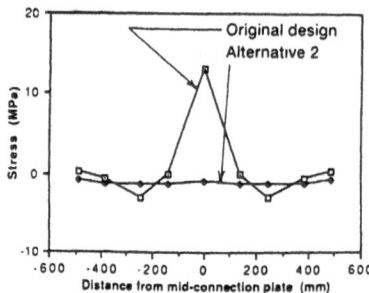

Figure 3.29 The variation of transverse bending stresses σ_z along the longitudinal x-direction at the level of the top of the stiffener gap (Alternative 2 compared to original design)

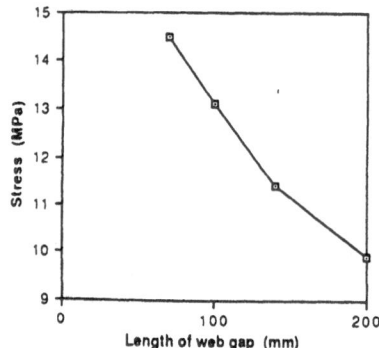

Figure 3.30 Influence of gap length on the maximum transverse bending stresses σ_z at the top of stiffener gap

Conclusions. The addition of six horizontal truss bars in Alternative 1 significantly improved structural behaviour with respect to stress concentrations in the web at the gap region. The design changes of Alternatives 2-3 had the same beneficial effects.

The result of Alternative 4 shows that when the gap length increased, transverse bending stresses decreased. However, the reduction in stress levels was small, so substantially increasing the gap length cannot be recommended. The simplest remedy is to attach the web stiffener to the flange, Alternative 3. To minimize any reduction in fatigue strength of the tension flange, the connection could use high strength friction grip bolts.

The design of the stiffener-to-flange interface without welding the stiffener end to the flanges was regarded as good engineering practice at the time of construction. After a few years of service, however, it was found to be inappropriate for the local lateral deformations of the web plate at the stiffener gap. By trying to avoid one fatigue problem in the flange, a design leading to fatigue cracks in the web was created. A cause of primary fatigue crack initiation was inhibited, but secondary effects caused another type of cracking. This type of out-of-plane displacement-induced fatigue cracking in these bridges can be difficult to foresee.

3.3 Fatigue Damage in Steel Retrofitting Works
Case Studies from Japan

The bullet train systems are the most important trunk lines in Japan. For example, the Tokaido Shinkansen, the first bullet train system completed in 1964, covers a distance of 515 km between Tokyo and Osaka. Nearly 250 trains run daily at the maximum operating speed of 270 km/h. The structures of bullet train systems have been supporting the operation without any fatal accident. But they are showing some signs of deterioration. In particular, the fatigue damage in welding joints of steel bridges attract our close attention. However, no fatal accidents have taken place yet because inspections of these bridges have been routinely carried out, and fatigue cracks discovered were properly repaired. Fatigue damage have developed only in the secondary members not incorporating any fatigue design [Miki and Ichikawa, 1997].

(a) Observed Fatigue Damage. The structures of bullet train systems suffered from very severe loading condition caused by high-speed train operation and highly repetitive frequencies.

Concerning steel bridges, several types of fatigue damage have been observed as shown in Fig. 3.31 [Isoura, 1989]. These types of damage began to be observed in about 8 years after the opening of service. Some types of fatigue cracks are comparatively rare in the conventional railway system. They are in many cases caused by stress concentrations due to structural details of members, by out-of plane displacements occurring between perpendicularly crossing members such as main girders and cross beams or cross beams and stringers, and by vibration due to distortion under high-speed operation of trains peculiar to bullet train system.

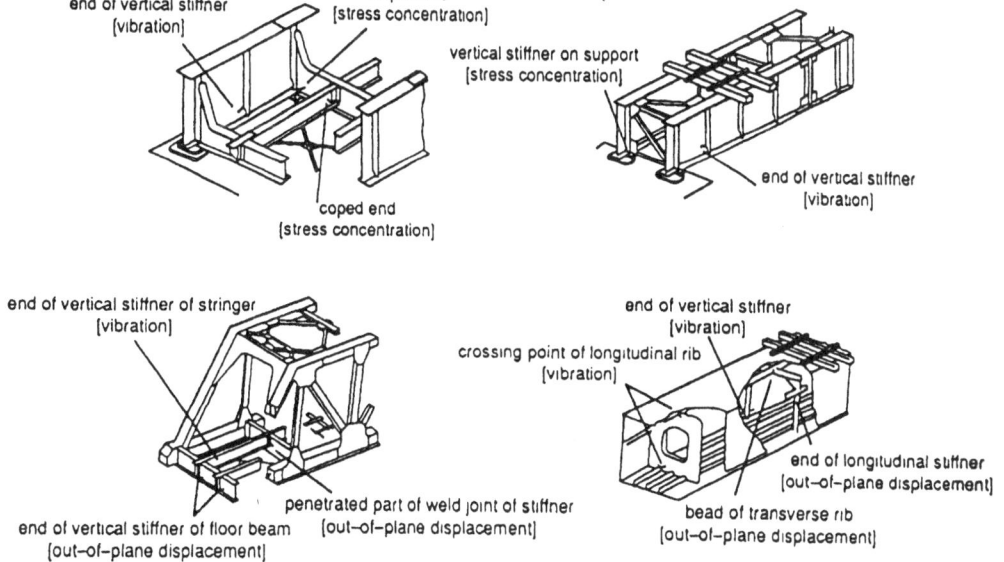

Figure 3.31 Fatigue damage in steel bridges

A fatigue crack has not yet been detected at important portion of main members, which lead to catastrophic failure where fatigue was assessed in the design stage. Fatigue cracks are, however, often discovered at such as secondary members as side walks, connections of attached facilities and diaphragms or secondary local portions of the main members. These types of damage were already repaired and almost of details which had the possibility of occurrence of the same kind of damage were also retrofitted.

(b) Coped Cross Beam of Through Plate Girder. The damage was observed at the coped end of the web plate of cross beam. Many cases of damage were observed in end cross beams. The crack usually develops obliquely from the corner of coped end of web plate to the inside of web plate. This type of damage was discovered in about 8 years after the opening of service.

Measurements of actual bridges, structural analysis and fatigue tests were carried out in order to study causes and retrofitting methods. The results are as follows;

1. The main cause of this fatigue crack was that the end of lower flange of the cross beam was coped to connect with main girder and this induced stress concentration.
2. The measured stress becomes higher in the case of normal shoe seat than in the case of damaged shoe seat. Thus, the settlement of supporting point was also one of the causes.
3. In the retrofitting, additional plates were applied to web plate in order to increase the loading capacity. A sufficient reinforcing effect was obtained.
4. The details of the currently designed bridge was improved as shown in Fig. 3.32.

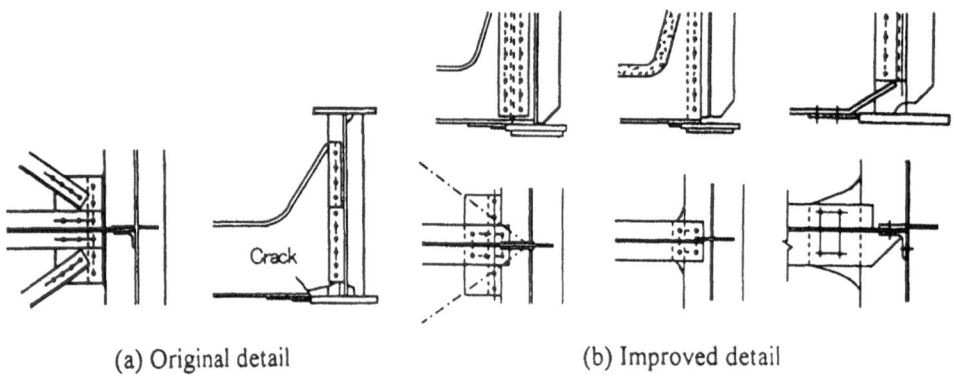

(a) Original detail   (b) Improved detail

Figure 3.32 Improvement of details of cross beam

(c) Coped Stringer of Open floor Type Bridge. A crack occurred at the stringer web plate of through plate girder or through truss, as shown in Fig. 3.33. This type of crack was discovered in about 10 years after the opening of service.

The stringer is a member which is directly subjected to the train load, and to the great lateral force and impact by high speed train. It is necessary to avoid a local stress concentration by improving the lateral rigidity or keeping the stress flow as continuous as possible. However, in some bridges the lower flanges of stringer were not connected to the cross beam. This caused the stress concentration at the cope combined with it and led to cracking. With regard to the repair, the lower flange was connected with the cross beam web plate by extending it to the stringer end.

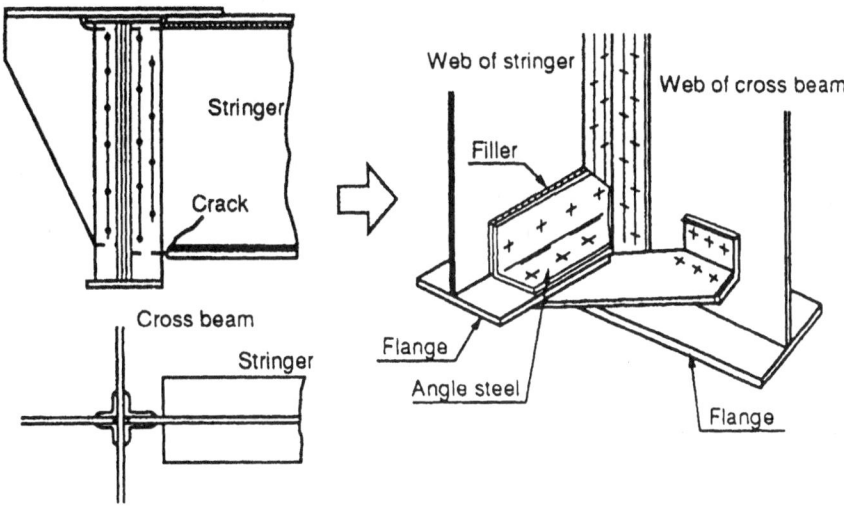

Figure 3.33 Damage of coped stringer

(d) Intermediate Diaphragm in Box Section Deck Plate Girder. In the box section deck plate girder, intermediate diaphragms are provided at intervals of 5 to 6 m in order to improve the torsional rigidity. Diaphragms are connected by welding with flange plates or longitudinal ribs. Fatigue cracks were observed at this detail, on the surface of diaphragm at the toe of fillet weld as shown in Fig. 3.34.

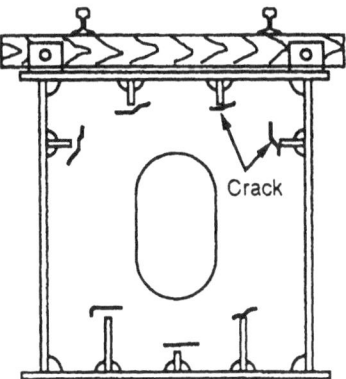

Figure 3.34 Damage of intermediate diaphragm

This type of crack is caused mainly by the stress concentration at the toe of weld and the high structural constraint. Diaphragms with such a structural detail are affected by an out-of plane vibration with the passage of a high speed train and subjected to considerable fatigue at the restrained weld.

This damage was repaired by drilling a stop hole at the crack tip, or rewelding after gouging and the tungsten inert gas arc remelting (TIG-melting) applied at the toe of fillet welds.

(e) End of Vertical Stiffener in Web Plate. Many fatigue cracks were observed at the lower ends of the vertical stiffeners attached to the web plates of stringers of truss girders or box-section deck girders. This type of fatigue damage was discovered in about 10 years after the opening of service, being hardly observed in the bridges in the conventional railway systems. The crack originates primarily at the toe or the root of fillet weld around the lower end of the vertical stiffener and develops horizontally into the base metal of the web plate. Such a crack may sometimes propagate along the weld toe, and progress in the direction horizontal to the base metal with some length. On the other hand, in the fatigue tests of beam specimen under in-plane bending, cracks usually progress in the perpendicular direction along the toe of weld as cruciform joints. These are shown in Fig. 3.35.

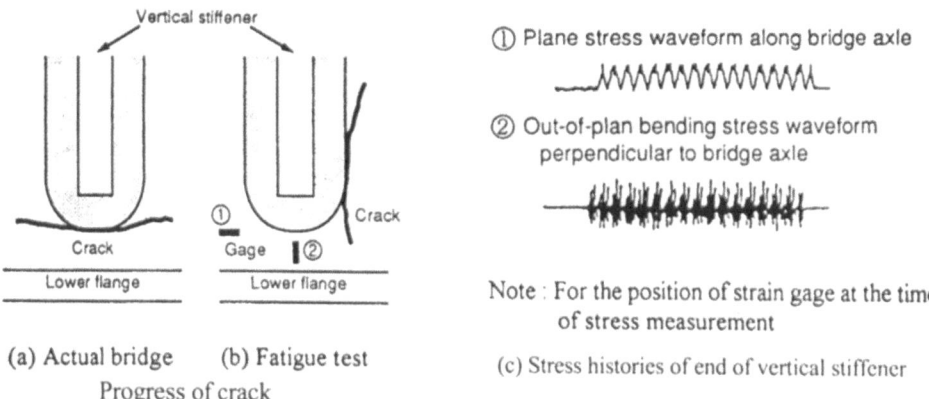

(a) Actual bridge   (b) Fatigue test
        Progress of crack

(c) Stress histories of end of vertical stiffener

Figure 3.35 End of vertical stiffener in web plate

Fig. 3.35.c shows a comparison between the in-plane stress histories along the bridge axle of lower end of stiffener and out-of-plane bending stress histories of that. This shows that the former waveform is almost the same as anticipated in the design, while the component of vibration is contained in the latter waveform. Thus, It seems that this type of crack in the actual bridge is caused mainly by the increasing of out-of plane vibration on web plate due to deflection of sleepers and those due to distortion induced by a high speed operation of trains, in addition to the plane stress in the design [Sakamoto et al., 1990]. This damage is repaired by applying additional plates to the web plate in order to prevent the deformation after gouging and rewelding, and furthermore, applying the TIG-melting at the toe of fillet welds. This type of fatigue damage is, however, increasing and there are many more points likely to suffer such damage.

(f) End Corner of Girder Reducing Section. Regarding the through plate girder and the deck late girder, there is a type of reducing the height of girder near the bearing. As shown in Fig. 3.36, a crack has been observed along the weld at the corner since several years ago [Sakai et al., 1994]. It was revealed that the stress components perpendicular to the weld bead is superior. The small radius of the corner, groove weld with full penetration not being used, and out-of plane vibration due to the high speed train operation are conducive. .

This damage was repaired by applying additional plates to the web plate in order to decrease the stress after gouging and rewelding the fatigue crack.

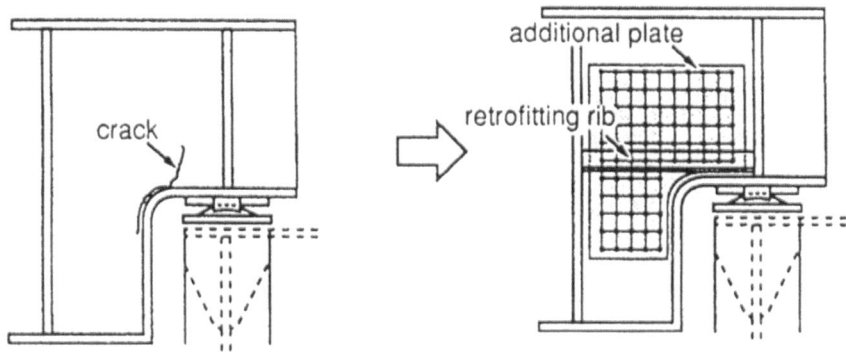

Figure 3.36 Damage of end corner of girder reducing section

(g) Sole Plate of Box Section Deck Plate Girder. A sole plate is usually attached to the lower flange plate with high strength bolts in railway bridges, but in the box section plate girders of long span, fillet welding is used together with bolt joint.

A crack has been observed since several years ago, as shown Fig. 3.37. This crack is observed on both the transverse welds and the longitudinal welds. The crack initiates from the root of weld, propagates along the bead of weld, and enters the lower flange plate [Sakagami et al., 1993].

Stress measurements of actual bridges, structural analysis and fatigue tests were carried out in order to study causes and retrofitting methods. The results are as follows;

1. A poor condition of movable bearing increases the stress around the sole plate.
2. A gap between the sole plate and the flange plate increases the stress.
3. In the retrofitting, a function of movable bearing must be improved, and sufficient reinforcing effect is obtained by exchanging the sole plate for a new one with bigger size after gouging and rewelding the crack.

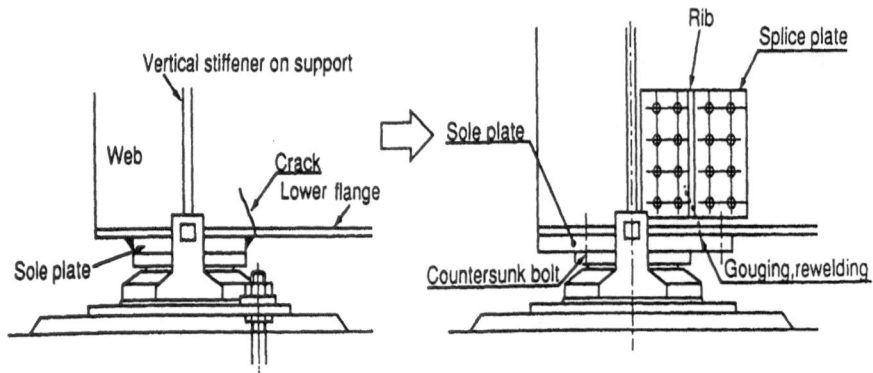

Figure 3.37 Damage of sole plate

3.4 Fatigue of Orthotropic Steel Decks
Case studies from the Netherlands

Structures of orthotropic steel decks, where the deck plate with open or closed stiffeners, is supported by and establishes an integrated structure with the deck and crossbeams, show to be susceptible to traffic induced fatigue [Leendertz et al., 1997].

Most bridges with orthotropic steel bridge decks have been built in the period from 30 to 10 years ago. Several of these structures show fatigue induced cracks in the locations where the stresses are governed by the traffic loads. Although the orthotropic steel decks have become relatively expensive solutions for bridges with shorter spans, they are still used in bridges with longer spans, where the dead weight must be low and for upgrading of existing bridges.

The amount of approximately 300 existing bridges with these structures in the Netherlands, controlled by the Ministry of Transport, causes the need for further investigations in order to obtain a good insight in the behaviour, the fatigue strength and the critical locations of these structures. As the critical locations for fatigue do not always coincide with the critical locations for the ultimate limit state, inspections can be more adequate and limited to specific locations.

(a) Orthotropic steel deck types. The earlier orthotropic steel decks were stiffened by flats and bulbs (see Fig. 3.38.a), thus allowing for spans of approximately 2.0m.

The connection to the crossbeams have been featured in two ways:

1. fitted between the crossbeams;
2. continuous stiffeners passing through special cut outs eventually with additional cope holes in the crossbeams.

The rather small stiffness and strength of these stiffeners caused the need for many crossbeams. A number of these crossbeams, the secondary crossbeams were supported by additional main girders, the secondary main girders. The latter were supported by the primary crossbeams that transmitted the loads to the main girders. The fabrication of these structures with many cut outs and welds was laborious and subsequently expensive. This caused the need to develop structures with less welded connections.

The introduction of the closed V-shaped, U-shaped and trapezoidal stiffeners as shown in Fig. 3.38.b, was a large improvement, which alloyed spans of approximately 4.0m. The secondary crossbeams and main girders were no longer needed. The amount of work involved reduced as well as the costs.

Like the open stiffeners the closed stiffeners could be fitted between or as continuous elements passing through the crossbeams. In this case the cut-out can be of close fit or being featured with an additional oval shaped cope hole or other like the "Haibach" cut out. All details showed to be susceptible to fatigue induced cracks, which resulted in world wide research to the fatigue strength of the details in order to develop data that can be used for the design of new bridges and the repair of existing structures.

Fig. 3.38.c shows a complete overview of all types of closed stiffeners and stiffener to crossbeam connections used in the Netherlands.

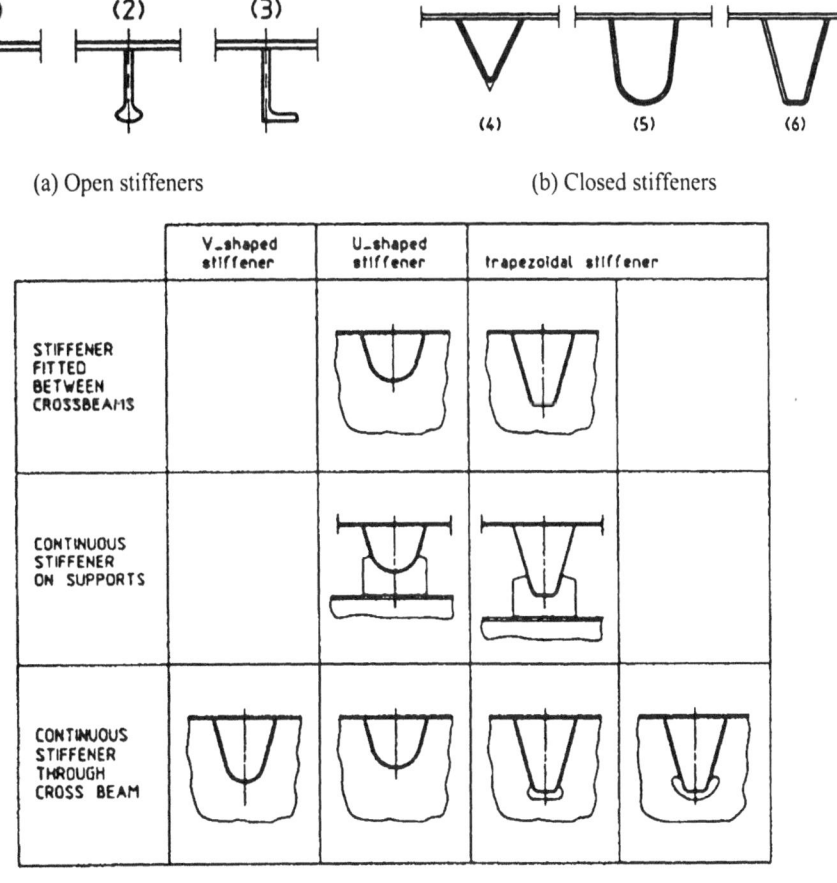

(c) Types of trough to crossbeam connection

Figure 3.38 Orthotropic steel deck types

(b) Structural Bridge Types.

In the orthotropic steel deck structures the deck plate acts together with the longitudinal stiffener. This system transmits the loads to the crossbeams. The latter transmit the loads to the main load carrying system, which can be constructed as plate girders or a box girder. The main load carrying system may be integrated in one of higher rank, such as an arch, a cable stayed structure or a suspension structure.

(bi) Structural systems. Fig. 3.39 shows four load carrying systems. For simplicity only systems with continuous closed stiffeners are shown.

Type I is a plate girder bridge with crossbeams that consists of a truss and a top cord acting as a continuous beam with short spans. The continuous beam transfers the loads by bending and shear to the truss nodes. The truss is supported by the main girders. Type II is a box girder bridge. The diaphragm of the box acts as a deep crossbeam that transfers the loads to the inclined and vertical webs of the box girder. Type III is a conventional crossbeam with cantilevers that are supported by the main girders. Type IV is an I-shaped crossbeam that receives the loads from the supports of the closed stiffeners. The shear connections in the cantilever sections cause a rotational spring using the axial stiffness of the deck plate and the I-beam and a lever arm. This is called the "Floating Deck Structure" and is used in a few bridges in the Netherlands [Ypey, 1972]. It has been developed for its easy of assembly.

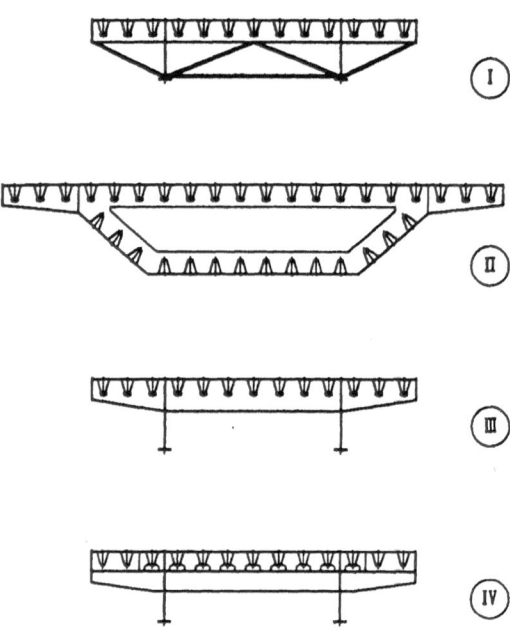

Figure 3.39 Structural types of bridges

(bii) Structural behaviour. In addition to the in plane shear and bending that is generated by the loading of the crossbeams, all structures are subjected to out of plane rotations, caused by the deflection of the stiffeners. In the combination effect the contribution from "in plane" and "out of plane" behaviour differs from type to type and depends strongly on the structural features.

(c) Stiffener to crossbeam connections

(ci) Open stiffeners. Open stiffeners are fitted between the crossbeams or are continuous. The first type is sensitive for eccentricities related to the continuous stiffener and the welded connection to the crossbeam.

Because the influence lines show short distances between the zero-crossings a rather unfavourable connection is submitted to many cycles caused by the wheel loads. Fig. 3.40.a shows the connections of continuous open stiffeners to the crossbeams. In the connections shown, cope holes have been used for fitting stiffeners purposes. The cope hole causes a discontinuity in the crossbeam. In Fig. 3.40.a type (a) and (b) it causes a local stress concentration, but type (c) a "Vierendeel" effect with additional bending is to be expected. This effect will be explained later in conjunction with the closed stiffener connections.

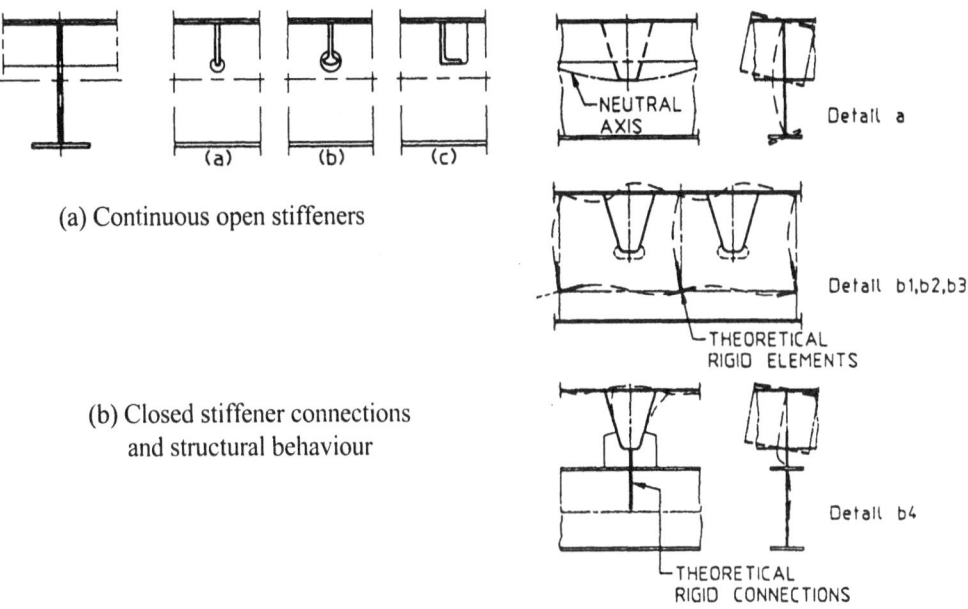

(a) Continuous open stiffeners

(b) Closed stiffener connections
    and structural behaviour

Figure 3.40 Crossbeam connections

(cii) Closed stiffeners. Closed stiffeners are sometimes fitted between the crossbeams, but more often they are continuous. In the latter case they may be welded all around, or passing through cut outs, eventually with additional cope holes for fitting purposes. In the past the connections with cope holes have been investigated extensively in order to analyse the fatigue strength and to optimise the shape [Bruls et al., 1991] [Bruls et al., 1995].

Fig. 3.40.b shows both the connections and their structural behaviour. Two groups are distinguished:

a. Troughs fitted between the crossbeams
b. Continuous troughs passing through cut outs with cope holes

Usually Group "a" connections are applied in structures with shallow crossbeams, where cut outs cause a too low shear capacity of the crossbeam. In the past, the detail with fillet welds showed many fatigue cracks. The details with full penetration welds show a much better fatigue performance.

The Group "b" connections are applied in structures with deeper crossbeams, diaphragms of Box Girder Bridges, "Floating Deck" structures. The following subdivision can be made:

"b1": Continuous troughs passing through a cut out with close fit and welded around with fillet welds;

"b2": Continuous trough passing through a cut out with an oval shape or similar;

"b3": Continuous trough passing through a cut out with additional cope holes with varying radius, the so-called "Haibach cut out" [Haibach and Plasil, 1983] or a similar shape;

"b4": Continuous trough supported by a counterfitted support plate, welded around the bottom of the trough. Further the support is welded to an I-beam.

(ciii) Mechanical behaviour of the connections. In the application of the detail "a", a discontinuity in the stiffener exists with the possibility of eccentricities. The crossbeam section remains practically unchanged. The stiffener rotations cause "out of plane" rotations in the web of the crossbeam.

The structural behaviour of the details "b1 ", "b2" and "b3" is the same, with minor differences.

In plane, the cut out causes a "Vierendeel Effect" if the depth of the cut out is substantially, compared to the depth of the crossbeam or diaphragm [Falke, 1983] [Leendertz and Kolstein, 1995a] [Leendertz and Kolstein, 1995b]. This is likely to occur if the detail is applied in Crossbeams Type I and III bridge structures (see Fig. 3.40.b). Further all details are subjected to locally applied forces and a contraction effect of the bottom of the stiffener caused by bending moments at the stiffener support [Wolchuk and Ostapenko, 1992].

Out of plane rotations are transmitted to the we of the crossbeam or diaphragm. The detail "bl" acts more rigid than the details "b2 and b3".

The detail "b4" does not participate substantially in the in plane load carrying behaviour of the crossbeam, as the shear connection between deck and I-beam caused by the trough is flexible. The out of plane rotations of the stiffeners are transmitted by bending in the support plate to the I-beam which will rotate and translate out of plane in line with the horizontal and torsional stiffness of the I-beam.

(d) Crossbeam in-plane behaviour
In crossbeams with connections "b2" a significant part of the web has been removed.

Consequently an "in plane" Vierendeel behaviour is generated (see Fig. 3.41.a). The part between the troughs, often called the "tooth" acts as a post clamped in a continuous T-beam upside down. Below the cut outs the T-beam remains as the bottom cord. Between the cut outs the web is fully intact. Features like the presence or absence of different shaped cope holes do not change the behaviour significantly [Leendertz and Kolstein, 1995b].

Due to crossbeam bending the locations "L" and "R" translate in horizontal direction. Depending on the neutral axis of the system an elongation or compression of the distance between them, is generated. Shear forces cause relative vertical displacements and rotations in the locations "L" and "R". In [Haibach and Plasil, 1983] these phenomena have been reported for detail "b2". Further, nominal stresses for a set of beams have been calculated. These results have been combined for the locations in the beam where the interaction between shear and bending effects reaches a maximum. For easy comparison the external load introduction is ignored in these results.

Fig. 3.41.b shows the results of FE-analyses with for a test specimen [Bruls et al., 1995] [Leendertzand and Kolstein, 1995b] [Kolstein et al., 1995] the principal stresses around the cut out for crossbeam to trough nr. 2 (detail "b2") connection and a fully welded around crossbeam to trough nr. 7 (detail "b 1 ") connection under the same but symmetrical bending and shear loads. The model consists of shell elements, which ignores the effect of the plate and weld dimensions in the neighbourhood of the welds. The arrows show the direction of the principal stresses at a specific location. Near to the welds the stress levels reach approximately the same level, but the direction with respect to the weld is completely different. Fatigue tests on a true scale specimen [Bruls et al., 1995] showed a better performance for the connection of trough nr.7.

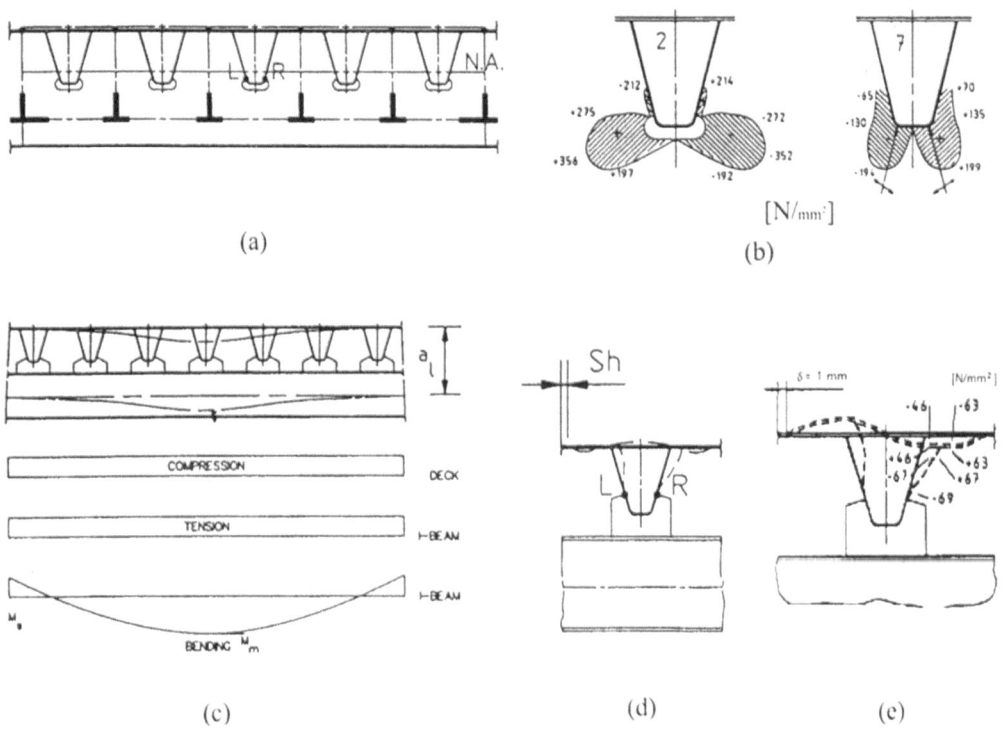

Figure 3.41 Crossbeam in-plane behaviour

In the "Floating Deck" structure as shown in Fig. 3.41.c [Ypey, 1972] [Leendertz and Kolstein, 1996], the end of the I-beam is restrained by the lever system, which generates a compression in e deck and a tension in the I-beam. The horizontal compression and tension forces at a distance a_1 are balancing the bending moment M_s at the support. The normal forces in the deck and the I-beam, the rotations in the I-beam and the shear deformation in the I-beam cause the deck shifting over a distance S_h with respect to the I-beam as shown in Fig. 3.41.d. Asymmetrical loads cause in some locations larger shifts S_h [Leendertz and Kolstein, 1996].

The shift S_h generates normal forces and bending moments in the trough web. Fig. 3.41.e shows the nominal stresses in the trough web and the deckplate caused by the shift S_h of 1 mm. The shifts S_h for a set of beams of with depths are shown in [Leendertz and Kolstein, 1996]. Realistic values of S_h under maximum crossbeam loading vary from 0.1 - 3.3 mm. In real structures these stresses must be multiplied with a stress concentration factor in order to find the Hot Spot Stresses, which are relevant for fatigue.

It is obvious that these stresses, which mainly are governed by bending effects in the deck plate and trough web, can not be neglected. The stress concentration factors related to the bottom connection are assumed to be higher than the stress concentration factors related to the trough to deck connection. If however the stresses due to the wheel loading on the deck are added, high stress amplitudes may occur due to the combination of both effects.

(e) Crossbeam out-of-plane behaviour

Passing vehicles generate bending and shear in the stiffeners, which deflect subsequently and makes the supports of the stiffeners rotate (see Fig. 3.40.b). The rotation of this connection causes an out of plane movement of the crossbeam web. This phenomenon takes place in all types of stiffeners. In [Bruls et al., 1991] [Bruls et al., 1995] the fatigue behaviour has been reported for various types of stiffeners and cut outs. The fatigue strength of details "b2" has been investigated under simultaneous vertical forces with out of plane bending in the web plate. Fig. 3.42 shows for three test specimens the stress results of FE analyses under equal vertical load and out of plane rotation.

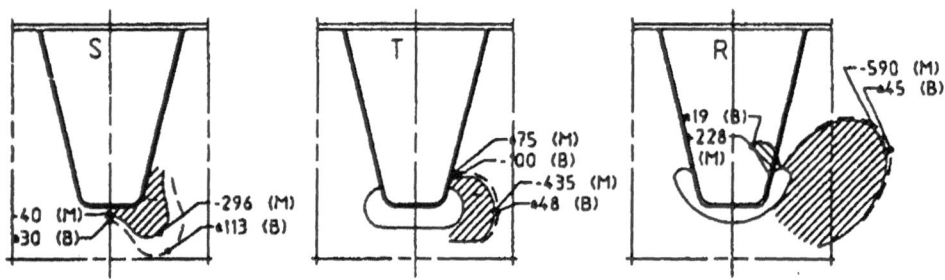

Figure 3.42 Stress distributions under vertical load in combination with out of plane bending (N/mm²)

The stresses shown are the membrane stresses (M) and the out of plane bending stresses (B) for test specimens at the Stevin Laboratory (NL) as reported in [Bruls et al., 1991]. The models consisted of shell elements, which mean that the stresses in the neighbourhood of the trough to crossbeam connection do not include the weld and plate dimension effects [Leendertz and Kolstein, 1996]. Nevertheless it is obvious that in this case, the rigid support not far below the trough to crossbeam web connection causes in the detail "bl" (S) much higher bending stresses than in the details "b2" (T) and "b3" (R). The types "bl";"b2" and "b3" refer to the detail categories of Section (cii). The tests in the Stevin laboratory showed a better fatigue strength for the type "S", if compared to the types "T" and "R". The V-shaped stiffener connection tested at TRL (UK) however showed a lower fatigue strength, the results have been reported in [Bruls et al., 1991].

3.5 Distortional Problems of Cross-Sections of Bridges

The influence of the deformation of the cross section on stresses has been analyzed by the Advanced Theory of Bending, Torsion and Distortion [Tesar, 1977] [Iványi et al., 1990] which is a simple method of analysis but precise enough to study the three-dimensional behaviour of the bridge.

Besides the axial deformation v_1, the two displacements v_2 and v_3 and the rotation of cross section v_4 displacement elements further sectional deformations develop in accord with the cross-section's transverse stiffness. These are equivalent to the relative position change of prismatic plates.

In case of open sections the number of displacement degree of freedom (v) equals to the number of joint-hinges (n). With the formation of the joint-hinges the cross sections' frame-stiffness ceases. To secure the kinematical stability of this "released" section n additional bars are needed. (Fig.3.43.)

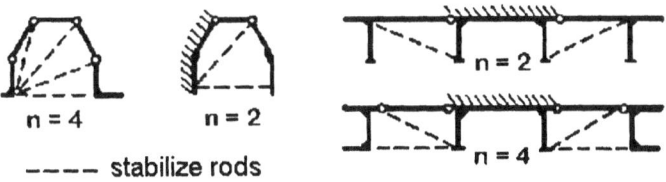

Figure 3.43 Displacement degree of freedom in case of open sections
(*n* is the number of the hinges that breaks up the frame-stiffness of the section,
v is the number of stabilizer bars)

In case of closed cross section (Fig.3.44.) the further number of displacement degree of freedom (v) are to be calculated according to the following formula:

$$2n-(s+3)=v \qquad \qquad /3.1/$$

where n is the number of the hinges that breaks up the frame-stiffness of the section, and s is the number of the section's prism-components (elements).

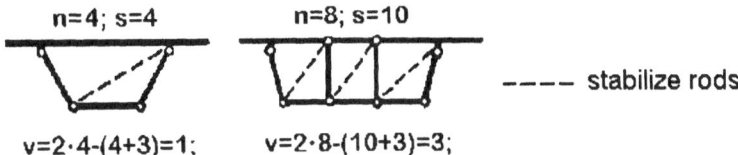

Figure 3.44 Displacement degree of freedom in case of closed cross sections
(*n* is the number of the hinges that breaks up the frame-stiffness of the section,
s is the number of the section's prism-components (elements),
v is the number of stabilizer bars)

The complete deformation of the cross section can be determinated, as the linear combination of the independent elements of the displacement of distortion of cross-section v_p (p>4). We will get these components (elements) if we progressively operate suitable chosen unitary $\vartheta=1$ deflections on the section that is released by hinges, while there is always a stabilizer bar released (see Fig.3.45).

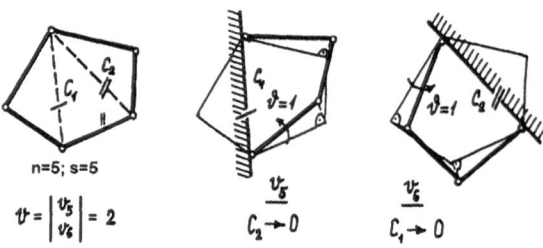

$$n=5; s=5$$

$$\vartheta = \left| \begin{matrix} v_5 \\ v_6 \end{matrix} \right| = 2$$

Figure 3.45 Kinematics of cross-sectional degrees of freedom

So the $\widetilde{\mathbf{v}}$ displacement vector:

$$\widetilde{\mathbf{v}} = \begin{bmatrix} \widetilde{\mathbf{v}}_t \\ \cdots \\ \widetilde{\mathbf{v}}_p \end{bmatrix} = \begin{bmatrix} \widetilde{v}_1 \\ \widetilde{v}_2 \\ \widetilde{v}_3 \\ \widetilde{v}_4 \\ \cdots \\ \widetilde{v}_5 \\ \widetilde{v}_6 \\ \vdots \\ \widetilde{v}_n \end{bmatrix} = \begin{bmatrix} \xi \\ \eta \\ \zeta \\ \vartheta \\ \cdots \\ \widetilde{v}_5 \\ \widetilde{v}_6 \\ \vdots \\ \widetilde{v}_n \end{bmatrix} \qquad /3.2/$$

of which has $\widetilde{\mathbf{v}}_t$ rigid body modes and $\widetilde{\mathbf{v}}_p$ distortional modes (p subscript refers to the displacement of the distortional cross-section).

Moreover we assert the basic principle of bending-torsion, which says that all the section's prism-components keep their shapes. So the Bernoulli-Navier hypothesis is valid for each prism.

According to the previous conception, the formula /3.2/ can be extended, which gives the tangential component of the strain:

$$\frac{df}{dx} = f' = f_t' + f_p' = \eta' \cdot \cos\alpha + \zeta' \cdot \sin\alpha + \vartheta \cdot r_M +$$

$$+ \widetilde{v}_5' \cdot \widetilde{r}_5 + \widetilde{v}_6' \cdot \widetilde{r}_6 + \ldots + \widetilde{v}_n' \cdot \widetilde{r}_n ,$$

So

$$f' = (\widetilde{\mathbf{v}}')^{\mathrm{T}} \cdot \widetilde{\mathbf{r}} \qquad /3.3/$$

Where

$$\tilde{r} = \begin{bmatrix} \mathbf{r}_t \\ \hdots \\ \tilde{\mathbf{r}}_p \end{bmatrix} = \begin{bmatrix} 0 \\ \cos\alpha \\ \sin\alpha \\ \hdots \\ r_M \\ \tilde{r}_5 \\ \tilde{r}_6 \\ \vdots \\ \tilde{r}_n \end{bmatrix}, \qquad \tilde{\mathbf{v}}' = \begin{bmatrix} \mathbf{v}_t' \\ \hdots \\ \tilde{\mathbf{v}}_p' \end{bmatrix} = \begin{bmatrix} \xi' \\ \eta' \\ \zeta' \\ \vartheta' \\ \hdots \\ \tilde{v}_5' \\ \tilde{v}_6' \\ \vdots \\ \tilde{v}_n' \end{bmatrix} ;$$

and $\tilde{r}_5, \tilde{r}_6, \dots, \tilde{r}_n$ are the referring sectional prism's force arms to the rotation-centres.
The axial component of the displacement of thin-walled section centre-line:

$$\tilde{u} = -\int (\tilde{\mathbf{v}}')^{\mathrm{T}} \cdot \tilde{r} \, ds = -(\tilde{\mathbf{v}}')^{\mathrm{T}} \cdot \int \tilde{r} \, ds = -(\tilde{\mathbf{v}}')^{\mathrm{T}} \cdot \tilde{\mathbf{w}} \qquad /3.4/$$

Where $\tilde{\mathbf{w}}$ is the warping vector.
Values of $\tilde{\mathbf{w}}$ for open sections (Fig.3.46)

$$\tilde{\mathbf{w}} = \begin{bmatrix} \mathbf{w}_t \\ \hdots \\ \tilde{\mathbf{w}}_p \end{bmatrix} = \int \tilde{r} \, ds = \begin{bmatrix} 1 \\ y \\ z \\ \omega_M \\ \hdots \\ \tilde{w}_5 \\ \tilde{w}_6 \\ \vdots \\ \tilde{w}_n \end{bmatrix} \qquad /3.5/$$

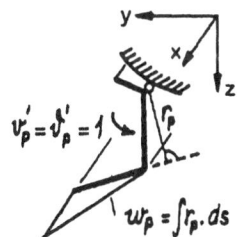

Figure 3.46 The axial component of displacement of the thin-walled section centre-line

In case of closed cross sections we extract the open "statically determinate" basic system first, with suitable chosen sections.

We can take into account the condition for continuity for the hollow section through the extension of $\dfrac{\tilde{\psi}}{t} = \dfrac{\tilde{\psi}_p}{t}$ vector.

We get an analogue equation system with /3.5/ for the vector components:

$$-\int_{j-1}^{j} \frac{\tilde{\psi}_{p,j-1}}{t}\,ds + \oint_{j} \frac{\tilde{\psi}_{p,j}}{t}\,ds - \int_{j}^{j+1} \frac{\tilde{\psi}_{p,j+1}}{t}\,ds = \int_{j} \tilde{r}_p\,ds$$

The warping vector is:

$$\tilde{\mathbf{w}} = \begin{bmatrix} \mathbf{w}_t \\ \hdots \\ \tilde{\mathbf{w}}_p \end{bmatrix} = \int \left(\tilde{\mathbf{r}} - \frac{\tilde{\psi}_p}{t} \right) ds =$$

$$= \int \left(\begin{bmatrix} 0 \\ \cos\alpha \\ \sin\alpha \\ r_M \\ \hdots \\ \tilde{r}_5 \\ \tilde{r}_6 \\ \vdots \\ \tilde{r}_n \end{bmatrix} - \begin{bmatrix} 0 \\ 0 \\ 0 \\ \psi_4/t \\ \hdots \\ \tilde{\psi}_5/t \\ \tilde{\psi}_6/t \\ \vdots \\ \tilde{\psi}_n/t \end{bmatrix} \right) ds = \begin{bmatrix} 1 \\ y \\ z \\ \omega_M \\ \hdots \\ \tilde{w}_5 \\ \tilde{w}_6 \\ \vdots \\ \tilde{w}_n \end{bmatrix} \qquad /3.6/$$

The normal stress of cross section centre line:

$$\tilde{\sigma} = -E \cdot (\tilde{\mathbf{v}}'')^{\mathrm{T}} \cdot \tilde{\mathbf{w}} \qquad\qquad /3.7/$$

The primary shear stress is:

$$\tilde{\tau}_1 = G \cdot (\tilde{\mathbf{v}}')^{\mathrm{T}} \cdot \frac{\tilde{\psi}_p}{t} \qquad\qquad /3.8/$$

Where $\tilde{\mathbf{v}}$ is the section's global deflection vector according to /3.2/. The relation between the rotation vector of the section's prisms $\underline{\tilde{\vartheta}}_p = \left\{ \tilde{\vartheta}_{p,m} \right\}$ and the section's $\tilde{\mathbf{v}}_p$ formal-change vector is the following:

$$\underline{\tilde{\vartheta}}'_p = \tilde{\mathbf{F}}_p \cdot \tilde{\mathbf{v}}'_p$$

When $\widetilde{\mathbf{F}}_p$ matrix has to be compiled by taking into account the chosen $\widetilde{\mathbf{v}}_p$ basic formal-changes (Fig. 3.45).

So the torsion moment vector referring to the sectional prisms' free torsion is the following:

$$\mathbf{T} = G \cdot \mathbf{K} \cdot \widetilde{\mathbf{F}} \cdot \widetilde{\mathbf{v}}' = G \cdot \widetilde{\mathbf{F}}^* \cdot \widetilde{\mathbf{v}}'$$

Where

$$\widetilde{\mathbf{v}}' = \begin{bmatrix} \mathbf{v}'_t \\ \cdots \\ \widetilde{\mathbf{v}}'_p \end{bmatrix} \quad \text{and} \quad \widetilde{\mathbf{F}}^* = \begin{bmatrix} \widetilde{\mathbf{F}}^*_t \\ \cdots \\ \widetilde{\mathbf{F}}_p \end{bmatrix}, \text{where} \quad \widetilde{\mathbf{F}}^* = \mathbf{K} \cdot \widetilde{\mathbf{F}} \cdot \qquad \qquad \text{/3.9/}$$

We have to determine the relation between the transverse moments in the section's prismatic element and the section's $\widetilde{\mathbf{v}}_p$ formal-change vector. So we compile the prism's common relative rotations at the hinges ε_{im}, in the common deflection vector of the section-walls $\widetilde{\mathbf{e}}$ (Fig. 3.47):

$$\widetilde{\mathbf{e}} = \{\widetilde{\varepsilon}_i\} = \{\Delta\widetilde{\vartheta}_i\} = \Delta\widetilde{\mathbf{F}} \cdot \widetilde{\mathbf{v}} \qquad \qquad \text{/3.10/}$$

Where

$$\widetilde{\varepsilon}_i = \widetilde{\vartheta}_{m+1} - \widetilde{\vartheta}_m = \Delta\widetilde{\vartheta}_i$$

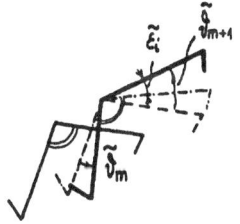

Figure 3.47 Relative rotations of walls of cross-section: $\varepsilon_i = \Delta\vartheta_i$

The rows of matrix $\Delta\widetilde{\mathbf{F}}$ come from matrix $\widetilde{\mathbf{F}}$ matching rows' difference. Further on we can suppose that the plate-stiffness of the wall can be substituted by the bending stiffness of infinitely many and infinitely broad frames that are close to each other. So the plates' torsion moments, and longitudinal moments in the section-wall are neglected.

If EI_r the bending stiffness of the unitary wide sectional frame, than M_{ri}, transversal moment at hinge i and which is a function of $\widetilde{\varepsilon}_j$, the common rotations of the section-walls at j places, comes from this formula:

$$\widetilde{M}_{ri} = \sum \beta_{ij} \cdot \widetilde{\varepsilon}_j \qquad \text{and} \qquad \widetilde{\mathbf{M}}_r = \{\widetilde{M}_{ri}\} = \mathbf{B} \cdot \widetilde{\mathbf{e}}$$

In case of a simply open section of which centre line is developable to a continuous line, the coefficients in matrix $\mathbf{B} = \{\beta_{ij}\}$ come from the inverse of matrix $\mathbf{D} = \{\delta_{ij}\}$, where δ_{ij} elements are unit factors of a statically determinated basic system according to the compatibility method (Fig. 3.48).

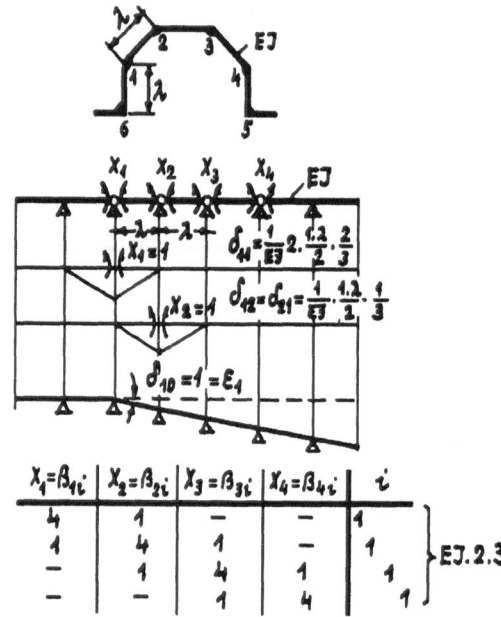

Figure 3.48 Determination of the β elements as the elements of inverse of matrix $\mathbf{D} = \{\delta_{ij}\}$ of continuous beam

If we express the rotation vectors of the section-walls, according to /3.10/ we'll get:

$$\widetilde{\mathbf{M}}_r = \mathbf{B} \cdot \Delta\widetilde{\mathbf{F}} \cdot \widetilde{\mathbf{v}} = \Delta\widetilde{\mathbf{F}}^* \cdot \widetilde{\mathbf{v}} \qquad /3.11/$$

Where $\Delta\widetilde{\mathbf{F}}^* = \mathbf{B} \cdot \Delta\widetilde{\mathbf{F}}$.

We depart from the principle of virtual work. The work of the internal forces comes from the extended formula bellow:

$$\delta\Pi_{internal} = \int_0^L \left[\int_A (\sigma \cdot \delta\varepsilon + \tau_1 \cdot \delta\gamma_1) \cdot dA + \mathbf{T}^T \cdot \delta\underline{\widetilde{\vartheta}}' + \mathbf{M}_r^T \cdot \delta\widetilde{\mathbf{e}} \right] \cdot dx,$$

$$\delta\Pi_{\text{internal}} = \int_0^L [E(\widetilde{v}'')^T \int_A \widetilde{w} \cdot \widetilde{w}^T dA \cdot \delta\widetilde{v}'' + G(\widetilde{v}')^T (\int_A \frac{\widetilde{\psi}_p}{t} \cdot \frac{\widetilde{\psi}_p^T}{t} dA +$$

$$+ (\widetilde{F}^*)^T \cdot F) \cdot \delta\widetilde{v}' + \widetilde{v}^T (\Delta\widetilde{F}^*)^T \cdot \Delta\widetilde{F}) \cdot \delta\widetilde{v}] dx \qquad /3.12/$$

If we do the examination of the energy balance, that means we use

$$\delta\Pi = \delta\Pi_{\text{internal}} - \delta\Pi_{\text{external}} = 0$$

condition, and do the partial integration, by taking into account the following extended stiffness matrixes:

$$\widetilde{I} = \left\{\widetilde{I}_{w_i w_k}\right\} = \int_A \widetilde{w} \cdot \widetilde{w}^T \, dA \qquad /3.13/$$

as the complex warping stiffness matrix,

$$\widetilde{K} = \left\{\widetilde{K}_{ik}\right\} = \int_A \frac{\widetilde{\psi}}{t} \cdot \frac{\widetilde{\psi}^T}{t} \, dA + (\widetilde{F}^*)^T \cdot \widetilde{F} \qquad /3.14/$$

as the simple torsional stiffness matrix

$$\widetilde{R} = \left\{\widetilde{R}_{ik}\right\} = (\Delta\widetilde{F}^*)^T \cdot \Delta\widetilde{F} \qquad /3.15/$$

and as the transverse bending stiffness matrix, then we get the extended simultaneous equilibrium system of differential equations:

$$\left|E\widetilde{I} \cdot \widetilde{v}'''' - G\widetilde{K} \cdot \widetilde{v}'' + \widetilde{R} \cdot \widetilde{v} - p \cdot \widetilde{r} - \int_A n' \cdot \widetilde{w} \cdot dA = 0\right|$$

$$\left[-E\widetilde{I} \cdot \widetilde{v}''' + G\widetilde{K} \cdot \widetilde{v}' + \int_A n \cdot \widetilde{w} \cdot dA - P \cdot \widetilde{r}\right]_0^L = 0 \qquad /3.16/$$

$$\left[E\widetilde{I} \cdot \widetilde{v}'' + N \cdot \widetilde{w}\right]_0^L = 0$$

These equations give besides $\widetilde{v}$ global deflection vector, the effect of the section centre-line change as well.

When solving open sections (Fig. 3.49) in case of ordinary plate slenderness of bridge systems, the torsional stiffness of the walls (prisms) can be neglected.

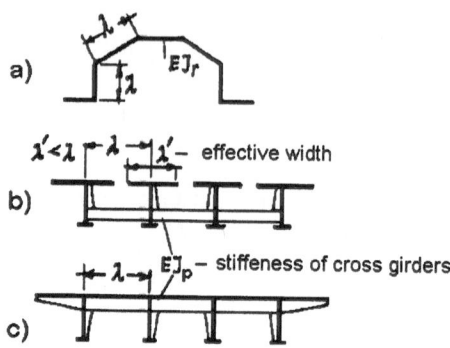

Figure 3.49 Different shapes of open sections

In case of <u>open sections</u> if we diagonalize matrix $\bar{I}$ and $\bar{R}$, it is obvious, that we can separate all the four deflection components of the shape-keeping section (ξ, η, ζ and ϑ).

After the normalization we get the independent difference equation system with I and R diagonal stiffness matrixes, where every element of v global deflection vector are equivalent to a model of a continuously elastic supported bar with R_i and with EI_i bending stiffness, and which is loaded by $p \cdot r_i$ load intensity in cross direction (see Fig. 3.50).

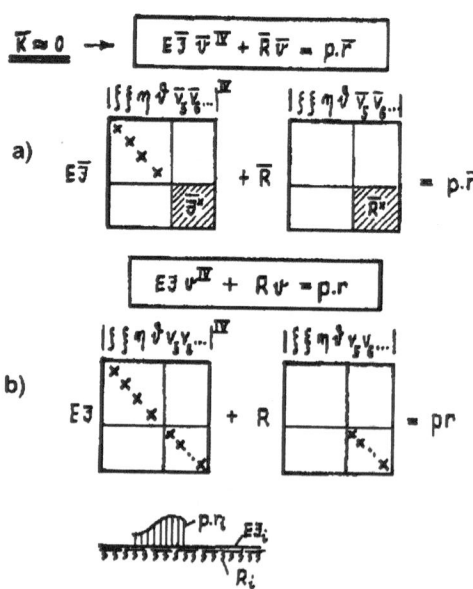

Figure 3.50 The normal function in case of open sections
a) The partially normalised simultaneous differential equations from the first phase to determine the eigenvectors in case of open section
b) Totally normalized independent differential equations in case of open section –analogue of the continuously supported transversally loaded bar

In case of <u>closed cross-sections</u> (Fig. 3.51) a relevant torsional stiffness comes because of the $\underline{\psi}_p / t$ vector. So it is suitable to neglect the cross directional stiffness of the section, which means that we assume that the section is a closed hinged mechanism (Fig.3.52). The fault of this assumption, that the fourth component of the deflection (9) is not independent.

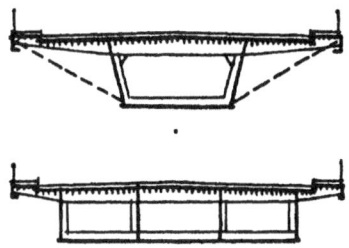

Figure 3.51 Shapes of closed sections

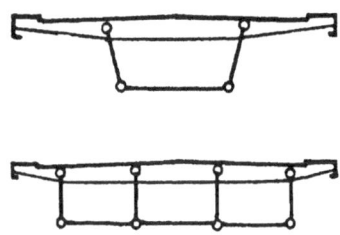

Figure 3.52 Kinematically determinated closed hinged mechanisms

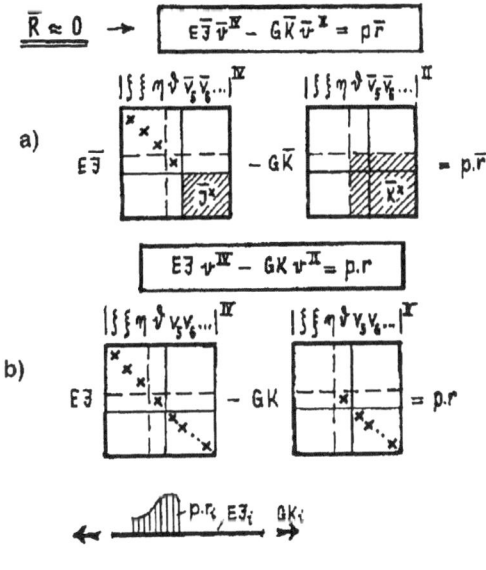

Figure 3.53 The normal function in case of open sections
a) The partially normalised simultaneous differential equations from the first phase to determine the eigenvectors in case of closed section
b) Totally normalized independent differential equations in case of open section –analogue of the transversally loaded and tensioned bar

After the normalisation we get an

independent system of differential equations with I and K diagonal stiffness matrixes, where each component of v global deflection vector equals to a bar with EI_i bending stiffness, with a transverse load of $p \cdot r_i$, and with GK_i fictive axial load at the ends. (see Fig. 3.53)

3.5.1 Traffic Data of the Examined Bridges.
Case studies from German

Lately, an increasing number of crack has been apparent in several steel highway bridges built between 1950 and 1970. Most affected are sway bracings, diaphragms, web stiffeners, components of orthotropic decks and crossing connections These problems are not confined to German bridges [Nather, 1991]. Earlier reports from the USA describe similar damage, and for some time fatigue damage detected in Japanese bridges has been reported too [Nather, 1991].

The main cause of the increasing number of damages is the development of traffic. Most bridges suffer more severe loading than was envisaged during their design due to increased traffic and higher axle loads. Other causes include inadequate fatigue design of structural details and fabrication defects. Another reason for fatigue cracks in steel bridges is inadmissible simplification of statical systems, mainly because the three-dimensional behaviour of bridges and secondary stresses are usually neglected in design practice.

Data of traffic flow and vehicle composition is necessary for bridge rehabilitation and investigation of damage. Earlier forecasts of lorry traffic on these motorways underestimated current traffic conditions. For instance, lorry traffic over the Haseltal bridge increased from 5700 lorries per day in 1978, to 9000 lorries per day in 1988. Over the period from 1980 to 1989, the number of applications for permission of special transports almost tripled [Nather, 1991].

Because of the predominant local traffic, the portion of vehicles with four and more axles did not exceed 55%. Further measurements indicated that the rate of capacity utilization was low and that the impact factor depends to a high degree on the velocity of vehicles. In assessing remaining fatigue life, development of future traffic must be considered. For instance with the European Community liberalization beginning in 1993, permissible axle loads and maximum weight of commercial vehicles will be increased.

3.5.2 Cracks in Connections of Cross Beams and Stiffeners.
Fatigue cracks have been detected:

- in fillet welds between the cross beam flange and the web or flange of the inside or outside main girder stiffener (Fig. 3.54, Detail No. 4)
- in the seam between web and flange of the cross beam (Fig. 3.54, Detail No. 2)
- in fillet welds between cap and connection plate of transverse stiffeners (Fig. 3.55, Type I, II, III) or sway bracings (Fig. 3.55, Type II, VI)
- in fillet welds connecting the connection plate to the main girder web (Fig. 6, Type III, IV)
- in the butt weld between connection plate and transverse stiffener (Fig. 6, Type V, VI), and
- in the connection between longitudinal and transverse web stiffeners.

Depending on the position of axle loads that may be located either between the main girders or on the cantilever, the haunches receive bending moments of changing signs. These moments are usually not considered during design of the connection. On curved bridges, the influence of local loads and forces due to the curvature are superimposed. On the Haseltal bridge (Fig. 3.54) the outside vertical stiffeners have been torn down in some spans. This led to a drop of the safety factor for lateral buckling of the bottom chord in compression. Fatigue assessment showed that fatigue cracking of the haunches could have been foreseen.

The Danube bridge near Sinzing (Fig. 3.55) has a superstructure for each lane. The main lane is located between the main girders, and every passage of a lorry results in bending stresses in the stiffeners. Field measurements and structural analyses revealed stresses larger than the yield stress even under service loading in welding outlets of nodes V and VI. Consequently, 17 of a total of 18 nodes of type VI and 47 of a total of 62 nodes of type V were cracked. The main reason for this damage was an inadequate statical system assumed during bridge design, i.e. the support points of the cross beams were assumed to be hinges.

On Fig. 3.55 node III, cracks are indicated in welded connections between vertical stiffener and cap plate. These cracks have been detected at an earlier date and repaired in 1977. Although 384 of a total of 424 welding showed cracks, no investigation of the causes of these cracks was conducted. Additionally, reinforcing plates have been mistakingly welded on top of the flanges of transverse stiffeners, and cap plates were also welded on flanges of cross beams. Thus, the cause of cracks was not eliminated, and new cracks similar to those presented in Fig. 3.55 developed. During repair work, 1990 diagonals had been fastened between transverse stiffeners and cross beams, as indicated on Fig. 3.56.

The influence of the deformation of the cross section on stresses in various structural details of the Danube bridge near Sinzing has been analyzed using the Advanced Theory of Bending, Torsion and Distorsion, which is a simple method of analysis but precise enough to study the three-dimensional behaviour of the bridge. Structural analysis indicated that the bending stresses in vertical web stiffeners reached values larger than those allowed (Figures 3.57 and 3.58).

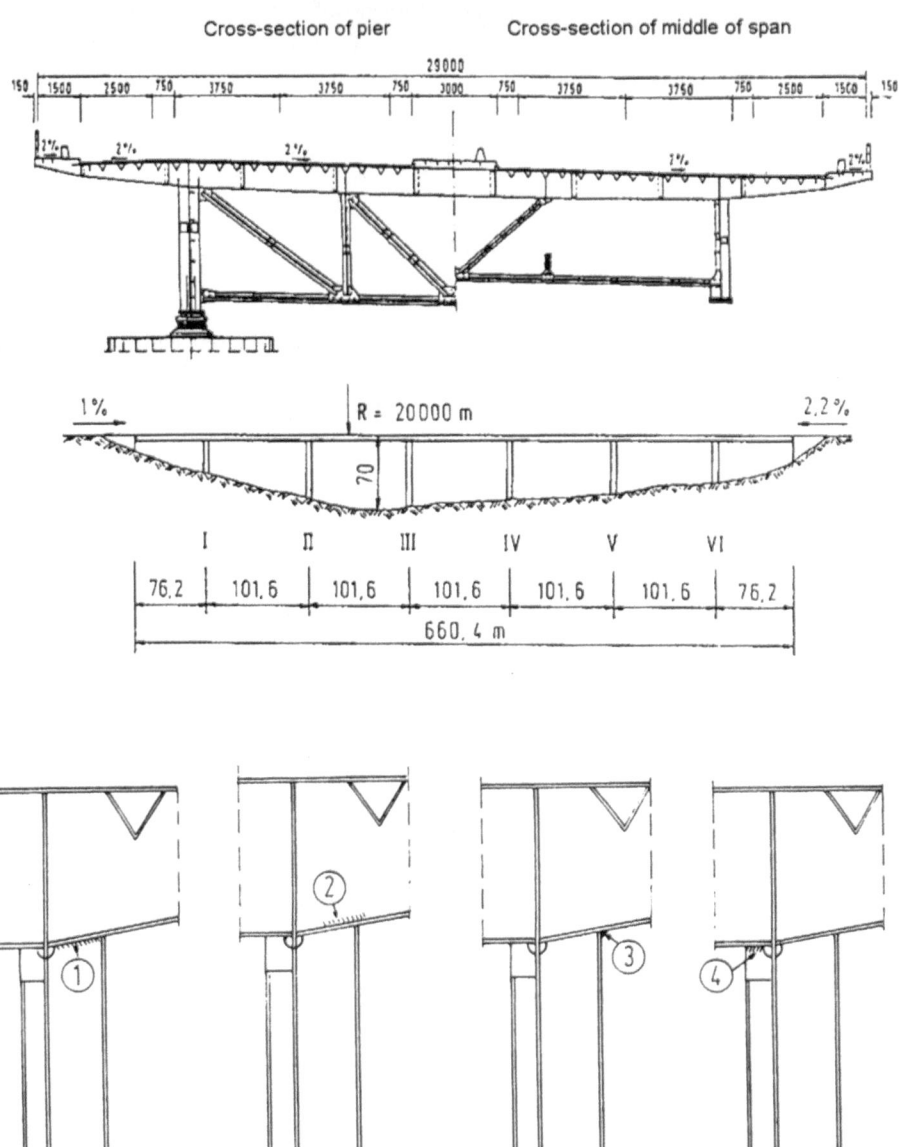

Figure 3.54 Haseltal bridge

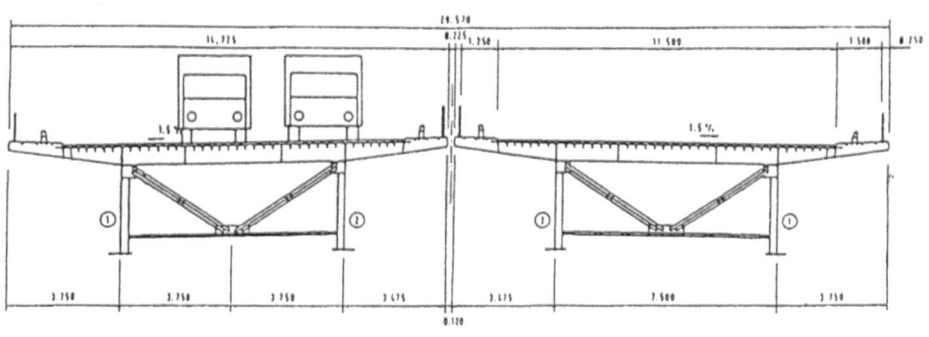

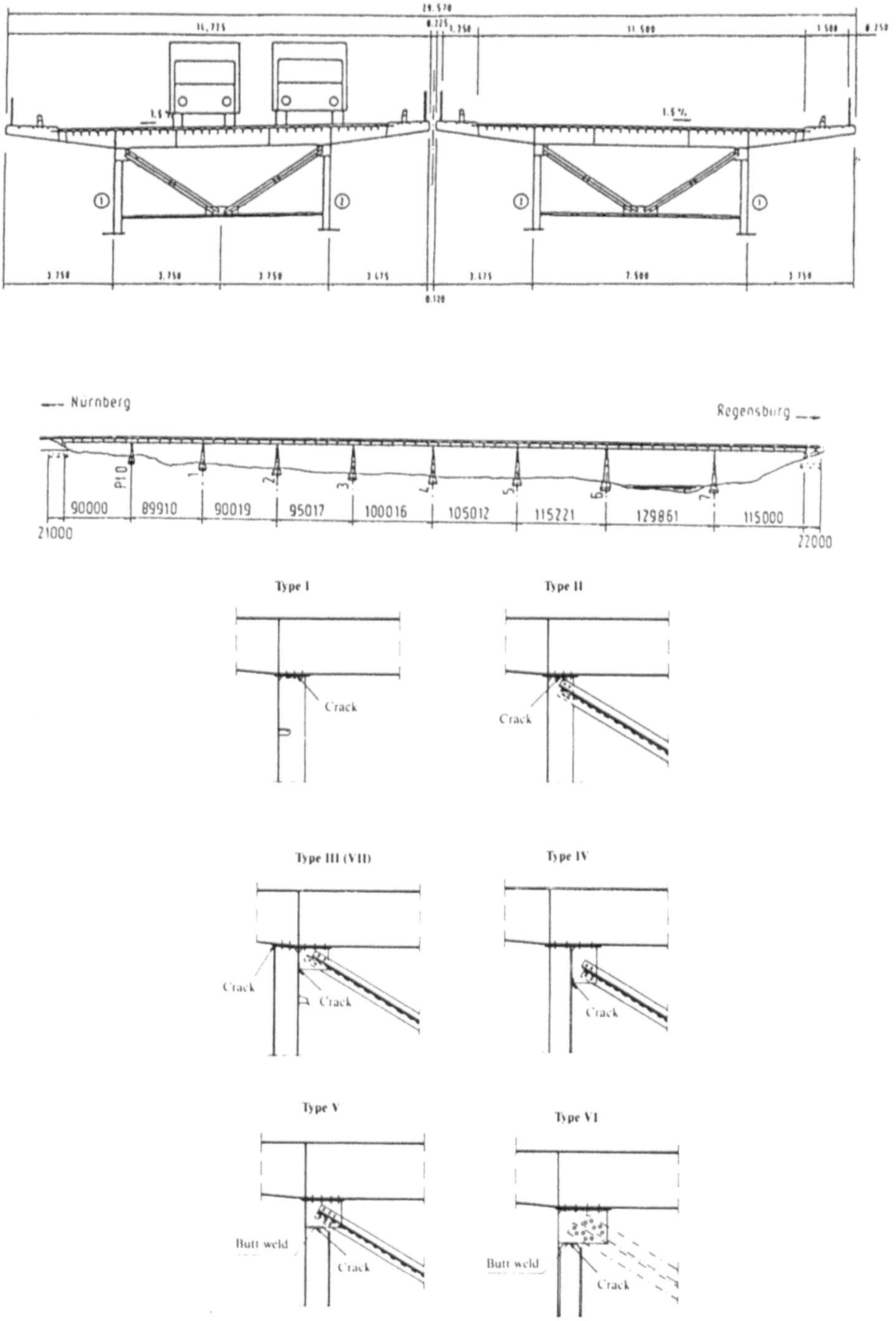

Figure 3.55 Danube bridge near Sinzing

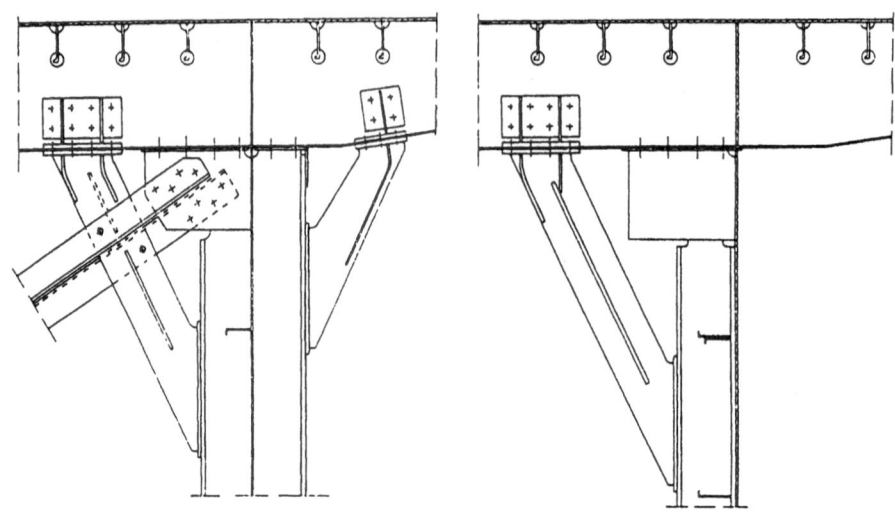

Figure 3.56 Danube bridge near Sinzing: Reconstruction of node type III (VII)

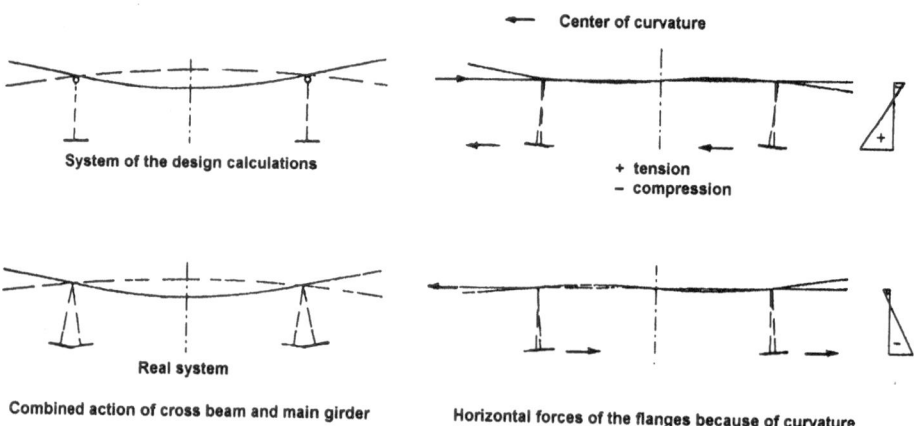

Figure 3.57 Haseltal bridge: Influences in the design calculation

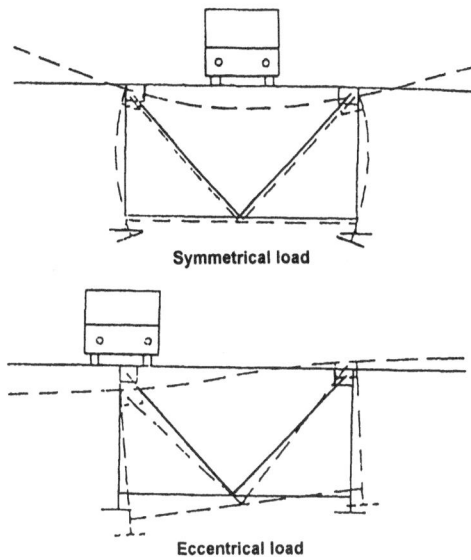

Symmetrical load

Eccentrical load

Figure 3.58 Danube bridge near Sinzing: Reconstruction of node type III (VII)

3.5.3 Cracks in Orthotropic Decks. Between 1960 and 1973, 25 steel bridges have been built in Germany with orthotropic deck and ribs of «Y»-(«wine glass») or «V»-shape. Ribs are cut into the floor beams, and their «V»-part is buttwelded with bevel groove welds to the floor beam webs. Outside Germany, only one bridge with «Y»-ribs and one with «V»-ribs has been built, i.e. the Bosporus Bridge (with continuous ribs) and the Komatsugawa Bridge (with backing strips).

The ideal case of a cross joint of the longitudinal V-stiffener, shown in Fig. 3.59, does not often occur. The axis of «V»-ribs tend to misalign, and an incomplete penetration may lead to fatigue cracks that are not visible from the outside. These fatigue cracks may even penetrate the cross beam web. For instance, the cracks in the rib-to-floor intersections of the Haseltal bridge could not be detected from the outside.

The ribs are directly stressed by the wheel loads. Structural analysis showed that a single passage of a vehicle causes several stress cycles of relatively high amplitude. For repair, short reinforcing plates have been welded first to the cap plate and then to both the deck plate and the «V»-rib by fillet welds (Fig. 3.60). Finally, the cap plates on both sides of the cross-beam web have been connected using high strength bolts. A fatigue assessment has been made for this connection to adapt the fatigue life of the ribs to the design life of other important structural members, i.e. 50 years. The detail category of 36 according to [ECCS, 1985] and a partial safety factor of 1.0 were chosen. From this, section modulus of the strengthened joint of 611 cm³ has been determined, while for the original joint it was 290 cm³.

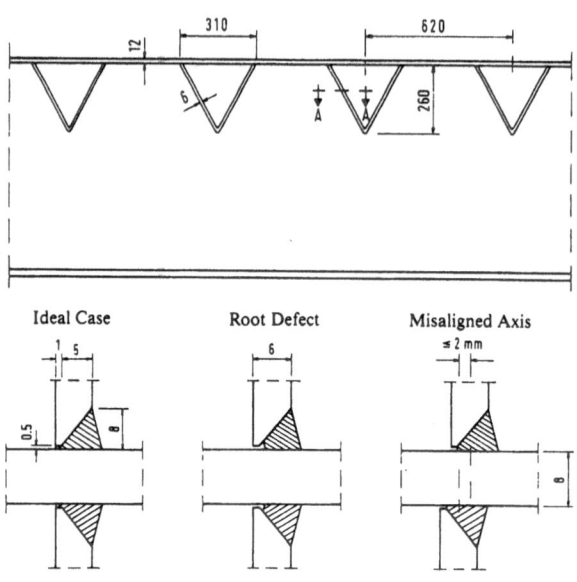

Figure 3.59 Cross joint of the longitudinal stiffener

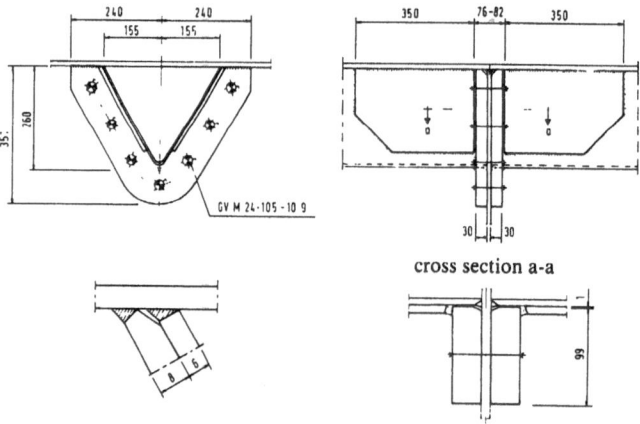

cross section a-a

Figure 3.60 Haseltal bridge: Reconstruction of the connection of V-stiffener to cross beam

Three different designs of the intersection between «Y»-ribs and floor beams are known (Fig. 3.61). The tee-portions of structural type A are spliced by a formed piece which runs continuously through circular cutouts in the floor beams. This detail has been used in the Sinntal bridge. The tee-portions of design B and C run continuously through the cut-outs and are welded, either single-sided or on both sides, to the floor-beam web by fillet welds. Design B for example, was used for the Rhine bridge near Leverkusen. In the deck of this bridge, a multitude of cracks developed from the lower end crater of the fillet weld, and some of them run through the inclined weld to the cover plate.

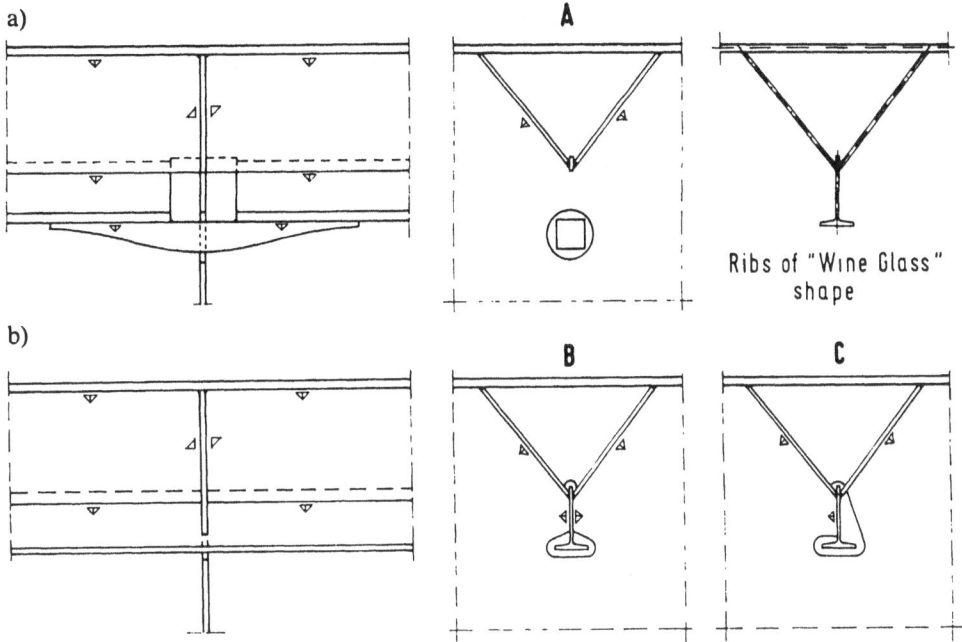

a) Splice profiles run through cutouts of the floor beam
b) T-portions run continuously through cutouts of the floor beam

Figure 3.61 Rib-to-floor-beam junctions

The numerous welds in the intersection of design A represent various possible locations for fatigue cracks. In a first phase of rehabilitation of the Sinntal bridge, cracks have been found in welds of the inclined plates of the «Y»-ribs. At some locations, the ribs were almost completely torn down. As a consequence, the grid effect of the deck was reduced which resulted in a propagation of damage similar to the opening of a zip-fastener. The accumulation of cracks in the last four spans near the southern abutment was explained by high dynamic stresses in the bridge deck caused by the traffic. It could be observed that trucks overtake one another even though no other vehicles were on the bridge.

In the second phase of rehabilitation, many new cracks have been detected which could not be attributed to fatigue. A considerable quantity of cracks may have developed as a result of the Sinntal bridge reached 115 °C on the side of the base.

The damage in the rib-to-floor intersection is typical for this type of structural detail. Cracks are not expected where the ribs with hollow sections run continuously through the cut-outs of the floor beams. Insufficient fatigue strength of rib cuts welded to floor-beam webs, has already been described in 1961.

3.5.4 Strengthening of Main Girders under Traffic. Increasing traffic loads, the high probability of occasional overstressing and a high utilization of the material without considering fatigue, are the main causes that may require strengthening of bridges [Nather, 1991]. The possibilities of strengthening and broadening of bridges, even under traffic, is the main advantage of steel structures. Two methods of strengthening are described next:

1. To reinforce the Haseltal bridge, two lattice girders were mounted between the existing main girders as shown in Fig. 3.62. Transverse truss bracings have been built-in, every 9.24 m, to form a grillage with the main girders. At the centre, pier hinges were built in the lattice girders to avoid additional loading of the cross bracings at the support which could result in replacement of the existing support bracings. The support reactions of the additional lattice girders are taken over by additional support bracings, from where they are led off to the main girders and finally to the bearings.
2. Another suggestion, shown in Fig. 3.63, is to build an additional lattice girder in the centre line of the bridge together with a torsional bracing that is fixed above the bottom chords of the existing main girders. The floor-beam cantilevers are braced by diagonals to the stanchions of the torsional bracings. These stanchions and diagonals form, together with the cross beams and web stiffeners, transverse diaphragms to keep the cross sections elastic. All joints are bolted using high-strength bolts. To reduce the stress-resultant components, additional transverse bracings may be built-in. In case removal or strengthening of the existing support bracing is not possible, the new lattice girder may also be supported spatially by diagonal bracings that are directly supported by the bearings.

3.5.5 Final Remarks. Experiences derived from the rehabilitation of highway steel bridges have been described. Fatigue cracks in orthotropic plates and in connections of cross beams and stiffeners have been detected, and repaired, on three steel road bridges. The main cause of the increasing number of damages, is the more severe loading of bridges due to increased traffic and higher axle loads. Other causes include inadequate fatigue design of structural details and fabrication defects. Repair and strengthening of steel bridges can be conducted relatively easily under traffic. Additionally, inadmissible simplification of statical systems is a further reason for damage on steel bridges.

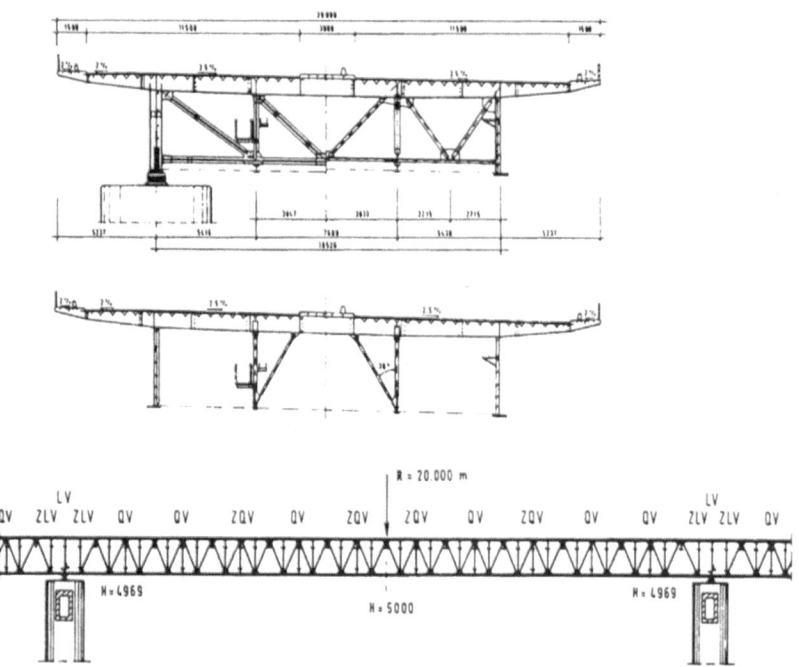

Transverse bracing above the support
Additional transverse bracing next to the support
Truss girder in the center span
/ = Transverse bracing over the pier (old), QV = Transverse bracing (old, reconstructed),
)V = Additional transverse bracing (new), ZLV = Additional transverse bracing beside
› supports (new)

Figure 3.62 Strengthening of the Haseltal bridge

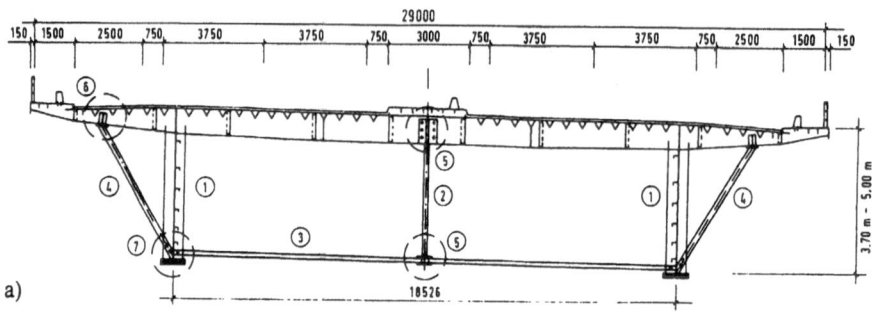

section B-B

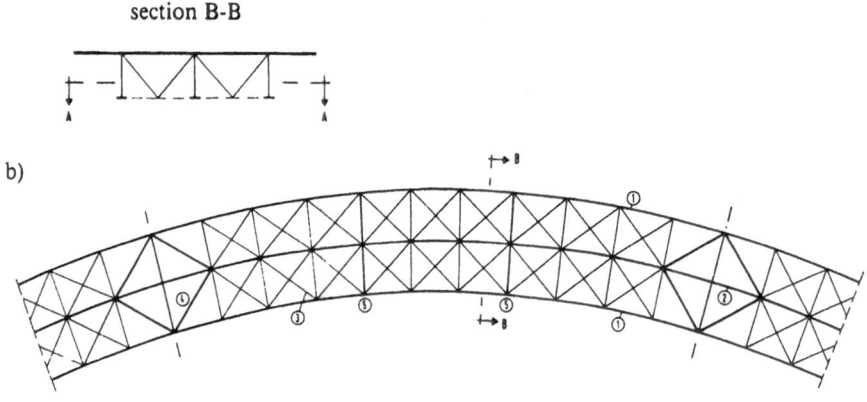

Strengthening of steel road bridges - suggestion:

a) Cross section   b) Horizontal projection

1: Plate girder	*5: Bottom chord member (new)*
2: Truss girder (new)	*6: Longitudinal girder (new, for load distri bution)*
3: Torsional web system (new)	*7: Supporting of the truss girder (new)*
4: Diagonal (new)	*8: Transverse bracing (alternative)*

Figure 3.63 Strengthening of Main Girders under Traffic

4 Strengthening of Steel Bridges

4.1 General Considerations

During service, bridges are subject to wear; in addition, the initial volume of traffic has increased, particularly over the last 50 years. Many bridges therefore require strengthening.

It is first necessary to assess the condition of the bridge, and then, if necessary, to have a scheme for strengthening. The inspection should consider:

- the age of the bridge and any repairs;
- the extent and location of any defects: cracks, local deformations, corrosion, etc.
- in situ data on steel grade, stress and strain at different points, etc.

The assessment should include a feasibility study to demonstrate the cost-benefit of strengthening. It must be emphasised that strengthening may extend the life of a bridge by about 20-40 years. It does not create a new bridge, and hence is only a realistic proposition if the cost is less than 40% of a replacement bridge [Bancila and Bondariuc, 1996], [Iványi, 1998].

4.2 Methods of Strengthening

4.2.1 Direct Strengthening (Strengthening with the enlargement of the cross-sections)
This is used to overcome local defects and involves fitting additional elements. There are several possibilities:

- - Existing flush surfaces to which new elements are fitted directly (Figure 4.1a).
- - Surfaces which require some initial preparation, e.g. by removing the heads of rivets (Figure 4.1b) to provide a flush face to which new elements can be fitted.
- - Strengthening of the longitudinal girders (Figure 4.1 c), reinforcing of the longitudinal girder connections (Figure 4.1 d) and strengthening of the cross girders (Figure 4.1 e)
- - Reinforcing of the splice of main girders (Figure 4.1 f)
- - Reinforcing of the connections of truss element (Figure 4.1 g)

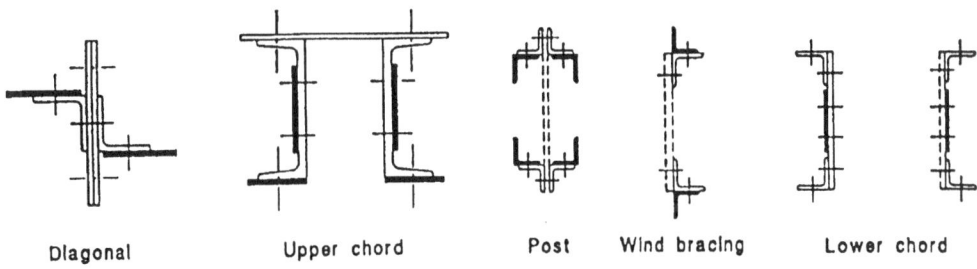

Diagonal   Upper chord   Post   Wind bracing   Lower chord

Figure 4.1a Direct reinforcement with additional elements applied on flush surface

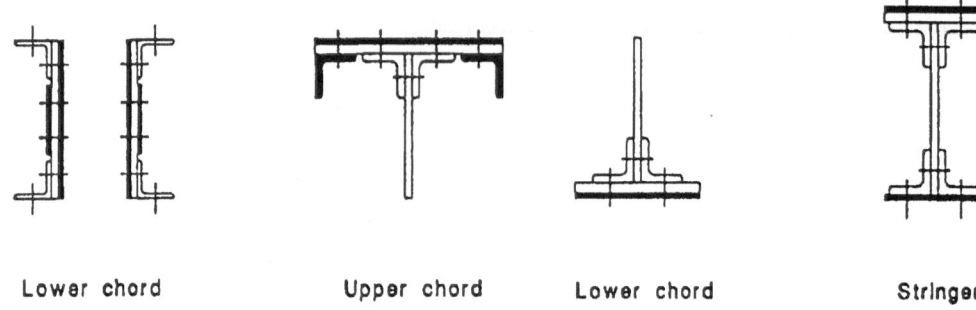

Lower chord   Upper chord   Lower chord   Stringer

Figure 4.1b Direct reinforcement after surface preparation, e.g. by removing rivet heads

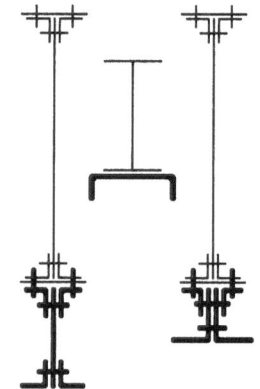

Figure 4.1c Strengthening of the longitudinal girders

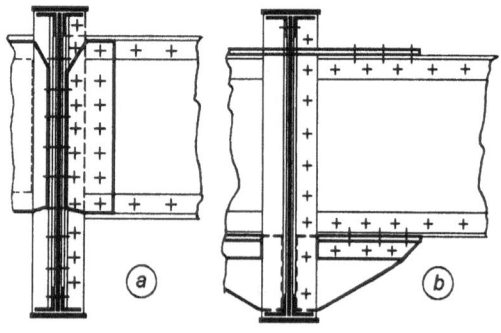

Figure 4.1d Reinforcing of the longitudinal girder connections

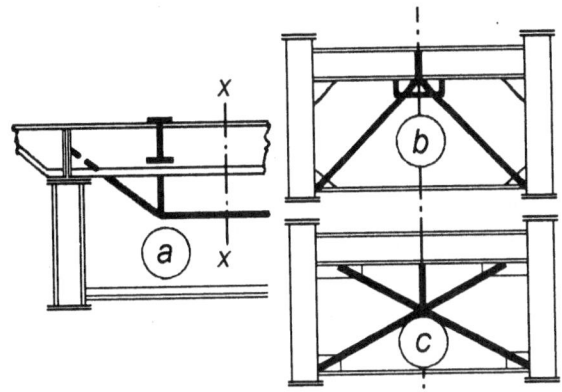

Figure 4.1e Strengthening of the cross girders

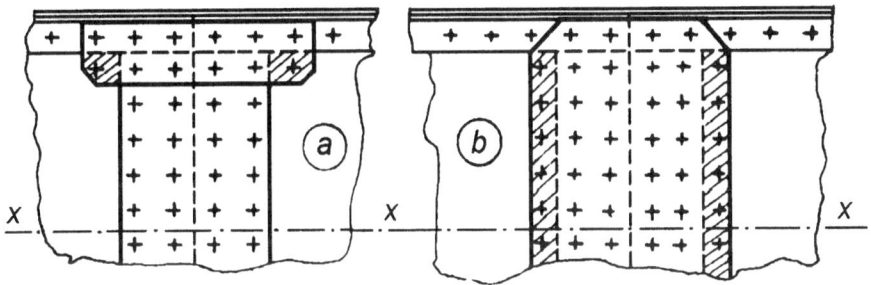

Figure 4.1f Strengthening of the splice of main girders Strengthening of the cross girders

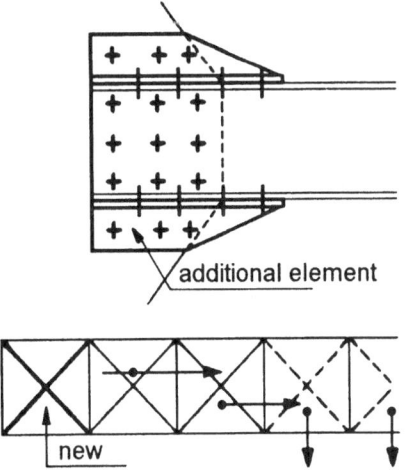

additional element

new

Figure 4.1g Reinforcing of the connections of truss element

4.2.2 Indirect strengthening (Strengthening with changing the statical behaviour of structures)
In this case independent elements are inserted within the structure. There are many types. Some
may allow initial pre-stressing forces, thereby increasing the efficiency of the reinforcement.
    The principal methods of providing indirect strengthening are:

Strengthening with tendons
The tendons can be a simple bar or a built-up section placed at the level of tension chord; it may
be prestressed and is typically used to strengthen bridges of lattice girder or truss construction
(Figure 4.2 and Figure 4.3).

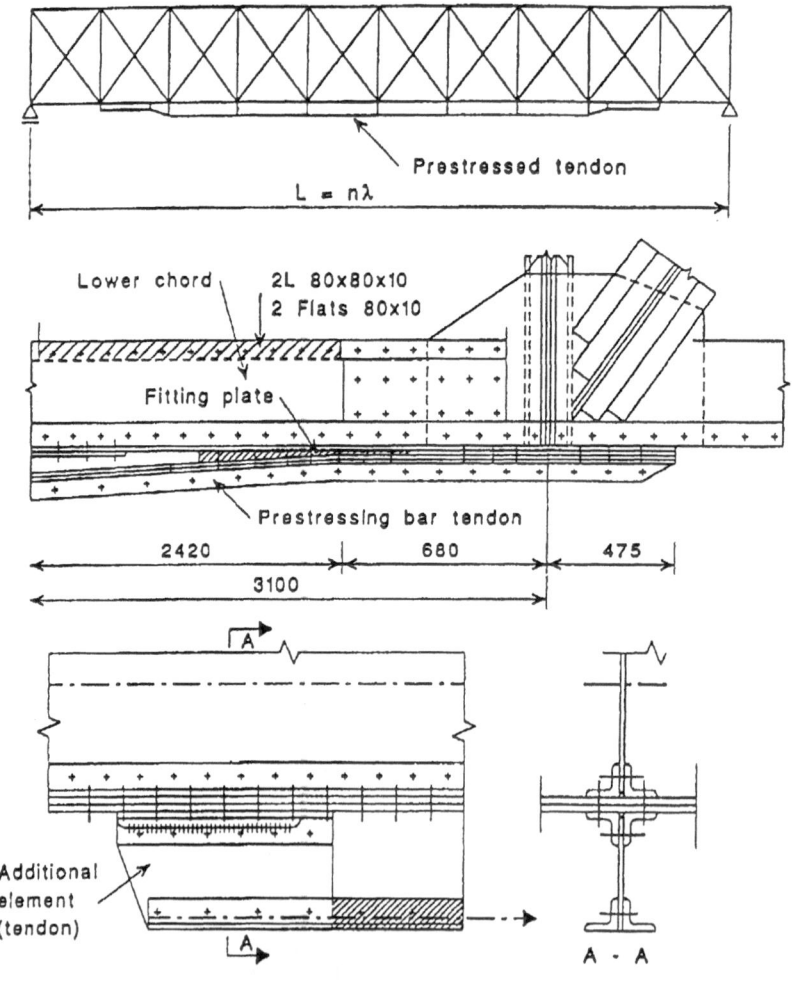

Figure 4.2 Reinforcement with tendons

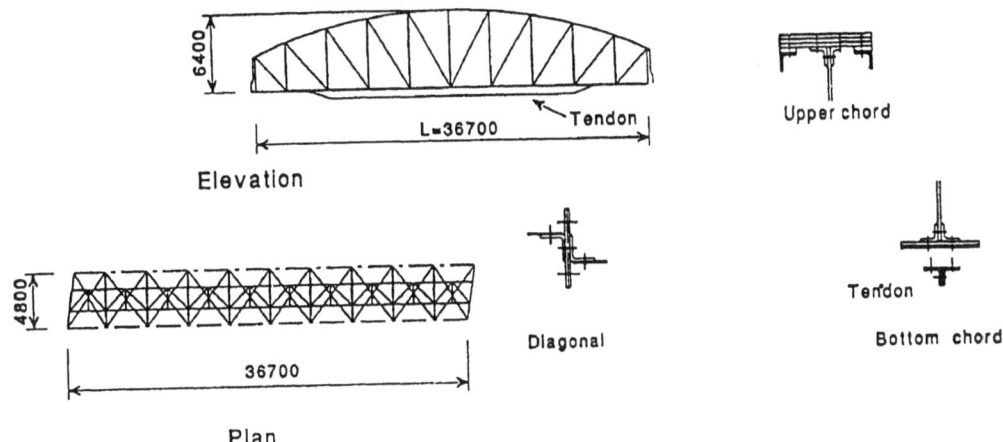

Figure 4.3 Reinforcement with tendons to bottom chord and applied elements to upper chord and diagonals

Strengthening by cables

Prestressed cables can be used in a similar way, fitted alongside or within existing tension elements.

For the example shown in Figure 4.4, three cables were introduced at the centre of the lower tension chord; as a result of the pre-stressing force in the cables this chord became almost completely unstressed under dead load. To increase the load carrying resistance of the complete trusses, the upper chord and the lateral diagonals were also reinforced but using direct strengthening measures.

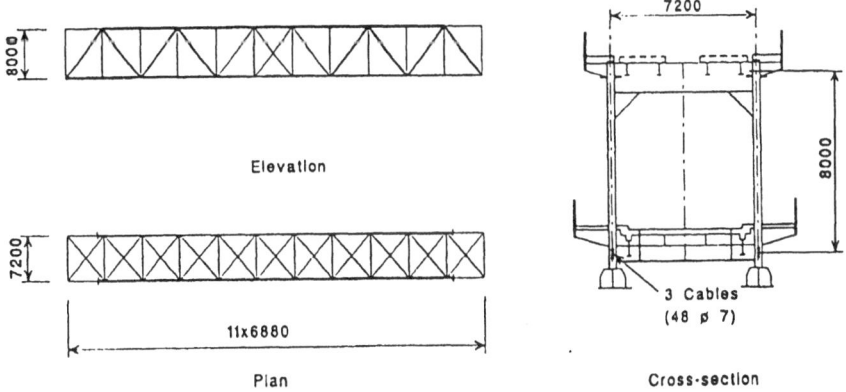

Figure 4.4 Direct reinforcement with cables

Reinforcement by additional trusses

Deck bridges can be strengthened by adding a third main girder connected to the existing structure in order to increases its load carrying resistance (Figure 4.5).

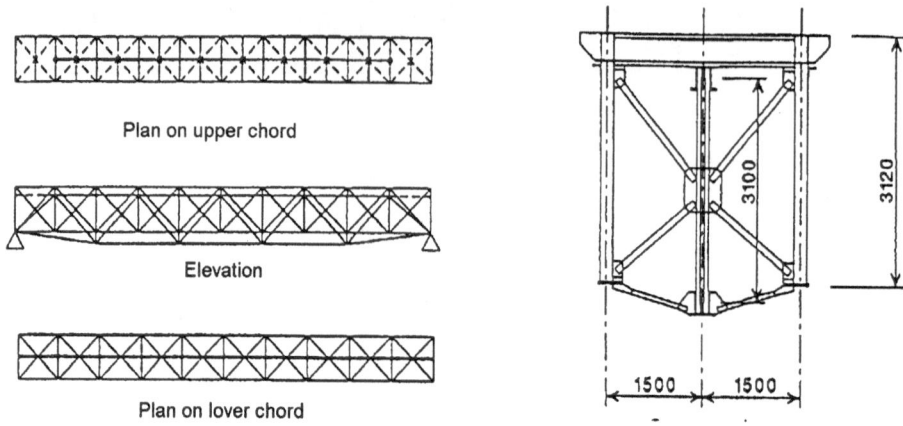

Figure 4.5 Reinforcement by introduction of the third truss-girder (deck bridges)

Increasing the effective depth.

Reinforcement by a new chord connected to that existing by a triangulated arrangement of members, effectively increasing the height of the main truss (Figure 4.6).

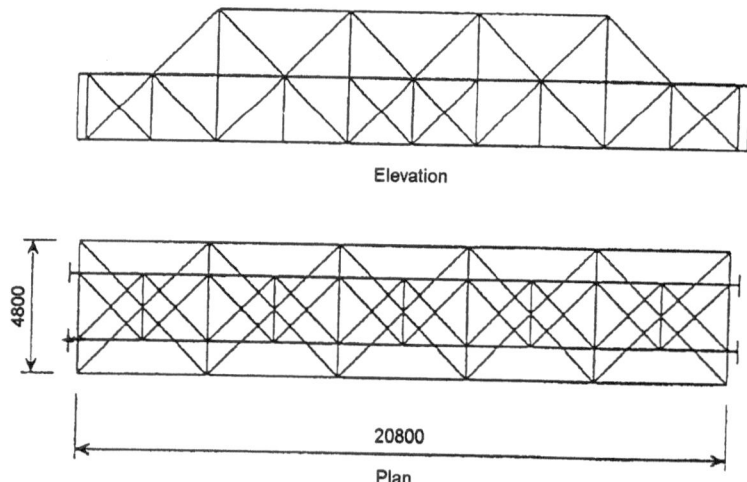

Figure 4.6 Reinforcement by introduction of the third truss-girder (deck bridges)

Strengthening plate girders by transformation into composite sections

The concrete slab replaces the sleepers; this maintains the original constructional depth. The concrete slab will act as an integral part of the compression flanges of the stringers and floor beams, Figure 4.7.

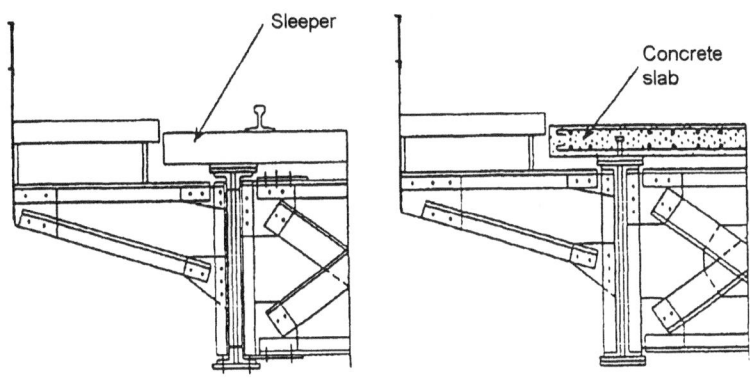

Figure 4.7 Strengthening by transformation into a composite structure

Reinforcement of the transverse frame

For semi - through trusses and through trusses where the wind-bracing has been removed during electrification of the line, it may be necessary to increase the rigidity of the transverse frame. For this purpose a tendon is introduced into the cross-section (Figure 4.8).

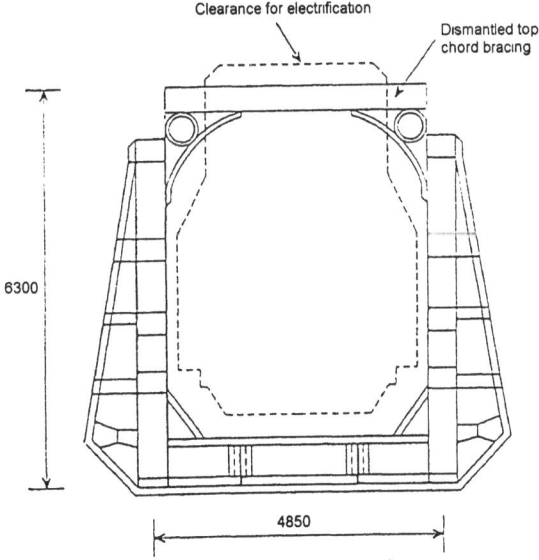

Figure 4.8 Strengthening the transverse frame

Support of the intermediate point of truss members: A new cross girder fitted to the half way of the old cross girder, and this new cross girder hung up with secondary truss elements to the diagonal (Ujpest Danube Bridge, 1932-35), see Figure 4.9 [Korányi, 1934].

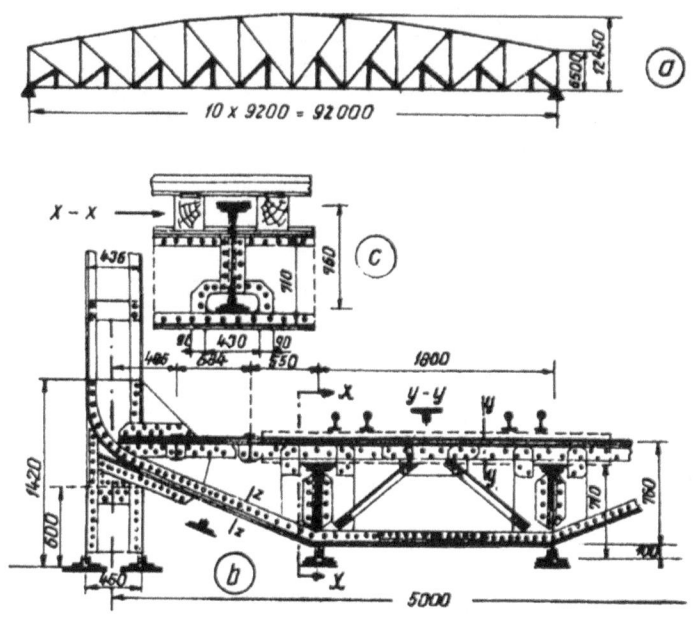

Figure 4.9 Strengthening of the diagonals with secondary truss elements
(Újpest Danube Bridge, 1932-35)

5 Bridges on the Danube
History, Design, Construction and Refurbishment

5.1 The River Danube

The river Danube is a unique international waterway flowing 2857 km across Europe. It flows from the heights of the Schwarzwald Massif down to its estuary in the Black Sea delta. In its passage, the river crosses 22 geographical longitudes joining 8 countries: Germany, Austria, Slovakia, Hungary, Yugoslavia, Romania, Bulgaria and Ukraine.

It is largely these factors, together with the importance of the management of its water resources, which historically encouraged civilisations and cultures to develop along the banks of the river.

The River Danube drainage basin includes glacier-covered mountains, mid-mountain chains clad with forests, karst formations devoid of growth, highlands and uplands, table lands, plateaus with deeply carved river valleys, and wide plains and depressions.

The Danube with a total length of 2857 km is listed immediately after the river Volga (length 3740 km), as the second biggest river in Europe. In terms of length it is listed as the 21st biggest river in the world, in terms of drainage area it ranks as 25th with a drainage area of 817000 square km.

The Danube basin extends in a westerly direction from the Black Sea into central and southern Europe. The maximum length of the river basin is 1630 km.

Twelve countries share the Danube catchments, though more than 70 % of the catchment lies within four countries. The smallest, almost insignificant parts of the catchment area are in Switzerland, Albania, Italy and Poland.

The sources of the Danube tributaries adjoin the source of the Rhine tributaries in the west and Northwest, the Weser, Labe, Odra and Visla river basins in the north, the Dnester river basin in the Northeast and the basin of the rivers flowing into the Adriatic and Aegean Sea in the south.

The rich, fertile riparian Danube territories attracted the interest of nations pursuing trade, as well as conquerors and nomads since earliest times. Early in the 8th to 7th century B. C. Phoenicians, Egyptians, and Greeks penetrated the Danube estuary, entering into trade contacts with the local population.

In the 6th century B. C. Persian Monarch Darius the First had tried to occupy the territories adjacent to the lower Danube basin, as did Alexander the Macedonian in 334 B. C. In the first century A. D. the upper Danube formed a part of the Roman Empire. In the years 101-106 the Roman Emperor Traianus defeated the Dacian tribes settled in the lower Danube basin. He had built the Traian's Bridge across the Danube at Turnu Severin in the Iron Gate Gorge, and on the right bank he constructed the Traian's Road, the existence of which is documented by the Tabula Traiana.

To investigate the origin of the name "Donau" (Danube) one has to go back to the Celtic tribes who lived in the upper Danube basin. The word "Danu" is of Celtic origin and signifies "swift, rapid, violent, undisciplined". Emperor Caesar in his work "De Bello Gallico" had named the stream Danubius. Phoenicians and Greeks had explored the river from the estuary upstream but knew only the lower course, known as Istros or Ister.

The name Danubius can be found in the works of Aristoteles, Ovid, Strabo, Pliny the Elder. "Istros" is mentioned in works by Herodotus and Virgil, when referring to the lower Danube, and Histerus is mentioned by Cicero. Other nations migrating along the stream called the river Donau, Dunaj, Duna, Dunav, Dunarea, the name being transformed and modified according to the language of the people.

In the age of the Roman Empire the Danube had an important strategic significance and this had persisted to the present time. Considerable migration of nations took place along the Danube, including the crusaders, and later the river became a strategic waterway and an artery of life influencing the development of European history. At the same time the economic significance of this mighty river had been growing.

5.2 Types of Danube Bridges

The main function of a bridge is to carry traffic over a crossing, which may be a river, a canyon, or another line of traffic. Besides serving its specific purpose safely and economically, a bridge should be designed aesthetically so that to will fit into and often enhance the beauty of its surroundings.

A good bridge layout must take into consideration the geographical and geological conditions of the site. Clearance requirements, erection procedures, and the method of foundation construction will affect the type and span of the superstructure. The designer should consider all these factors during analysis and detailing.

Of the thousands of bridges existing throughout the world, no two bridges are identical. They differ at least in some details. However, they may be classified into a number of main types.

Bridges can be classified either according to the service they perform or according to the structural arrangement they possess. The majority of bridges are either highway or railway bridges. There are also bridges carrying a combination of traffic, such as a highway bridge with streetcars or pedestrian sidewalks, or a railway bridge carrying highway traffic at the same time. Occasionally there are bridges for pedestrians only, or bridges carrying canals and pipelines. Some bridges are moveable, they can be opened either vertically or horizontally so, as to permit river traffic to pass under them.

Classified according to the cross section of the bridge, a deck bridge is one that has its floor resisting on top of all the main load carrying members, so that there is no bracing over the top of the traffic. If the floor is connected to the lower portion of the main load carrying members so that the bracing goes over the traffic, it is called a through bridge. If there is no overhead bracing but the main load carrying members project above the floor level, it is called a semi-through or a pony truss bridge. A double-deck bridge is one, in which there are decks on two different levels, both of which can be through decks, or one can be a through and the other·an open deck. For economy in bracing, either the deck or the through type may have a triangular section where the roadway is supported by two inclined frames.

Classified according to the make-up of the main load carrying members, an I-beam bridge has rolled I beams as the main load carrying members. When spans exceed a certain limit, built-up plate girders are used and they are known as plate girder bridges. For still longer spans, truss bridges are usually more economical. Occasionally, the Vierendeel truss with quadrangular framing instead of the usual triangular framing is used. For very long spans, the suspension bridge is found to be economical, with high-tensile-strength steel cables carrying the main loads. Suspension bridges are usually stiffened with girders to obtain rigidity. In all these cases, the number of trusses or cables may vary from two to three or more, depending on the economy of layout. Single truss bridges have been occasionally advocated, though seldom built as yet.

Classified according or the structural layout of the main load carrying members, most of truss, girder, and beam bridges can have the following various arrangements. One common arrangement is the simple span type where the main carrying members span from one support to another, but are discontinuous over the piers. It is sometimes economical to make the spans continuous, so as to reduce the maximum positive moments. Such a layout is slightly more difficult to analyse and may be objectionable if the foundations are likely to settle unevenly, thus producing settlement stresses in the members. Some engineers prefer the cantilever layout to the continuous, possessing favourable moments along its length but not subject to settlement stresses and also easier to analyse. However, a cantilever layout requires special hinge arrangements and is less rigid that a continuous one. Besides, the continuous layout has a higher ultimate load carrying capacity, as indicated by the theory of limit design.

The arch layout is considered to be more aesthetic than the simple spans. The arch itself can be made of girders or trusses as may suit the case.

The most common design is to provide two hinges, one at each support - rendering the arch stresses statically indeterminate to the first degree. But the three-hinge, one-hinge, and hingeless types are sometimes employed. The horizontal component of the arch reaction force may be resisted by the abutments and piers, or by ties along the roadway.

Classified according to the type of connections, the great majority of steel bridges are riveted, although welded and bolted bridges are being designed and constructed more and more.

Taking into account the classification categories mentioned above, the structural system of the main girders of the Danube bridges was summarised in tables, including the traffic conditions as well.

Truss Girder Steel Railway Bridges
Truss Girder Steel Highway Bridges
Truss Girder Steel Railway and Highway Bridges
Plate Girder Steel Bridges
Box Girder Steel Bridges
Reinforced Concrete Bridges
Arch Bridges
Cable Stayed Girder
Chain Suspension Bridge
Cable Suspension Bridges
Pipe – Line Bridges

5.3 About the CD-ROM

By the proposal of the Hungarian National Committee of the International Association of Bridge and Structural Engineering (IABSE), the Department of Steel Structures, Technical University of Budapest, with the active cooperation of similar departments of the countries along the Danube, first of all in Bratislava, Vienna and Belgrade, in 1990 began to organise the International Conference "BRIDGES ON THE DANUBE" for September 7 - 11, 1992 held in Vienna - Bratislava - Budapest.

On the occasion of the Conference - for subsequent publication - collection of material for a catalogue of the Danube bridges had been initiated, including all kinds of technical and historical data.

For this task a Committee was formed including members from Germany, Austria, Slovakia, Hungary, Yugoslavia and Romania.

The collected material can be originated from different sources.

Generally used references were:

ALBUM DES PONTS (Description des conditions de passage des batiments sour les ponts sitnes sur le Danube), Commision du Danube, Budapest, 1967.

INDICATEUR KILOMETRIQUE DU DANUBE Commision du Danube, Budapest, 1992.

In Romania a detailed collection work was carried out, supervised by Prof. Radu Bancila. Their material contained data for the Bulgarian-Romanian and Yugoslavian-Romanian bridges. Very efficient contribution has been given by Dipl.-Eng. Sabin Florea, President - Executive Director, Stock Company VIACONS S. A.

In Yugoslavia Bratislav Stipanic summarized the main important data of the bridges in a table, it was the basis of the further ones. Figures were taken from the general references mentioned above or they were made on the basis of the figures of the Proceedings papers from Yugoslavia. In this third version a more completed material can be found about the Yugoslavian bridges thanking to his efforts.

In Hungary the basic materials for the catalogue were the study of UVATERV Bridge Desing Office, prepared with the collaboration of the Department of Steel Structures, Technical University, Budapest in 1985-86, titled "Tanulmány a magyarországi Duna- és Tisza-hidak fejlesztésére" and the summarising studies of Jenő Hargitay. A great number of data was presented by László Szabó museologist (Hungarian Museum of Transport).

In Slovakia Prof. Eugen Chladny and Prof. Zoltán Agócs were the leaders in the data collection work concerning the bridges of the Slovakian Danube reach.

In Austria Prof. Günter Ramberger was responsible for this task. The fact, that in the frame of her diploma work in 1991-1992 Ms. Maximiliane Pisecker made the collection and compilation, was a very important guidance.

In Germany Prof. Herbert Kupfer and Prof. Friedrich Nather were the reponsible persons for the catalogue. Along the Danube in Germany there are bridges in a very high number, the present collection contains the German bridges up to Regensburg. Ms. Pisecker's diploma work was an important tool even for the Bavarian Danube bridges.

Collection of the data of bridges above Regensburg was leaded by Ministerialrat Dipl.-Ing. Jürgen Weber with the contribution of Dipl.-Ing. Hans Grassl (TU Munich).

In the Catalogue data are presented in an adjusted, unified form.

The compilation work was carried out at the Department of Steel Structures, TU Budapest. The undersigned got very valuable assistance during it from the late Sen. Assist. Prof. Mihály Farkas and Assist. Prof. László Horváth staff members, and from Gabriella Verci, civil engineer.

In the name of the Catalogue Committee of the Conference BRIDGES ON THE DANUBE we are grateful to all of our colleagues for their activity.

In 1993 some new data and corrections were presented by Assoc. Prof. Ferenc Szépe and Dr. Herbert Träger.

A new revision of the Catalogue was carried out for the occasion of the second international conference "Bridges over the Danube" being held in Bucharest in September 1995. This version contains three new bridges additionally and some further corrections, additions to the previous material.

The CD version has been preparred for the third international conference "Bridges on the Danube" being held in Regensburg in October 1998, with the valuable contribution of the Local Committee (K. Zileh, G. Albrecht, A. Swaczyna, J. Weber)

Most of the bridges are illustrated by colour photos. The colour photos for the bridges of the Hungarian, Slovakian, Austrian and German Danube reach have been made by Dr. László Kristóf.

Compilation of the CD ROM has been made by Moare Ltd. (László Tóth, Tamás Bolberitz, Balázs Fekete, Balázs Juhász) in Budapest, Hungary. Send your feedbacks to tocsi@mail.datanet.hu

The compilation of the CD-ROM was in October 1998.

Dr. László Hegeds   Prof. Dr. Miklós Iványi
Scientific Secretary   Chairman
of the Catalogue Committee

5.4 Case Study on the Danube (Sample)

1645.300 km "Szabadság" Bridge (Liberty Bridge)
H/6 highway bridge with tramway

5.4.1 Location and name of the bridge

Name of the bridge:	"Szabadság" bridge (Fig. 5.1)
Distance:	1645+300 km
Country:	Hungary
City/town:	Budapest
Year of building - completion:	1894-1896
Year of reconstruction:	1945-1946, 1979-80, 1985-86
Span lengths:	79.3 + 175.0 + 79.3 m
Roadway widths:	2.9 + 1.4 + 11.5 + 1.4 + 2.9 m.
Designer:	MÁVAG, on the basis of the original plans
Main contractor:	MÁVAG, Gusztáv Fáber
Construction cost:	6.164 million HUF

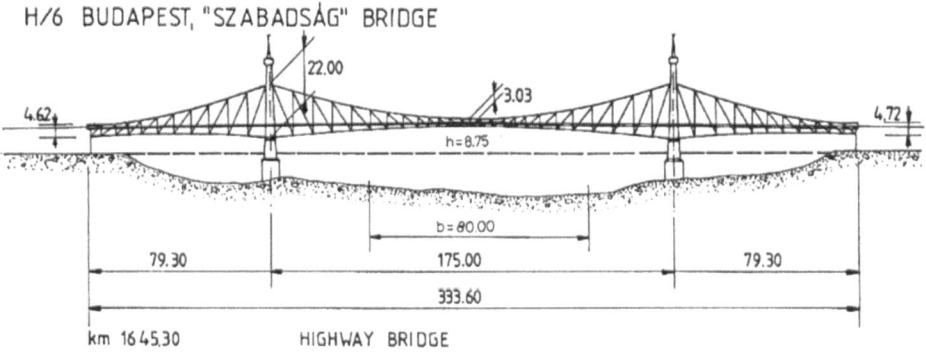

Figure 5.1 "Szabadság" Bridge

5.4.2 Traffic function of the bridge

Highway bridge carrying tramway line

Number of lanes:	2
Number of tramway tracks:	2

5.4.3 Antecedents; the history of the bridge

The construction of the Erzsébet ("Elizabeth") and Szabadság ("Liberty") bridges was ordered by law No. XIV. in 1893 on the occasion of the forthcoming millennium of Hungary in 1896, and a common tender was held with great success. All together 74 designs were presented, among them 53 for the Elisabeth bridge, and 21 for the Szabadság bridge. The members of the international jury were bridge engineers of the highest reputation of that time. The reporter of the jury was Antal Kherndl, the well-known professor of Technical University. 41 designs were solved with a single-span structure; 15 designs were Hungarian, 16 ones were American, several ones were Italian, Austrian, German and French.

The first prize was given to the cable-bridge of Julius Köbler from Germany for the later Elisabeth bridge; while the second prize winner plan of János Feketeházy (Fig.5.2) and the third prize winner plan of Róbert Totth prepared for the Liberty bridge.

The second prize winner plan of János Feketeházy was realized with small modifications. The details of the plan were worked out by István Gállik and József Beke (Fig.5.3 and Fig. 5.4), he portals, having an architectural importance, were designed by Virgil Nagy, professor of the Technical University, who was the architect specialist of the design team (Fig 5.5).

Figure 5.2 The Prize winner plan of János Feketeházy

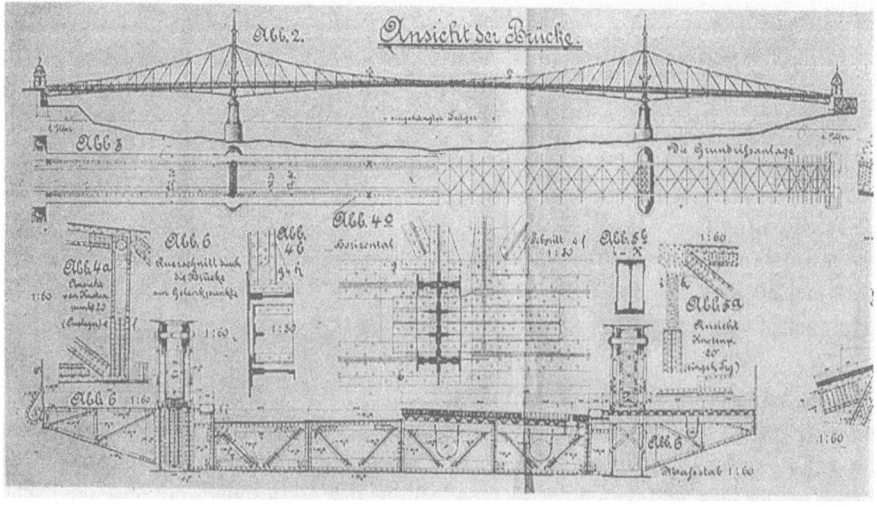

Figure 5.3 The general plan of the original structure

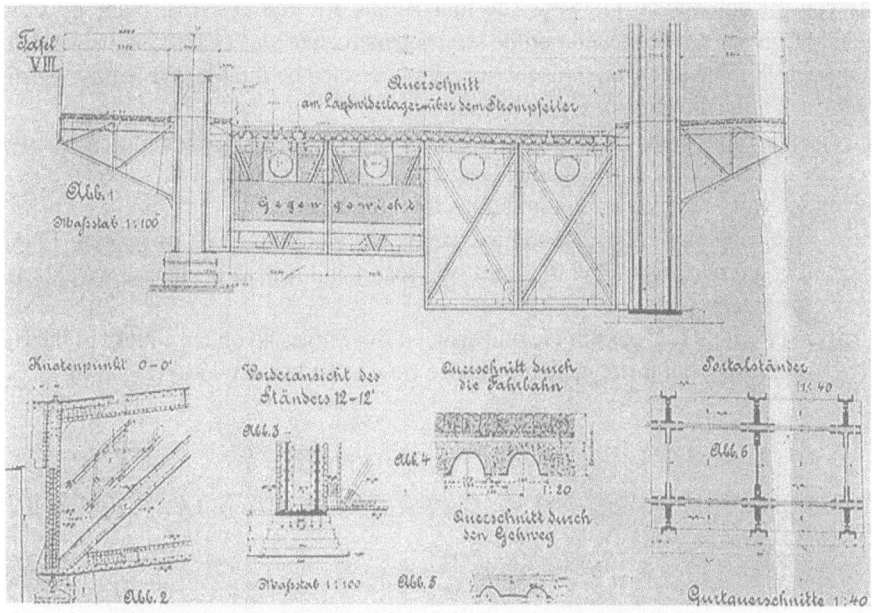

Figure 5.4 The cross-section and the floor system of the original structure

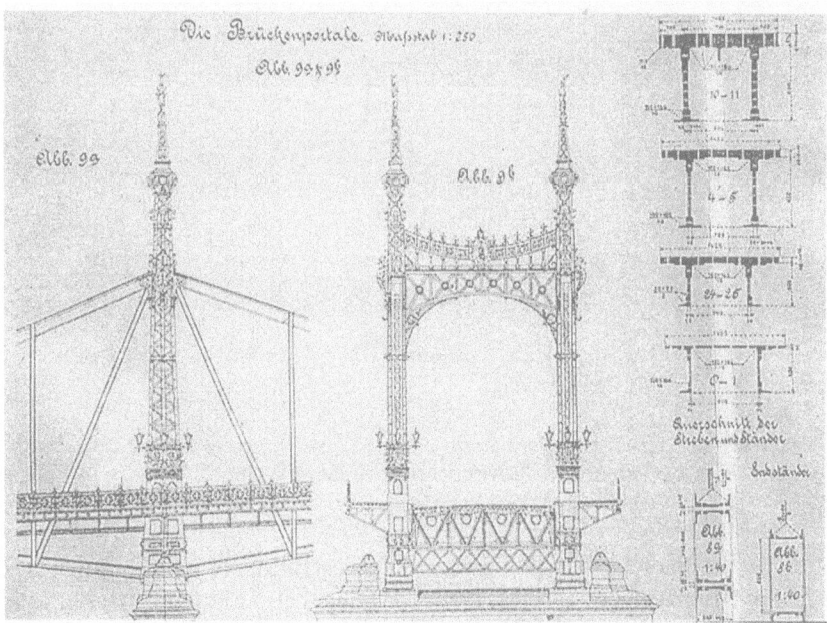

Figure 5.5 The portal and the cross-sections of the main bars

The construction started in 1894. The foundations are iron caissons, made by Gartner and Zsigmondy Company. The erection of the steel superstructure was carried out between July 1895 and August 1896. The steelwork was fabricated and erected by the Magyar Királyi Államvasutak Gépgyára, under the direction of Gyula Seefehlner.

The opening of the bridge took place on October 4th, 1896, during the millennium ceremony in the presence of Ferenc József, Hungarian King and Austrian Emperor.

The bridge was called Ferenc József bridge for it for decades.

The simple supported mid-span and the cantilevers were blown up in January 1945. During construction of a provisory bridge; the Buda side span fell down, as the ballast weights at the side support were not removed.

A provisory bridge was built in the mid-span on five barges an on the wrecks of the Buda span from 15. March 1945, but it was drifted by an ice-flow in 10 January 1946 (Fig. 5.6).

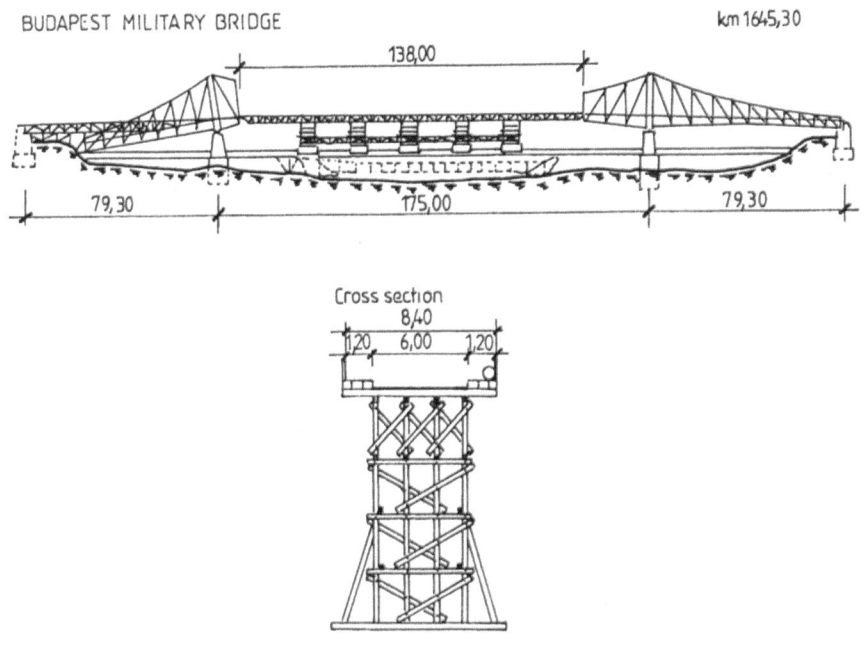

.**Figure 5.6** Provisory bridge in January 1946

As the bridge was the relatively less damaged one on Budapest, its reconstruction was the most realistic. For this material of the other destroyed bridges also was used. The midspan had to completely rebuilt, its material was produced in the iron-works of Diósgyr and Ózd. It was opened 20 August 1946.

The main girders of this steel bridge are of the cantilever type running over three spans (Fig. 5.7.) [Haviár, 1947].

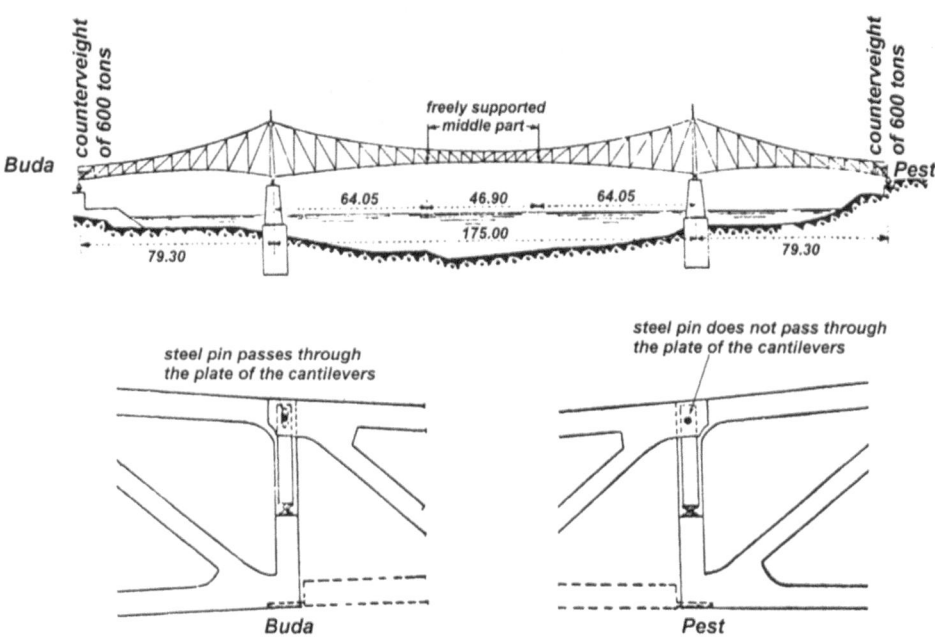

Figure 5.7 The cantilever type main girders

The reconstruction was carried out on the basis of the original plans. 2250 tons of steel were needed for the reconstruction - representing 38 % of the total steel material. About 550 tons of steel necessary for the less important cross girders were taken from the wrecks of two other Danube bridges, so that only an amount of 1700 tons had to be supplied by the rolling mills.

The reconstruction of the bridge was executed in three steps.

I. For raising the 730 tons weight of the anchor arm on the Buda side an enormous scaffolding was erected. The uplifting device worked by 16 hydraulic jacks of 100 tons each was placed on the top of the stage. The lifting forces of the jack were transmitted through thrust blocks and through huge suspending rods (bars) made of chain links of the Erzsébet-Bridge. The latter served also as straps for binding the sunk down ends of the main girder.

The bridging scheme as, planned by Chief-Engineer Ladislaus Lébényi was the following.

Two thrust blocks were supported directly by two hydraulic jacks (Fig. 5.8 and 5.8.a.) and the perforated lifting rods were passed through the slot. A high resistance steel pin was put across the bore of the rod and supported on the bearing on the top flange of the thrust block in order to transmit the force of the hydraulic jacks to the lifting rod. The rod was perforated at every 20 centimetres, the holes having a diameter of ti5 millimetres. The maximum throw of the piston of the hydraulic jacks was also 20 centimetres. When the pistons arrived at the stroke end the jacks had to be released and their pistons brought back to their initial position. For this purpose another pair of similar thrust blocks was placed underneath. The load of the lifting rod (chain) could be transferred temporarily to the pin passed through these additional cross blocks. For transmitting the load the jacks were emptied, the upper thrust blocks and pin were lowered down by 20 centimetres and the hydraulic jacks with pistons in the initial position could raise the structure by another 20 centimetres. The height of lift was altogether 9 metres 30 centimetres and by such steps of 20 centimetres all the work was done within four days.

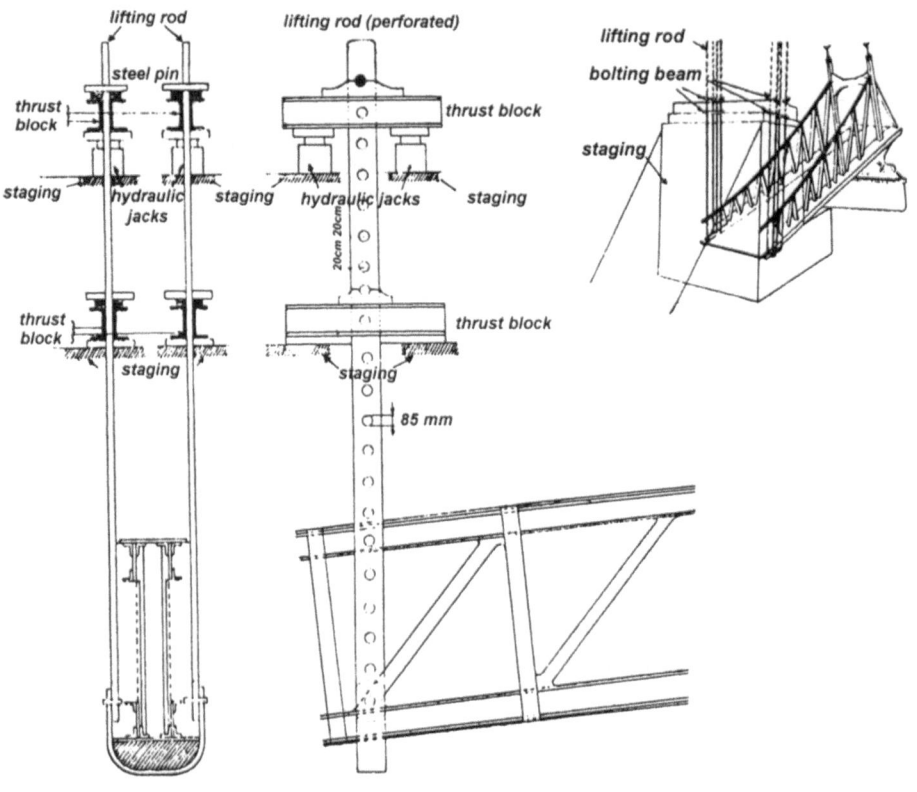

Figure 5.8 The bridging scheme

II. The wrecks of the 2,5 metres long broken part of the right anchor arm were removed and a newly fabricated steel structure was built. The new portion was adjusted from a stage erected under the side span. The uplifted part of the bridge was carried by the above mentioned chain links while the new bays were being riveted together with the old ones. When all this was done the steel super structure was let down on , its final supports by taking out the supporting wedges.

Apart from this 25 metres long section the other missing bays of the bridge in a length of about 140 metres were erected without any false-works partly because the wrecks of the blown bridge would have hampered piling and partly because of the great timber shortage.

The cantilever trusses of 64 metres each were built out by cantilever method piece by piece and so the main girders of the 46.9 metres freely suspended middle span carried by two floating derricks of 100 tons each in the middle of the river and placed directly on the hinges of the projecting ends of the cantilever arms.

The assembling of the cantilevers themselves was done with the aid of a derrick of 5 tons moving forward on each end of the existing structure, the lifting up of the heavier pieces was effected by means of a floating derrick of 100 tons.

Towards the ends of the cantilevers the procedure was simplified. The panels were smaller and lighter and their top chords could be tied to the derrick so that it was possible to assemble them on the embankment and thus raise them into position in one piece.

The freely suspended middle part of the cantilever bridge was 46.90 metres long and each main girder weighed 120 tons. They were assembled and riveted together on the embankment. The next step to be taken was to bind the two 100 ton floating derricks stiffly together with timber and wire cables. The structure to be lifted was bound to the lifting hooks of the floating derricks by huge clamps. Then a tug towed the derricks up to a distance of about 30 metres from the bridge where they cast anchor, hence the towing was done with the aid of winches bound to the anchors and driven by the motors of the floating derricks. By means of the anchorage chains the floating derricks got into the right position necessary for lifting and suspending the main girder in its final place. The procedure took about four hours for each main girder.

III. The third stage of work was the repair of the bomb-damaged parts of the old steel structure on the bridge itself. With a single exception these works were done without falseworks. The damaged vertical struts of the main girder were replaced one by one. At places where the load was greater, a special device was used. The hardest work which needed most preparation was the replacement of the two adjoining vertical struts of the main girder, marked 8 and 9. 'The lattice work of the verticals being shattered, some sections of the columns buckled sideways to an extent of 35 centimetres so that their length became shorter by 11.2 centimetres and 6.7 centimetres respectively and both the main girder and the deck sagged by about 10 centimetres.

As a consequence of the buckling of the two struts the stresses in the chords computed by Hook's formula exceeded the ultimate tensile strength of 3600 kg/cm^3 of the steel material. (Fig. 5.9.) Actual stresses, however, exceeded but slightly the yield point, because straining compensated and levelled the increase of the stresses. However, luckily the bridge did not collapse owing to the fact that the stiffness against bending was much greater in the chords than in the lattice work. In this way similarly to a continuous girder the chords transmitted the load over the damaged vertical columns to the other undamaged parts of the bridge. At the final reconstruction of the bridge it was impossible to insert columns which were shorter by 11.2 and 6.7 centimetres respectively because such great bending stresses would have remained in the lattice work that in case of live load the structure would have broken owing to these excessive stresses. The chords which were bent had to moved apart first. This could be done in two ways:

1. with a direct method by lifting up the damaged main girder.
2. with an indirect method by actually pressing the top and bottom chords apart by the aid of temporary steel struts worked by hydraulic jacks.

At the reconstruction both methods were resorted to.

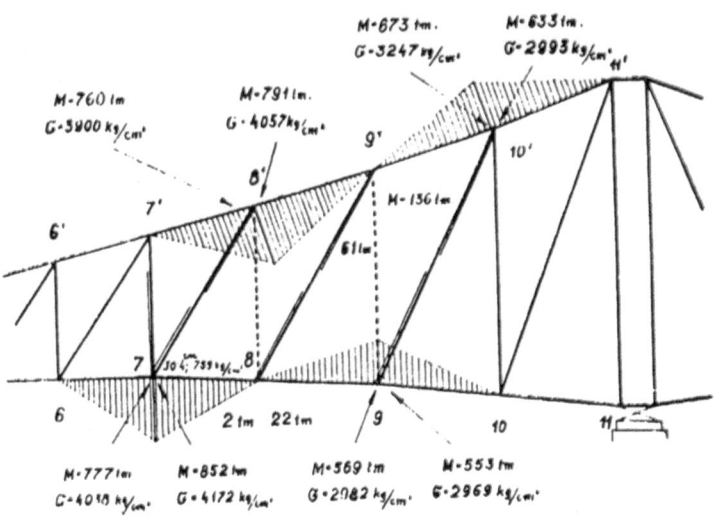

Figure 5.9 The computed stresses in the chords

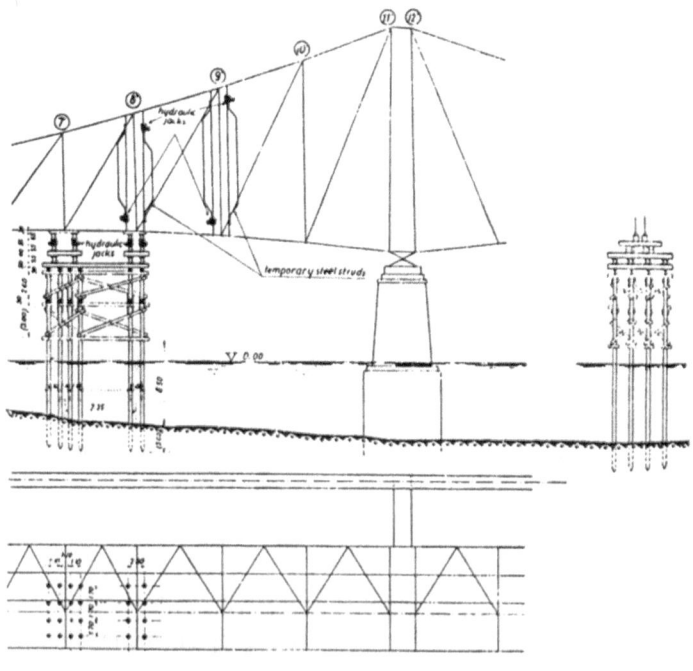

Figure 5.10 The stage for lifting

The stage needed for lifting was built on 24 wooden piles driven in the river bed at the knots of the main girder marked 7 and 8. (Fig. 5.10.) Beside each buckled strut two temporary steel columns were placed for pressing the chords asunder. On one end of these temporary columns hydraulic jacks were inserted.

In order to replace the buckled struts the following steps were taken.

1. The struts were released of stress before being cut off. 2. Their sprained ends were pushed asunder by the aid of the pressing device. 3. The new struts were then placed in and fixed between the old knots. 4. This being done the hydraulic Jacks of the pressing device were released.

The relieving of the struts was effected by slightly raising the main girders from the stage at two points. The degree of lift was calculated from the deformation of the main girder. The buckled struts were cut off with oxygen torch from two sides, for in this way eventual forces could be balanced without shaking the structure.

2. For pressing the shattered chords asunder the points of the main girder marked 7 and 8 were lifted. The raising forces $P_1 = 186$ t and $P_2 = 22$ t were calculated as supporting forces of the main girder (Fig. 11.) charged with the dead load of the bridge. The main girder was considered without the buckled columns, which had already been removed. By applying these raising forces the gusset plates were released of the moments because the main girder was balanced as a hinged structure. (Fig. 11.)

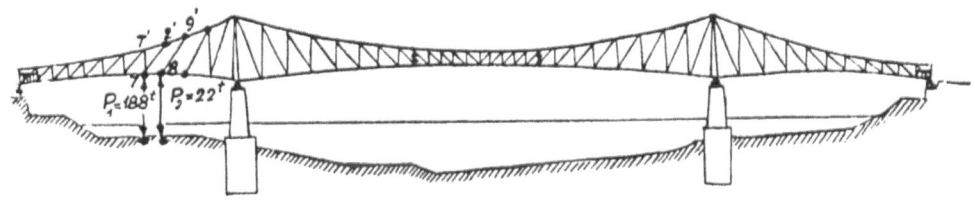

Figure 5.11 The raising forces P_1 and P_2 were calculated as supporting forces of the main girder

When the hydraulic jacks placed on the stage produced lifting; forces P_1 and P_2 indicated by manometers, the chords were pressed apart but not full to their original position because at the buckling of the columns the steel material of the gusset ,plates had passed the yield point and had undergone permanent deformation. In order to decrease permanent deformation chords were moved asunder by the application of further 400 tons with the aid of the hydraulic jacks of the pressing device. In this way the distance between the chords increased to such an extent that the new steel struts became only 4 centimetres shorter than the original ones previous to their deformation. With regard to possible overstressing of the new struts the actual length of these latter was determined only when the latter 400 ton expanding force had been released.

When the struts had been adjusted, the hydraulic jacks were released on the stage and both pressing device and scaffolding could be removed: At the load test of the reconstructed bridge the obtained results corresponded to those calculated. After testing the bridge was opened to traffic. The delicate work of reconstruction was completed in half a year.

Further reconstructions were made in 1979-80, when the roadway structure was changed to a composite one, and in 1985-86, when the pavements and their supporting structures were renewed. During this process, because of very heavy corrosion damages of some members of the main girder, a local collapse developed in a column of the main truss. After it the corroded parts of the main trusses were strengthened.

5.4.4 The technical data of the bridge

Structural system, span lengths, widths: The statical system of the main girder is a so called Gerber truss, having an approx. 50 m long simple supported part in the symmetry axis. Because of balancing the tension reaction forces from self weight, ballast weights of 609 t each were necessary at both bank supports.

The quality of the shape of the bridge was recognised by many [Merthens, 1898]; the end portal and the bridge is considered as one of the most important monuments of the turn of the century.

Deck: The deck was reconstructed in the original form and changed later into a composite construction..

5.4.5 Method of constriction/erection; joints

First the raising operations of the Buda span took place in 1946. Lifting of the structure was done with 16 hydraulic jacks of 100 t capacity each resting on piled pedestals to a height of 9 m. It was accomplished in five days.

After reconstruction of the two cantilevers the lift-in operation of the mid-span structure was realised with the floating crane of "József Attila" and "Ady Endre" having a lifting capacity of 100 t each. The total weight of the 46.9 m long piece of main was 240 t, so this operation needed special care.

5.4.6 Test loading(s); periodical assessment of serviceability

After 33 years of service the complete floor system and the suspension system of the counterweights had to be reconstructed in 1979. The next step of rehabilitation became necessary in 1985, when the replacement of the side walks and the repair of main elements was decided.

Vertical and diagonal members of the main trusses are led through the slab of the side walks, practically without any gap, which is the "result" of a previous repair. Dust and salted sand, thrown in winter time, caused very heavy corrosion. Cross section reduction of the diagonals was around 10 %, while that of verticals exceeded 40 % in some cases.

During the rehabilitation process, the one of the mostly damaged columns in its complete cross section broke and moved 15 mm downward and 35 mm sideways. The traffic was closed immediately, provisional fixing elements were built in by HSFG bolts.

Reconstruction should be started with the determination of the remaining force in the damaged member. Applying the trepanation method, strain gages were bonded to the elements of the column. Drilling pairs of hole near the ends of strain gages, the remaining stresses could be measured and the total force was calculated. The active force in the column was found as 871 kN instead of 1930 kN, calculated from the dead loads.

Special devices and technology were designed to achieve the necessary position of the column ends, to replace the damaged part of them and to induce the required force into the column. The complete process was continuously controlled and measured.

The renovation of the broken bar was carried out in the following steps (Fig. 5.12):

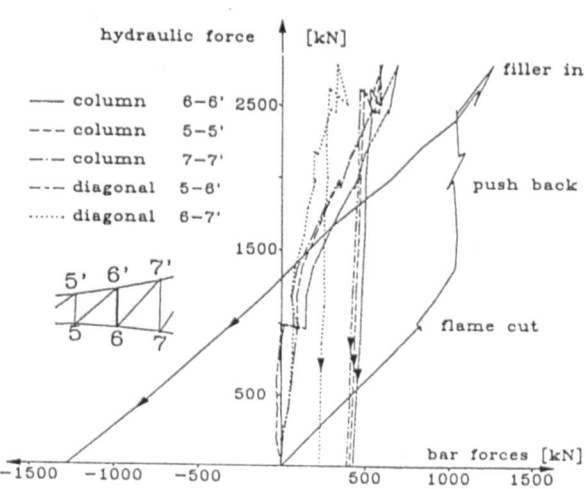

Figure 5.12 Bar forces vs. Hydraulic lifting force during the repair process

- (a) At 20 kN hoisting force the provisional fixing elements were removed.
- (b) The force was slowly increased until the compression force in the broken bar decreased to zero. This in fact happened at 1000 kN, because the raising force was partially taken by the neighbouring parts. At this point, the column was fully flame cut at the plane of the damage. No considerable effect of the cutting process was observed in any of the measuring points.
- (c) The force was increased further, up to 2000 kN. Here, the horizontally shifted bar end of the upper part was forced back to its proper place by two 50 kN jacks.
- (d) Increasing the force up to 2500 kN, the position of the broken bar ends were temporarily fixed. The defected sections were removed and the two opposite butt ends were cut plane. At a force of 2800 kN, the gap between the butt ends opened wide enough for a prefabricated element of 200 mm height to be put into it. Now decreasing the hoisting force back to 2500 kN, four new cover plates were welded to the flanges, then connected by precision bolts to 'the angles.' The bar ends were also welded to the filler in order to ensure the continuity of material and a more favourable transmission of force.
- (e) The repair was finished by unloading the hydraulic system, i.e., by transmitting the force from the trimming structure to the repaired column. While unloading, the transfer of the load to bar 6-6' was found to be linear. The reduction of the lifting force from 2610 kN to zero resulted in a measured compression force of 2548 kN in the upper part of the bar, and, in 2816 kN, in the lower.

To examine the behaviour of the repaired part of the main trusses local loading tests were carried out. The live load intensity in the column achieved the maximum design value. The difference between the actual and calculated behaviour was realistic.

At the end of the reconstruction work e complete bridge was tested again using 30 trucks of 20 t each (Fig. 5.13).

In the middle of the 1990's, it was found that the floor system of the bridge showed corrosion, which necessitates further rehabilitation. Heavy corrosion was also found in the stringers and the cross beams. The two bottom structural parts of the bridge has therefore been scaffolded. At the moment of the preparation of this material (summer 1998), the last phase of the work is being done.

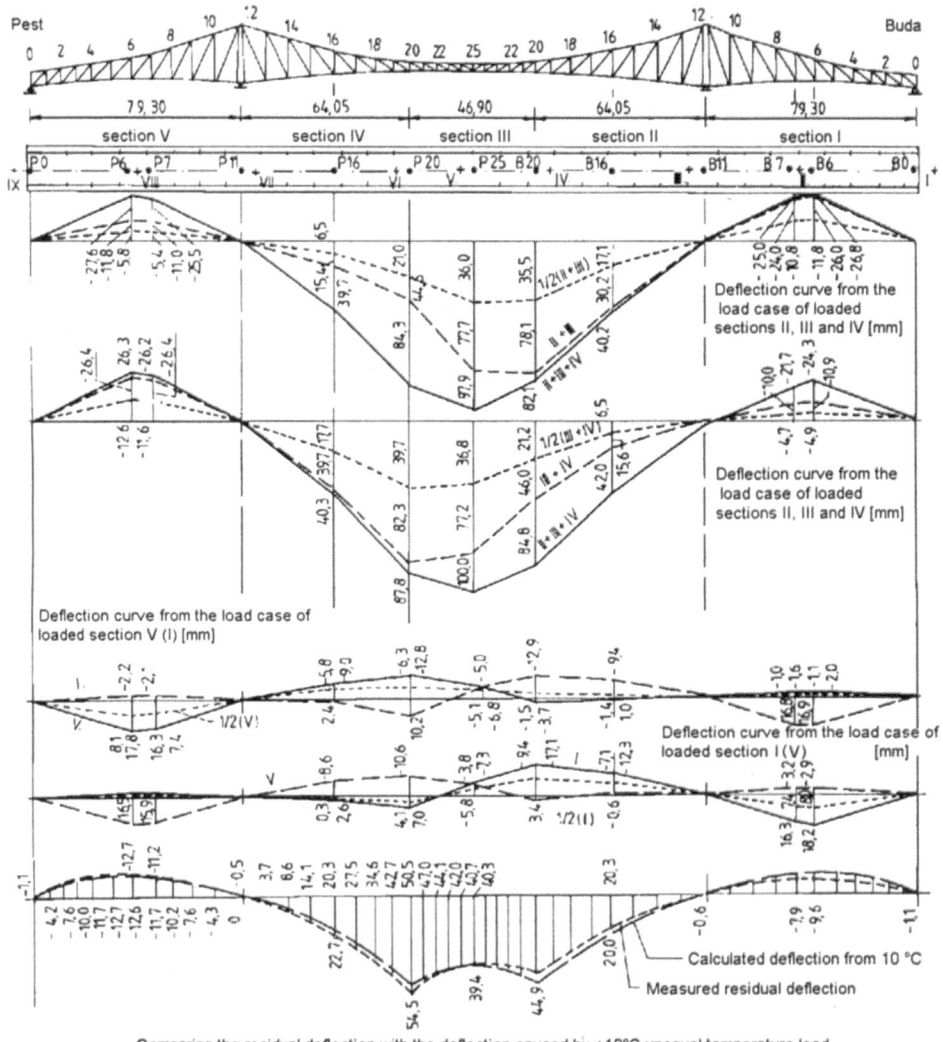

Figure 5.13 Deflections of the Budapest Liberty "Szabadság" Bridge

Summary:
The activities made on the "Szabadság" Bridge belong to different levels of the refurbishment.

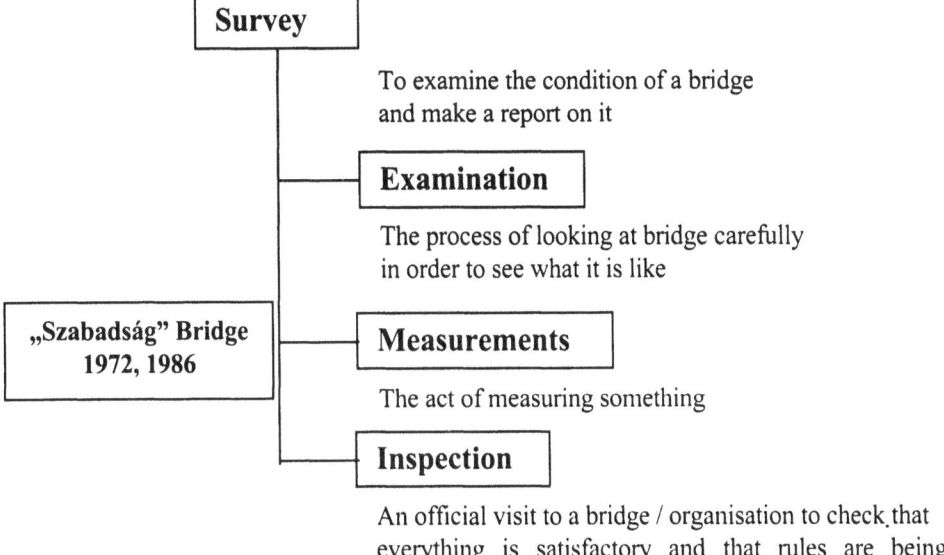

Survey
To examine the condition of a bridge
and make a report on it

Examination
The process of looking at bridge carefully
in order to see what it is like

"Szabadság" Bridge
1972, 1986

Measurements
The act of measuring something

Inspection
An official visit to a bridge / organisation to check that
everything is satisfactory and that rules are being
obeyed

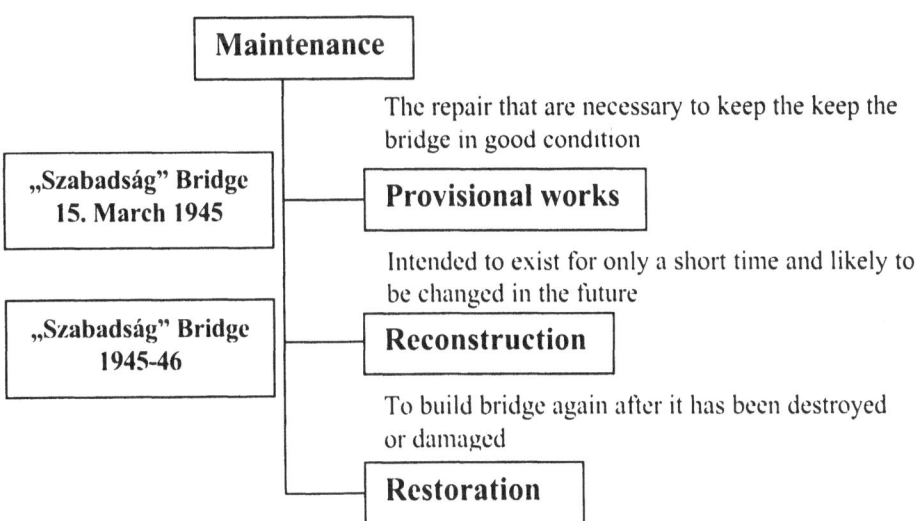

Maintenance
The repair that are necessary to keep the keep the
bridge in good condition

"Szabadság" Bridge
15. March 1945

Provisional works
Intended to exist for only a short time and likely to
be changed in the future

"Szabadság" Bridge
1945-46

Reconstruction
To build bridge again after it has been destroyed
or damaged

Restoration
The act of thoroughly repairing bridge so that it
looks the same as it did when it was first made

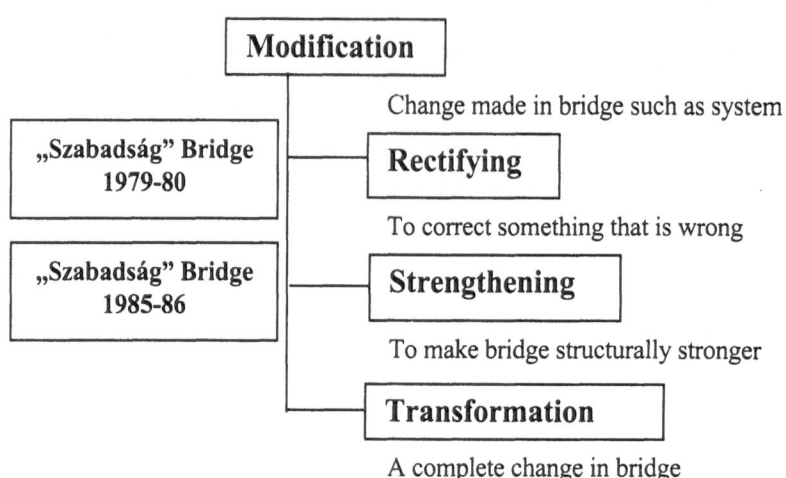

References

AASHTO (1989). Standard Specifications for High-way Bridges, In *The American Association of State Highway and Transportation Officials,* 14th Edition

Agócs, Z. (1996). Diagnostics of steel structures In Iványi, M. and Hegeds, L., eds., *Refurbishment of Structures. Report prepared for the first workshop of Tempus SJEP 09524,* Technical University of Budapest, 201–209.

Åkesson, B., Edlund, B. and Shen D. (1997). Fatigue Cracking in a Steel Railway Bridge. In *Structural Engineering International (IABSE),* Vol. 7., 118-120.

Bancila, R. (1996). Curriculum for Refurbishment Steelworks Engineering In Iványi, M. and Hegeds, L., eds., *Refurbishment of Structures, Report prepared for the first workshop of Tempus SJEP 09524,* Technical University of Budapest, 27–31.

Bancila, R. and Bondariuc, V. (1996). Some examples of refurbishment in industrial steel structures and bridges, In Iványi, M. and Hegeds, L., eds., *Refurbishment of Structures, Report prepared for the first workshop of Tempus SJEP 09524,* Technical University of Budapest, 97–115.

Basler, K. (1961). Strength of plate girders in shear. *ASCE St. 7,* 151-180.

Bruls, Beales, Bignonnet, Caramelli, Croce, Frolli, Jacob, Kolstein, Lehrke, Poleur and Sanpaolesi (1991). Measurments and Interpretation of Dynamic Loads on Bridges (3), *Commission of the European Communities,* Report, Brussels.

Bruls, Caramelli, Cunninghame, Jacob, Kolstein and Lehrke (1995). Measurements and Interpretation of Dynamic Loads on Bridges (4), *Commission of the European Communities,* Report, Brussels.

Castigliani, C. A., Fisher, J. W., Lee, J: J. and Yen, B. T. (1986). Field analysis of fatigue damage in two steel bridges. In *Construzioni Metalliche,* No. 4., 212-234.

Dowling, P. J. and Chatterjee, S. (1977). Design of box girder compression flanges. Designof webs of plate and box girders. In *2nd International Colloquium in Stability of Steel Structures, Introducory Report,* Liège, 153-208.

Dubas, P. and Gehri, E. (1986) Behaviour and Design of Steel Plated Structures. *ECCS Publication No. 44*

ECCS (1985). <<Recommendation for the Fatigue Design of Steel Structures.>> *European Convention for Constructional Steelwork,* Brussels, Publication No.43.

Falke, J. (1983). Zum Tragverhalten und Berechnung von Quertägern orthotroper Platten. Diss. Technische Universität Braunschweig.

Fisher, J. W. (1977). Bridge Fatigue Guide, *AISC*

Fisher, J. W. (1977). Bridge Fatigue Guide: Design and Details. American Institute of Steel Construction, New York

Fisher, J.W. (1984). Fatigue and Fracture in Steel Bridges, Case Studies. John Wiley and Sons, New York

Fisher, J.W. and Mertz, D: R. (1985). Hundreds of Bridges – Thousands of Cracks. In *Civil Engineering / ASCE*, April, 64-67.

Ghosh, U. K. (2000). Repair and Rehabilitation of Steel Bridges. A-A. Balkema, Rotterdam, Bookfield.

Gioncu, V. and Ivan, M. (1978). Instabilitatea structurilar din plci curbe substisi (in Romanian), Editura Academie Rep. Soc. Románia.

Haibach, E. and Plasil, I. (1983). Untersuchungen zur Betriebsfestigkeit von Stahlleichtfahrbahnen mit Trapezhohlsteifen im Eisenbahnbrückenbau. In *Der Stahlbau 9*, (in German)

Halász, O. (1977). Stability problems in national specifications. In *Final Report to the Regional Colloquium on Stability of steel Structures*, Budapest, 9-13.

Halász, O. and Iványi, M. (1985). Some lessons from tests with steel structures. In Periodica Polytechnica, Civil Engineering, Volume 29, No. 3-4, 113–122.

Haviár, Gy. (1947). Reconstruction of the blown Szabadság Bridge. In *Megyetemi Közlemények*, No. 2

Hegeds, L. (1997). Diagnostics In Iványi, M. and Hegeds, L., eds., *Refurbishment of Structures, Report prepared for the second workshop of Tempus SJEP 09524*, Technical University of Budapest, 167–169.

Isoura, K. (1989). Maintenance Program of Shinkansen Structures, *IABSE Symposium*, Lisbon

Iványi, M. (1979a). Yield mechanism curves for local buckling of axially compressed members. *Periodica Politechnica (Budapest)*, Civ. Eng., 23(3-4), 203-216.

Iványi, M. (1979b). Moment-rotation characteristics of locally buckling beams. *Periodica Politechnica (Budapest)*, Civ. Eng., 23(3-4), 217-230.

Iványi, M., Agócs, Z. Balaz, I. (1990). Torsion of Steel Beams (in Hungarian), Budapest-Bratislava, Mérnöki Továbbképz Intézet, BME.

Iványi, M. and Skaloud, M. (1995). Steel Plated Structures. *CISM Courses and Lectures No. 358*, Udine, Springer-Verlag, Wien

Iványi, M. (1998). Bridge Construction (in Hungarian), Megyetemi Kiadó, Budapest

Kármán, T. (1910). Festigkeitsprobleme im Maschinenbau. In *Enzykl. der Math. Wiss.* IV. 348-351.

Kármán, T., Sechler, E. E. and Donell, L. H. (1932). The strength of thin plates in compression. *Trans. Am. Soc. Mech. Eng.*, 54.

Kolstein, M. H., Wardenier, J. and Leendertz, J. S. (1995). Fatigue Performance of the Through to Crossbeam Connections in Orthotropic Steel Bridge Decks. Nordic Steel Conference, Malmö.

Korányi, I. (1934). Die Verstärkung der Fahrbahnträger der Eisenbahnbrücke über die Dunau bei Újpest (Ungarn), (in German), *Bautechnik*, Vol. 18., pp 19.

Leendertz, J. S. and Kolstein, M. N. (1995a). Numerical Analyses of the Through to Crossbeam Connections in Orthotropic Steel Bridge Decks. In *Proceedings Nordic Steel Conference*, Malmö.

Leendertz, J. S. and Kolstein, M. N. (1995b). The Behaviour of Through Stiffener to Crossbeam Connections in Orthotropic Steel Bridge Decks. In *Heron*, Volume 40, Delft.

Leendertz, J. S. and Kolstein, M. N. (1996). Structural Types and Endurance Aspects of Crossbeams in Steel Orthotropic Plate Girder Bridges. In *ICASS Proceedings*, Hong Kong.

Leendertz, J. S., v.d.Weijde, H. and Kolstein, H. (1997). Inspection of Bridges with Orthotropic Steel Decks with Particular Attention to Fatigue. . In *Evaluation of Existing Steel and Composite Bridges IABSE Workshop*, Lausanne, Vol. 76, 57-66

Madea, Y. and Okura,I. (1983). Influence of initial deflection of plate girder webs on fatigue crack initiation. *Eng. Struct.*, 58-66.

Madea, Y. and Okura,I. (1984). Analysis of deformation-induced fatigue of stiffened plate girder in bending. *Verba Volant, scripta manent, Vol. d'hommege au Prof. Ch. Massonet*, CERES, Liège., 255-264.

Maquoi, R. and Massonet, Ch. (1971) Theorie non-linearia de la résistance postcitique des brandes pouters au caissons raidies. *IABSE Publ.*, Zürich, 3-II

Maquoi, R. and Massonet, Ch. (1976) Interaction between local plate buckling and overall buckling in thinwalled compression members. *IUTAM Sym. Cambridge*, Springer

Marguerre, K. (1937). Die mittragende Breite des gedrükten Plattenstreifens. Luftfahrtforschung 14.

Massonnet, Ch. (1977). Design practices in Europe. In *Proceedings of the International Colloquium on Stability Structures*, Washington, 503-531.

Massonet et al (1977). Plate box girders. In Introductory Report 2nd International Colloquium on Stability of Steel Structures, Liège, 145-208

Mehrtens, G. C. (1898). Der Brückenbau sonst und jetzt. In *Zeitschrift für Architektur und Ingenieurwesen*, 18-55

Miki, C. and Ichikawa, A. (1997). Fatigue Assessment of Steel Bridges of the Bullet Train System. In *Evaluation of Existing Steel and Composite Bridges IABSE Workshop*, Lausanne, Vol. 76, 221-230, 1997.

Nather F. (1991). Rehabilitation and Strengthening of Steel Road Bridges. In *Structural Engineering International*, Volume 1, No.3, 24-30.

van der Neut, A. (1969). The interaction of local buckling and column failure of thin-walled compression members. In *Procedings 12th International Congress in Applied Mechanics*, Standford University, Spronger-Verlag

Okura, I., Yen, B. T. and Fisher, J. W. (1993). Fatigue of Thin-Walled plate Girders. In *Structural Engineering International (IABSE)*, Vol. 1., 39-44.

Reis, A. J. and Roorda, J. (1977). The interaction between lateral-torsional and local plate buckling in thin-walled beams. In *2nd International Colloquium on Stability of Steel Structures* Liège, 415-427.

Rockey, K. C. and Skaloud, M. (1972). The ultimate load behaviour of plate girders in shear. *The Structural Engineer*, 50(1).

Sakai, N., et-al. (1994). Consideration of Damage Occuring End Corner of Girder Reducing Section, In *Proceedings of the 49th Annual Conference of JSCE*, I-252, (in Japanese).

Sakagami, A. et al (1993). Study on Cause of Fatigue Crack Occurring Sole Plate of Railway Bridge. In *Proceedings of the 48th Annual Conference of JSCE*, I-223, (in Japanese).

Sakamoto, K. et al (1990). Vibration Fatigue of Steel Bridges on the Bullet Train System, *IABSE Workshop*, Lausanne, 157-166.

Skaloud, M. (1967). The limiting state of thin-walled columns with regard to the interaction of the deformation of the column as a whole with the buckling of their plate elements. *Acta Technica CSAV*, No. 8.

Skaloud, M. (1978). General report on plate and box girders. *Final Report Regional Colloquium of Steel Structures*, Budapest

Tesar, A. (1977), Kovové konštrukcie a mosty, *Moderné ocel'ové mosty, II. ast'* (in Slovak), ES SVŠT, Bratislava.

Thompson, J. M. T. and Hunt, G. W. (1973). *A general theory of elastic stability*, John Wiley and Sons

Ypey, E. (1972). New Developments in Dutch Steel Bridge Building. *IABSE Congress*, Amsterdam.

Wagner, H. (1922). Ebene Blechwandträger mit sehr dünnem Stegblech. *Zeitschr. f. Flugtechnik und Motorluftschiffart*, 20.

Wolchuk R. and Ostapenko, A. (1992). A secondary Stresses in Closed Orthotropic Deck Ribs at Floor Beams. In *ASCE Journal of Structural Engineering*, Volume 118, No.2.

CHAPTER 3

REFURBISHMENT OF SINGLE-STOREY BUILDINGS

H. Pasternak
Brandenburg Technical University Cottbus, Cottbus, Germany

Of course, one can build a building in a green meadow. Another possibility is to move into a 30 or 40 years old standard building. Both solutions would be cheaper. But, what about the clients? Would they like it? They usually would to buy in a city and not in a green meadow.
The factory (Fig. 1) was erected in 1877, later the building was modernized. The building was partially destroyed during the second world war. The reconstruction took place in 1946. All the time, there had always been a typical workshop for mechanical engineering. It is a threespan single-storey building. The middle span is 8,5 m and the heigth is also 8,5 m.

Figure 1. Workshop in Braunschweig.

A conversion (Fig. 2) was made from a workshop to a service enterprise. It's a view of 1991. Here they are selling computers, printers etc. Note that even a small stream and trees were incorporated.

The lattice columns and trusses haven't been changed. The steel construction was strong enough to carry additional load. The fibre-cement-plates have been kept, in addition heat insulation was installed. The heating was not situated - as is usual - under the truss. The reason was **not to disturb** the slender view of the truss. The heating is now situated on the top of the lower chord of the truss.

Even the overhead travelling crane was not changed (hand-operated) and it's still used for transport. The floor was carefully changed because of polution level.

Figure 2. Service enterprise in Braunschweig.

Another example is the famous central **market hall** of the city of Budapest (Fig. 3, 4). It was built from 1894 till 1897 [1] and it is a protected monument. The roof of the main span - it spans 20 m - has been carried out like an arch truss. Therefore the horizontal forces from the roof had to be introduced to the adjacent spans. Only a few steel members had to be removed.

Figures 3 and 4. Market hall in Budapest.

Most effort was put into the foundations (Fig. 5, 6). The brick foundations were reinforced by steel bands and then a concrete covering was given. There was a nice cap ceiling. It had a reduced load bearing capacity because of corrosion. Concrete was used to make the slab stronger (composite structure).

Figure 5. Market hall in Budapest.

Figure 6. Market Hall in Budapest (model).

Railway stations are - as architects say - cathedrals of modern times. At the end of 19th century when Amsterdams central station (Fig.7) [2] was built, the work of the architect and that of the structural engineer were strictly separated.

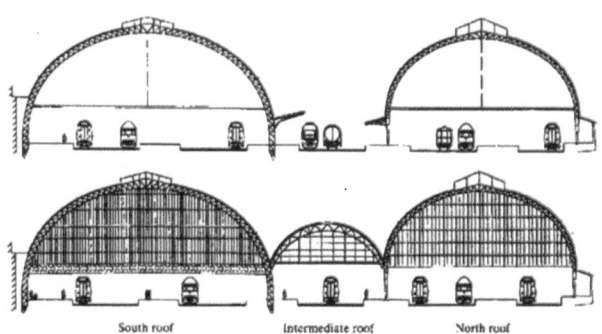

Figure 7. Central Station in Amsterdam.

The architect designed the station building, while the structural engineer concentrated on the roof structures. Nowadays, it is common practice that architects and structural engineers work (or should work) in close cooperation.

Until recently, three tracks within Amsterdam central station were not covered by a roof structure. Open canopies, cantilevering out from two existing roof structures covered the platforms along these tracks.

This situation was rather uncomfortable for passengers, and therefore it was decided to build an intermediate roof over the three uncovered tracks.

This roof would increase passenger comfort and accomodate the high-speed trains coming to Amsterdam.

Amsterdam Central station is a typical facade station, built at the end of 19th century. At that time, large important railway stations usually consisted of an impressive monumental building - let's say a gate to the city - with a roof structure behind the building.

The facade building was erected between 1881 and 84, the South Roof was added between 1884-89. It's a three-hinged arch, spanning 44.5m.

In 1922 , the North Roof spanning 34 m, was built. Again, a three-hinged arch. All structures are rivited.
The Intermediate Roof spans 19,5 m and is connected to the existing roofs. So it is possible to avoid seperate columns. The supports and the arches are tapered.
Because of their smooth apperance without sharp edges, tubular structures are a good choice from a maintenance point of view. The surface area to be painted is relatively small - in comparison with open sections.
From the point of structural design, tubular structures also offer good possibilities. The arch-like span (Fig. 8) consists of a 3D-truss with tubular braces and chords. Four longitudinal 3D-truss girders connected the arches.The lower two truss-girders transfer their loads directly to the supports. The arches are built up by welding.

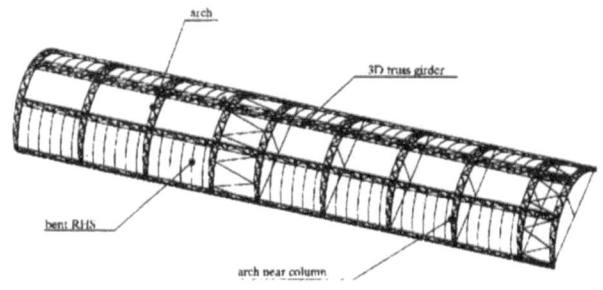

Figure 8. Intermediate roof.

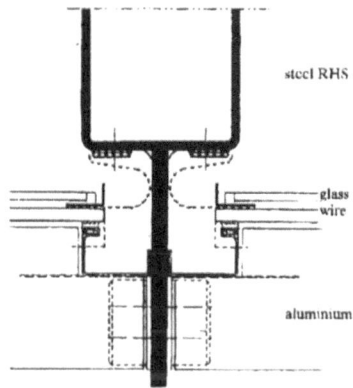

Figure 9. Aluminium glass-supporting structure carried by RHS.

Between the truss girders, bent rectangular hollow sections (RHS) are placed every 2m. These RHS carry the aluminium glass-supporting structures (Fig. 9).

A horizontal beam takes the horizontal arch reaction.
The supports of the arches are hinged, transferring the reactions to the adjoining roofs (Fig.10).
Brackets attached to the existing arches, carry the arches of the intermediate roof. The brackets were bolted to existing structures using M20 10.9 injection bolts, which replaced the rivets, that were removed.

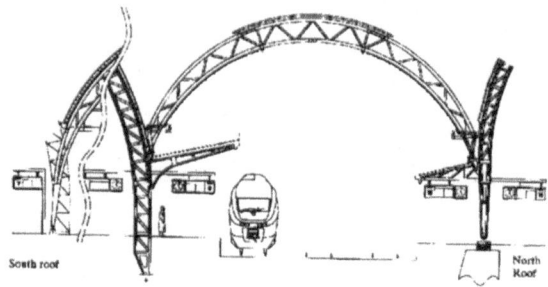

Figure 10. Intermediate Roof supported by brackets on adjacent roofs.

Fig.11 and 12 give the final view of the Amsterdam Central station.

Figure 12. Intermediate Roof between North and South Roofs.

Another case study: It is normal to use a truck crane to lift a roof.

But there are some cases - especially under rehabilition -, where it may be useful to heigthen a roof without the usage of a truck crane [3]. In a single-storey building lighting equipments are produced. It was decided to increase its production. The building was **not** provided with an overhead travelling crane. The machines used for production were serviced by special lifting gear able to move on air.

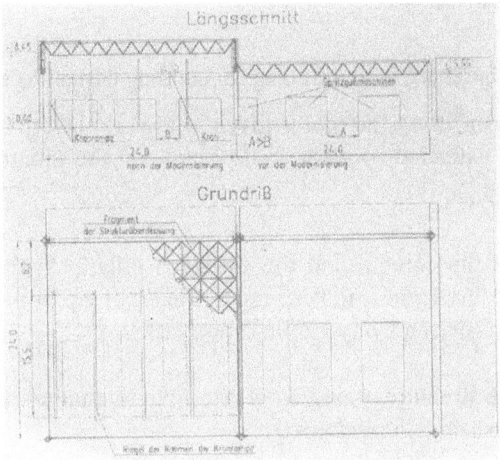

Figure 13. Modernisation concept of factory hall and its outfit.

The lifting gear required wide communication lanes, and those couldn´t be used by machines. An increase of production was intended to be achivied by installing some additional machines (Fig. 13). This made it necessary to equip the building with an overhead travelling crane to take over the job of the pneumatic lifting jack, used for routine maintenance of the machines. The idea is now, to use the communication lanes as additional production area. To carry out the project, it was necessary, to raise the heigth of the building by 3,0 m. The supporting structure was a space truss. The dimensions were 24 x 24 [m]. In Fig. 14 „1" represents the space truss, and „2" the truss girders embracing the space frame.

Figure 14. Load-bearing structure ot the roof supported on a central column.

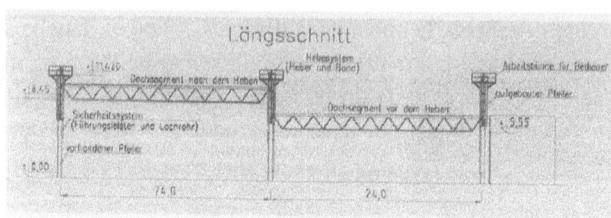

Figure 15. Schematic diagram of lifting the hall roof.

The longitudinal section of lifting the roof is given in Fig. 15: before the lifting and just after. The columns have been extended by additional pillars. On the pillars a lifting system and a safety device are mounted.

Fig 16 shows a view of the column and the extended pillar. On their tops lever arms are installed. The task of the lever arm is to keep the hoisted roof up by means of perforated steel tapes. The hoisting gear is driven by hydraulic lifting jacks.

It is possible to hoist the structure, going from one hole to another, step by step. The safety device consists again of perforated steel tape.

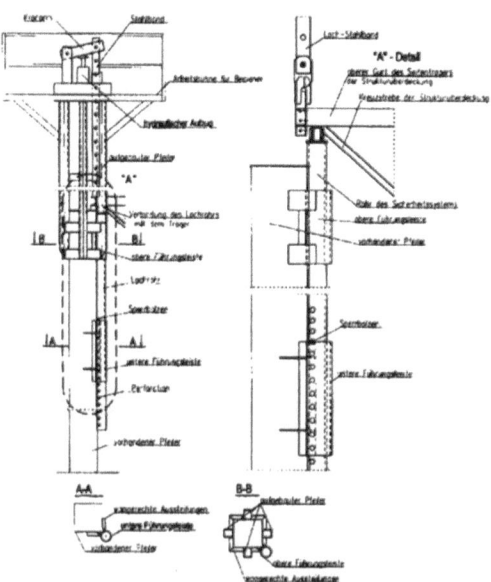

Figure 16. Lifting system and safety device to protect a section of the hall roof.

The lever arm is in the second hole (Fig. 17). The procedure of hoisting begins (Fig.18).

Figure 17. Lever attachment for lifting the hall roof installed on an assembly platform fixed the extended column.

Figure 18. Fragment of the hall roof section before the lifting operation.

The procedure of hoisting the roof is finished (Fig. 19, 20). The first roof raised up to its final position (Fig. 21, 22).

Figure 19. The lever attachment.

Figure 20. The lifting system of the hall roof after the placement of the first roof section to its new position.

Figure 21. The first roof section of the

Figure 22. View of the hightened hall raised up to its destined position hall after adding a curtain wall.

Sometimes single-storey building are supplied with overhead travelling cranes. In Fig. 23 one sees a top view of a particular craneway. The span of the craneway is 27.6 m, and the span of a craneway girder is 12 m.

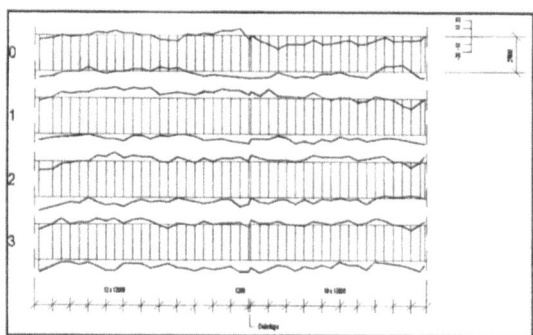

Figure 23. Horizontal position of a craneway.
0 - October 1975
1 - April   1976
2 - October 1976
3 - October 1978

Figure 24. A typical damage case.

The position of the imperfections differs from the planned position by plus/minus 20 mm. Such a difference is acceptable, but not 50 or 60 mm, and it changes with the time.

Two things are important for the behaviour of the crane are: the lateral imperfections of the craneway and the skewing (or oblique) position of the crane wheels.

In Fig. 24 a rail and a crane wheel is shown. This wheel has two flanges, one of them has been destroyed by the rail.

Figure 25. Craneway stubs.

In Fig. 25 one sees a top view of two craneway girders. The rails have been connected by fillet welding to the upper chord of this craneway girder.

The whole left rail, which lies on the left girder, is connected by welds. And the right rail is connected by welds to the craneway girder. The left part of rails above the right craneway girder is free.

What happens? Let´s consider the situation on the left.

Due to the lateral forces acting in the weak axis of the girder and columns, the girder will move to and from, it will oscillate.

The situation on the right side will be stable (no vibration). Even in case of normal imperfections you will hear a terrible noise (it reminds one of a shot).

Figure 26. Craneway stubs.

It is necessary to carry out a cut with oblique angle (Fig.26). „l" is the length with any welding and it should more than 500 mm.

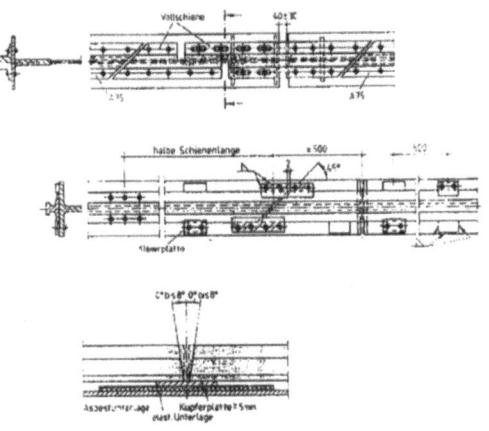

Figure 27. Craneway stubs.

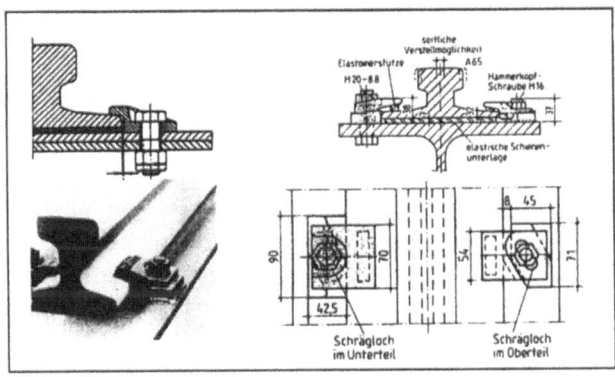

Figure 28. Craneway fasteners.

In case of large extensions, step joint may be recommened (Fig.27). This solution is to avoid moveable rails. It consists of copper plate, asbest lining and elastic lining. The rails are mounted to a craneway by clamping plates.

The rails are often a working part (Fig. 28). Working parts have a shorter lifetime. Then it may be nessecary to exchange them.

The Fig. 28 (left) shows a bolted clip between the rail and the upper chord of the craneway girder. The bolt should be put in from the upper side. A nut is necessary from the bottom. A difficult solution, because it requires the enclosure of the whole upper chord. There is no possibility to balance lateral imperfections.

Let´s consider the Fig.28 (right). Now we have two cross-section on the same figure - one is from the left, and one from the right. The rail clip consists of two parts.

On the left the bolt should be introduced through the upper flange. One has is the possibility to balance lateral imperfection by means of a slotted hole in the lower of the clip. The piece below the rail is made of elastomer.

On the right, the lower part of the rail clip is welded. The upper part has a slotted hole and it's bolted. A T-bolt (or a head bolt) is introduced to the welded lower part of the clip. The upper part is fixed by a nut (Fig. 28, photo).

Figure 29 and 30. Conventional overhead travelling cranes.

The disadvantage of the solution is, that we have high stress concentration on the welds.
Since 1873 conventional overhead travelling cranes have been constructed (Fig. 29, 30). The Norail crane concept [4] inverts the overhead crane principle (Fig. 31, 32).

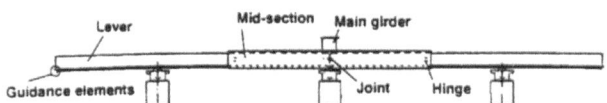

Figure 31. Construction of the NORAIL crane.

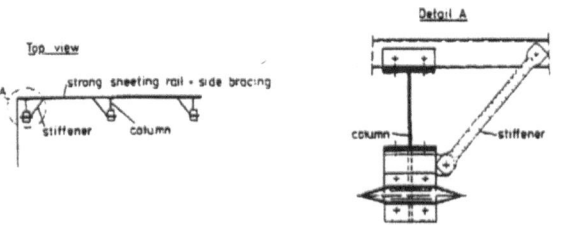

Figure 32. Absorbing buffer forces.

Whereas conventional cranes have wheels travelling on rails, the Norail crane has short rails attached to the bridge, travelling on wheels, mounted on the cantilevers of columns.

These rails run along a series of stationary wheels. The rails are designed to be somewhat longer than the maximum distance between three adjacent support points, so that the crane is always supported by at least two wheels on each side.

The economy of Norail cranes increase
- with increasing load capacity of the crane and
- with increasing length of the building.

Having a span of 21 m and a frame distance of 7.5 m, and a 20t Norail crane is 18 % lighter and 17% less expensive than the conventional solution.

What does it mean for rehabilitation? Due to this design, the long conventional craneway becomes unnessary. It would be possible to introduce a crane with increased load capacity (let's say 20 t-Norail crane instead of 16 t conventional crane) without stiffening the members.

A special solution for the transfer of buffer forces is necessary. The buffer load is transferred by a diagonal stiffener to a strong sheeting rail and to the side bracing and from there to the two adjacent columns. About 30 Norail cranes exist.

References

Nagy, G., Szelenyi, K.: Markthallen in Budapest von der Jahrhundertwende bis heute. F. Szelenyi Haus, *Veszprem* 1997

Snijder, H.H., Spoon, M.: Intermediate Roff Structure for Amsterdam Central Station. *Structural Engineering International*, Vol. 10, May 2000

Ziolko, J., et al.: Höhersetzen eine Hallendaches - Hubmontage ohne Autokran. *Bauingenieur* Vol. 75, May 2000

Goussinsky D., Scheer J., Pasternak H.: Design Aspects of Steel Buildings with Norail Bridge Cranes. *New Steel Construction*, February 1993

CHAPTER 3

3.2 SILOS AND TANKS

H. Pasternak

Brandenburg Technical University Cottbus, Germany

It is well-known that shell structures do not have postcritical reserves. Therefore, the consequences of incorrect assumptions in calculations and design of silos or tanks are more drastic than for other structures. So, it is not surprising that silos and tanks belong to those steel structures that have the highest frequency of failure.

There are several types of silos. We can distinguish between a few possible shapes of silos, that means rectangular or cylindrical cross sections.

More important for the calculation and the design of a silo, however, are the boundary conditions.

Figure 1 presents a silo with continuous support. The vertical cylindrical wall is supported on a continuous basement or a plate. So, the reaction forces at the bottom are equal around the perimeter. The rules for calculating such a silo are given, for example, in the German standard DIN 18800 "Stahlbauten" Part 4 "Stabilitätsfälle, Schalenbeulen" or in Eurocode 3 "Design of steel structures","Silos".

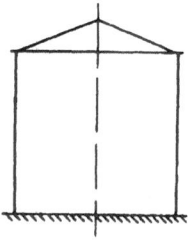

Figure 1. Silo with continuous support.

Figure 2 shows a silo construction which is supported on discrete columns. In this case, it is characteristic that the reaction loads and thus, which are local loads, in the shell are much higher than in the case of a continuously supported shell. This is also the main reason for a lot of failures.

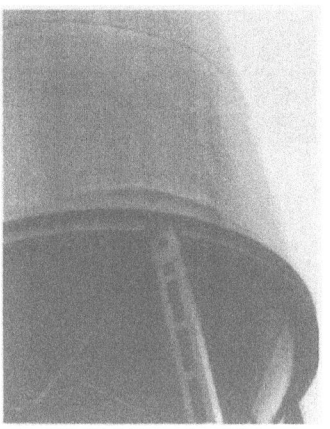

Figure 2. Silo with discrete support.

Figure 3 shows a silo with local deformations above the columns. This silo collapsed 21 years ago. The diameter of this silo was 6 metres and its volume 225 m³. The r/t ratio amounts to 375. The filling was cement. The construction was supported on four columns. At the bottom rim of the silo, a ring stiffener with I-section 120 was arranged. The deformation above the columns, a so-called "elephant foot", can clearly be seen. What was the reason for the failure?

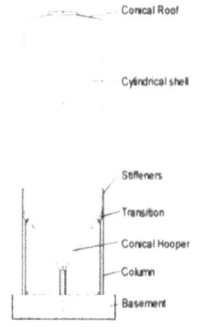

Figure 3. Collapsed silo

The stiffness of the ring beam at the bottom rim of the silo construction was too small to ensure an equal distribution of the longitudinal compression stresses along the lower rim of the cylindrical shell. So, the stresses above the supports were much higher than the buckling stresses, due to local buckling. In between two supports, however, the stresses were not that high.

We have analysed this special problem in detail. The analysis was carried out by using the finite-element method. The calculation was done geometrically and physically nonlinear. After the evaluation of the results, the same failure mode like in reality could be observed.

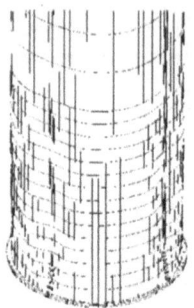

Figure 4. Results of the finite-element analysis.

According to [1] it is possible to do a safety check against local buckling.

$$\frac{\rho_{x,local}}{\varphi_{x,local} \cdot f_{y,d}} \leq 1,0$$

$$\bar{\lambda}_{x,local} = \frac{0.0283}{\bar{\lambda}_{x}} \cdot s_0/r + (0.77 - 0.56\bar{\lambda}_{x})$$

$$\alpha \leq \alpha_{x,local} \leq \alpha_x \cdot \frac{2}{\pi \cdot s_0} \leq 1.0$$

(steel grade Fe 360)

The parameters used are:

$f_{y,d}$	yield limit of the steel grade used
$\bar{\lambda}_{x}$	non-dimensional slenderness
S_0	support width
R	radius

If condition (1) is met, then the silo shell above the local support is safe from shell buckling. Figure 5 shows the trial of a reinforcement with plates of a thickness of 5 mm and a length of 2 metres on average.

Figure 5. Reinforcement of the silo.

Now the failure occured above the longitudinal stiffeners (figure 6). Obviously, the length of these stiffeners is too short to ensure a continuously vertical force transfer from the stiffeners into the shell.

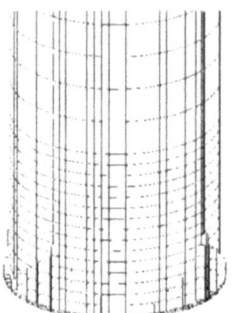

Figure 6. Failure mode of the reinforced silo.

The finite-element calculation shows the same failure mode, that is shell buckling at the end of the stiffeners. The reason for it was the high stress concentration. Figure 7 presents the von-Mises stress distribution. The appropriate deformations are shown in figure 8.

Figure 7. Von-Mises stress distribution.

Figure 8. Deformations in failure mode.

Figure 9 presents another case of damage. The silo failed 10 years ago due to local buckling above the discrete support. The shell was made of profiled sheet.

The reason for the failure was the same as in the last example. The stiffness of the ring beam was not large enough to ensure a continuously vertical load application in the bottom ring girder. So, a stress concentration above the support occured and the shell failed due to local buckling.

Figure 9. Failure mode of a silo with profiled sheet.

Figure 10. Failure mode of a silo with profiled sheet.

Figure 10 shows a silo, which also failed due to local buckling above the local supports. This silo was built in 1999.

Figure 11 shows a silo with a ventilation flap. The silo is equipped with a special tool for emptying it, so-called screw conveyors. In the upper part of the silo, you can see volume V1. In the lower part, you can see volume V2, caused by an arch of the material (bin arch) or by an "internal static system".

If the bin arch at the bottom breaks down, the whole air volume is now in the upper part of this silo. During the sudden break down of this bin arch, negative pressure in the upper part of the silo will occur. The ventilation flap is necessary to avoid this problem. So, fresh air can come into the silo and negative pressure cannot occur.

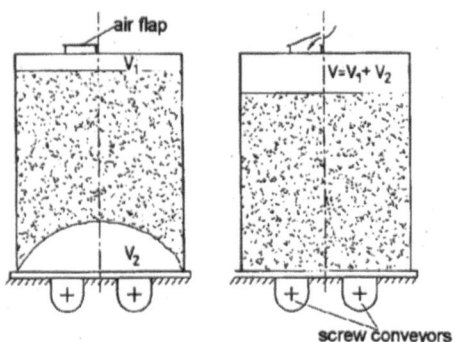

Figure 11. Low pressure due to a break down of the bin arch.

Figure 12 presents a silo with a volume of 2100 m^3 and a diameter of 11 metres. The filling was limestone meal, which is a cohesive material. The silo was filled pneumatically by using moist air. Therefore, the occurrence of bin archs is possible from time to time.

The bin arch in the lower part suddenly broke down and the ventilation flap was not large enough to let in enough additional air into the silo.

This was the reason for the global buckling of the silo shell in the upper part.

The reinforcement of the silo was done by arranging three additional ring stiffeners and vertical stiffeners in the upper part (figure 13).

The best way to avoid such damages is the installation of special emptying tools, like screw-conveyors, which can help to prevent such bin archs.

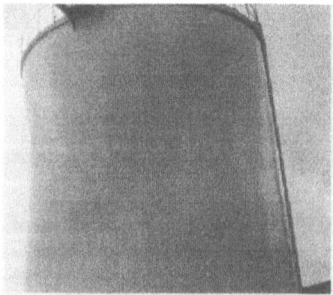

Figure 12. Global buckling due to negative pressure in the upper part.

Figure 13. Silos after reinforcement.

In the next case, the diameter of the silo was 7,8 metres and the volume 570 m³. The silo was filled with soya beans.

The cylinder was made of corrugated metal. The vertical stiffeners are necessary due to the small stiffness in vertical direction ("accordion"-effect).

The hopper of this silo was made of flat sheet with a thickness of 4 mm. It is a French product.

The friction coefficient between the corrugated metal and the bulk material is relatively high, approximately 0,6 to 0,7 (instead of 0,3 for flat sheet).

During the process of emptying, at first the light roof collapsed (not shown here) and, after that, parts of the cylindrical shell and the hopper pulled down (figure 14).

Figure 14. Collapsed silo with corrugated sheet.

The reason for the collapse was the break down of the bin arch. At first, the roof was damaged, and then, the joints between the cylinder and the vertical stiffeners failed due to the unequal distribution of the forces in the bolts.

Figure 15 shows the silo after its rehabilitation. It was necessary to remove the damaged parts of the corrugated sheet. The horizontal line marks the end of the cylinder and the beginning of the hopper.

Figure 15. Silo after rehabilitation.

The next picture (figure 16) presents a silo with a volume of 200 m³. The silo wall is made of corrugated sheet and the hopper of flat sheet. The filling of this silo was grain.

Figure 16. Failed silo with corrugated sheet.

Along the vertical shell wall, longitudinal stiffeners are arranged in small distances. They are on the inside of the cylinder. In the last example, the longitudinal stiffeners were on the outside. These longitudinal stiffeners carry the whole load, which appears between the shell wall and the bulk material. The longitudinal stiffeners are placed on a ring stiffener, which is supported on 4, 6 or 8 columns.

That is why the longitudinal stiffeners are not loaded equally.

If the stiffener is placed directly on the column, the load will increase. In contrast to this, the load will decrease if the longitudinal stiffener is placed in the middle of two supports. The reason for this is the unequal distribution of the compressive stresses along the ring beam at the lower rim.

So, the longitudinal stiffeners placed on the columns are more loaded than the adjoining stiffeners. This effect has to be considered during the process of designing.

Figure 17. Failed silo

Figure 17 presents a similar case of damage. The silo is made of corrugated sheet and the hopper of flat sheet.

At first, the wall sheet was damaged due to a failure of the bolts. Next, the hopper fell down and the bulk material came out. After that, the roof was destroyed by a sudden increase of negative pressure in the upper part of the silo.

Figure 18. Hopper of a coal dust silo.

Figure 18 shows the hopper of a coal dust silo in South Africa. The geometrical dimensions are presented in the longitudinal section in figure 19.

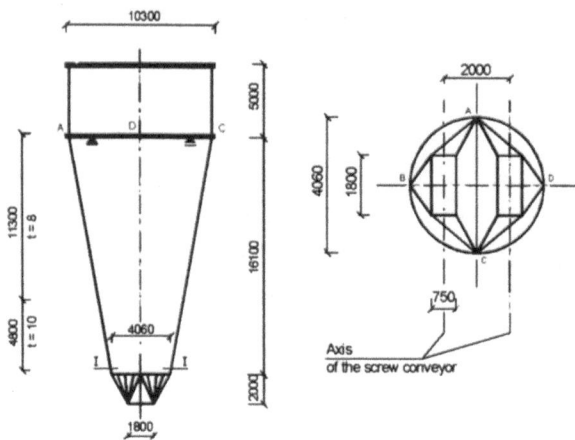

Figure 19. Geometrical dimensions.

The diameter of this silo was 10.3 metres and the volume 1100 m³. Compared with the height of the hopper the cylindrical part of this silo was very short. At the lower rim of the hopper, a screw-conveyor, a tool for transportation, was arranged.

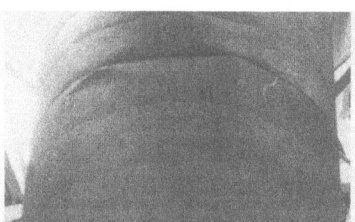

Figure 20. External view of the deformed hopper.

Figure 20 presents an external view of the deformed hopper. The deformations look like local buckling. But in fact, the hopper was loaded in tension. What was the reason for these deformations?
The view into the inside of the hopper (figure 21) gives a more precise picture.

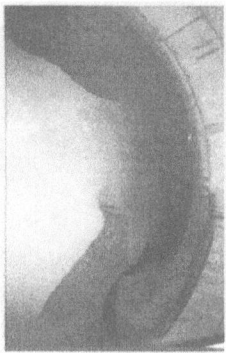

Figure 21. Internal view of the hopper

As mentioned above, the emptying is done by a screw-conveyor. Figure 21 shows that there are regions filled with material and regions without material.

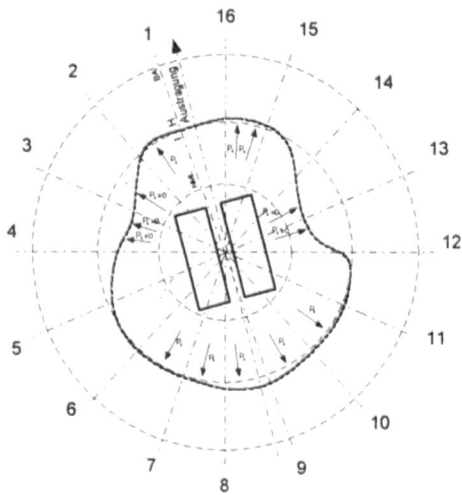

Figure 22. Internal pressure distribution.

Figure 22 presents the internal pressure distribution. It can be observed that the pressure is almost zero in regions without material. In regions filled with material, the full internal pressure appears. As a result, the emptying of this silo passed in an unequal manner. This led to an "out-of-roundness" deformation of the thin shell.

This "out-of-roundness" effect was even increased by welding deformations in this region.

Due to this increased effect the deformations look like local buckling.

In order to improve this situation and to refurbish the hopper it was found a solution by applying a special tool, which is improving the material flow during the process of emptying. After the installation of this tool, the hopper was working properly again.

Figure 23. Failed tank in Kuwait.

Figure 23 shows a gigantic tank situated in the desert in Kuwait. The tank was subjected to a wind load of 100 km per hour with hail. The height of this tank was 30 metres.
Wind pressure from the outside was superimposed by wind suction from the inside. The roof had not been installed yet. The storm took 10 minutes during working time. It was an extraordinary and unexpected situation. There was no time to fasten the shell wall by cables, so the deformation occured. The deformations were very impressive.

Figure 24. Gap at the lower rim of the tank.

As a result of this strong storm, the dent with an area of 400 m^2 occured and the mantle lifted off the foundation, so that a slit of 10 cm arised (figure 24).

Figure 25. Refurbishment of the tank.

The refurbishment was follow in this way, that by means of prime mover and jacks the dents were pulled out (Fig. 25).

The figure 26 shows a tank with a volume of 5000 m^3 for fuel oil.

Intensive emtying and deficiency of the safety valves caused low pressure and the shell was deformed.

Figure 26. Failed oil tank.

The strategy of repairing was to fill up this tank with water (figure 27) and then hydrostatic pressure was in the inside. By means of this hydrostatic pressure the deformations were pressed out. In the regions with large deformations leaks in the wall occured. These regions were cutted out and new sheets were built in. So the tank works again.

Figure 27. Repairing of the tank.

REFERENCES

[1]   Pasternak, H., Hotala, E.: Schäden an Stahlsilos-Ursachen und Beispiele.
        Bauingenieur 71 (1996) 223-228
[2]   Greiner, R. U. A.: Local loads in cylindrical structures. In: Final report ECCS
        Contract No. 7210-SA/208, Brüssel 1998
[3]   Bodarski, Z., Hotala, E., Pasternak, H.: Zur Beurteilung der Tragfähigkeit von
        Metallsilos. Bauingenieur 59 (1984) 49-52.
[4]   DIN 18 800, Teil 4 (11.90), Stahlbauten-Stabilitätsfälle, Schalenbeulen, Beuth-Verlag
[5]   Eurocode 3, part 4-1, Design of steel structures-silos, Brussels, 1997

Thanks from the autor to S. Komann

·CHAPTER 3

3.3 TOWERS AND MASTS

H. PASTERNAK

BRANDENBURG TECHNICAL UNIVERSITY COTTBUS. GERMANY

1 Introduction

The lecture informs about steel structures of this kind, which architects do not necessarily contribute to. It is only engineering design – towers and masts.

The difference between masts and towers lies in their static systems. Masts mostly have a pinned base and the mounting of them is done by guys. Towers, on the other hand, have several columns with separate supports. The number of columns of a tower depends on its cross-section, three columns at least.

Figure 1. High-tension line tower.

The first figure shows a high-tension line tower. The question is: Why did this steel tower fail? Answering this question, one can find two reasons – direct wind forces and vibrating cables – which are not mutually dependent. Both of them have to do with forces and loads and finally resulted in resonance. In this case, any kind of renovation is useless. After such a collapse, the only remaining task is the removal of the construction. However, the next examples of masts and towers will show the possibilities of restoration and renovation.

2 THE EIFFEL TOWER

The Eiffel Tower in Paris was finished in 1889. It was the symbol of that year's World Exhibition. Having a height of 307 metres, the Eiffel Tower was the highest building of the world until 1931. The tower is a truss frame steel construction. The amazing number of 2.5 million rivets have been used to connect the 15,000 elements. The whole steel structure weighs over 7,500 tonnes. Therefore, the tower is one of the first examples of mass production. The technology of riveting was used in France and Germany up to the earlier 1960ies. In Romania, it was used until seven years ago, and now this technology is still used in India. Using rivets is possible for the production in a workshop, but it is not longer possible when it comes to the rehabilitation of a tower in its vertical position. The strategy is to replace rivets by high strength bolts, that means bolt grade 8.8 or 10.9. The preloading of a bolt is achieved by controlled tightening, which can more or less be compared to the preheating of a rivet. This solution was used in the Eiffel Tower (figures 3 and 4).

Figure 2. Eiffel Tower.

Figure 3 and 4. Replace rivets by high-strength bolts.

3 THE HABICHTSBERG TOWER [1]

The tower is situated in the south of Berlin. It was built in 1936 right before the Olympic Games in Berlin (figure 5). The tower then had a height of 88 metres. It is a timber frame steel construction with angle bars and bolted connections. Originally, the tower was used as a television tower. Later on, in the time of East German government, its height was reduced to 56 metres. Five years ago, in 1996, the tower was renovated and rebuilt up to 66 metres. The result of the renovation can be seen in figure 6. Now, it is used as a basis station for a mobile phone company (e-plus), as, in this respect, the location of the tower is excellent. The surroundings of Berlin are rather flat; the tower, however, is situated on top of a hill (178 metres). The tower has not been put under a preservation order yet, which is quite untypical for Germany. Under these circumstances, the question occurs why there has not been built a new tower instead of the old one. The answer is that it is easier to get permission to modernise the tower than to build a completely new one.

Figure 5. Habichtsberg tower – old.

Figure 6. Habichtsberg tower – new.

The material of the elements is a non-weldable steel with a yielding stress of 240N/mm². Therefore, it is not possible to weld even the smallest detail. At first, investigations have been carried out to determine the degree of corrosion of the whole construction. It became obvious that the most corrosive details were the gusset plates and the adjacent bars. The figure 7 shows some examples of the situation during the inspection in December 1995 before the restoration. The degree of damage due to corrosion is a decisive factor for choosing the appropriate way of restoration. If the progress of corrosion is low, sand-blasting of the gusset plates is sufficient. If the progress of corrosion is high - that means the extent of damage is approximately 0.2 of the thickness of the plate - the gusset plate has to be replaced. Moreover, the bolt holes are critical spots for cracks.

Figure 7. Situation before restoration.

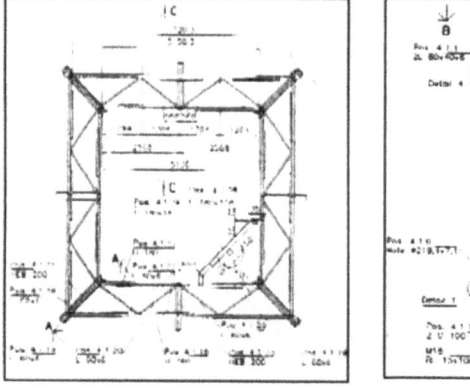 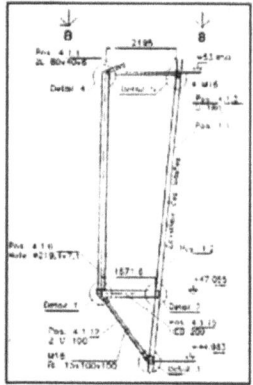

Figure 8. Construction details.

The structural details show the platform. In the left figure of figure 8 you can see the columns of the tower. The four tubular sections are used for putting up the aerials. However, a problem occurs in the fastening of the aerials. Under maximum wind loading, the deviation of the aerials from their vertical positions may only be 0.5 degrees. The structural analysis was performed by the computer program "SOFISTIK".

In the figure 9, you can see the three-dimensional model. The calculation of the originally planned structure led to the result of unacceptable deformations. In this case, the real angle was much larger than 0.5 degrees. So it became necessary to change the fastening of the aerials.

The platform was supported by inclined bars, above and below. The new calculation then led to the result of reduced deformations due to a better resistance against wind forces.

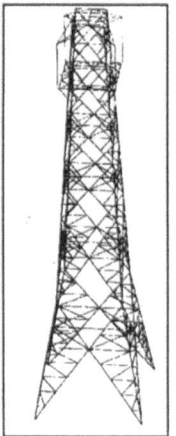

Figure 9. 3D-model.

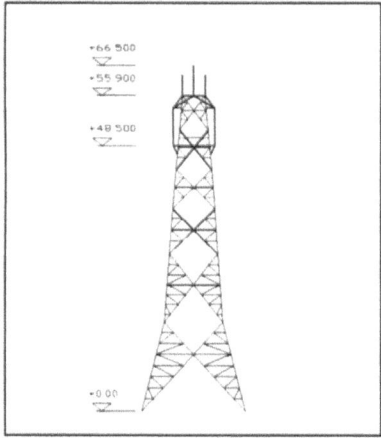

Figure 10. Structure of the Habichtsberg tower.

Figure 10 shows the original structure having a weight of 38 tonnes (thin lines), the new structural elements and strengthened bars (thick lines), as well as the dismantled structural elements (red lines).

Figure 11. Habichtsberg tower – old.

Figure 12. Habichtsberg tower – new.

Another reason for modernising the tower was the opportunity to improve the safety system. In the figure 11 of the old tower, you can see a safety ring around the ladder.

Figure 13. New access ladder.   **Figure 14.** Foundation.

The new tower has an access ladder, a special ladder with a rolling device. The rolling device prevents the person from accidentally rolling down.

Another problem was the concrete foundation. It was concrete without reinforcement. Inside the foundation some cracks appeared. The areas affected by the cracks had to be repaired by repair mortar. The concrete had to be partially removed and the upper side of the foundation was additionally concreted.

4 COLLAPSE OF MASTS

What are the typical features of masts? Masts are very slen- der structures, which are anchored by cable guys. Due to their high slenderness, masts belong to those structures who have a strong tendency to fail. There have been at least five masts (USA, Poland) with a height of more than 600 metres, but today only one of them – in the USA – still exists. In 1991, one mast in Poland collapsed when its cables were being exchanged [6]. Another mast failed after a crash with a helicopter. It seems in building masts there is a kind of limit to height which is 600 metres. Below are the results of a Finnish study (figure 15). It shows that the main reason for the collapse of masts in the 70ies was the iceload.

Ursache cause	USA/ Kanada	England	Finnland/ Schweden	Deutschland	andere * others	Σ
Eisansatz ice	17		3			20
Sturm, Gewitter storm, thunderstorm	10		2		3	15
Fremdeinwirkung ** extraneous cause	4	3	2	1	2	12
Montage installation	7		1	1		9
Querschwingung transverse vibration	2	2	1	1	2	8
Konstruktion construction	1	1		3	3	8
Sabotage sabotage		4				4
Mangel bei Wartung fault by maintainance	1					1
unbekannt unknown	19	1				20
Σ	61	11	9	6	10	97

Figure 15. Compilation about collapse of masts.

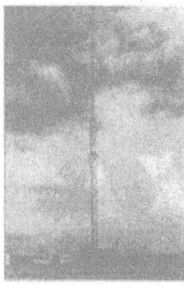

Figure 16. Mast in Nauen.   **Figure 17.** Collapse.

The list of collapses starts in 1911 with a mast that was situated west of Berlin near a town called Nauen. The poor quality of the photos can be attributed to the year of their taking (1911). At this time, the relation between wind forces and the height of a building or construction was still unknown. Therefore, the reason of the collapse of the 200m mast was the underestimation of the wind loading (Figure 16 and 17).

4.1 MAST IN THE TEUTOBURGER FOREST

Figure 18. Masts in the Teutoburger forest

In the Teutoburger Forest (west from Hanover), there were two masts of the TV-station WDR. One of the masts had a height of 60 metres, the other one was 300 metres high. The 300m mast had been built in 1970. It collapsed in 1985. There are many possible reasons for a collapse. Causes which could be excluded for this collapse are:
- an earthquake
- critical moves owing to air traffic
- sabotage
- ice and snow loading
- breaking of the cables due to low temperatures.

The collapse resulted from a combination of two causes. The first reason was vibrations due to wind forces, the second reason was the fatigue of the material caused by defective welds and the high-stressed steel-members. The course of the collapse of the mast can be described as follows: It was a winter night with a temperature of –20°C. The weather was bad, 12 to 15 hours of turbulent wind streaming had led to vibrations. A plate between cable and mast was demolished by these vibrations. The mast leaned over and finally cracked at the height of 160 metres. The upper part of the mast fell down to the ground and pulled down with it parts of the remaining mast of between 70 metres and 160 metres length. The falling elements cut the cables of the lower part of the mast and so the remaining 70-metre-part of the mast collapsed as well.

Figure 19. Collapse of the masts.

The details show the difference between the planned and the performed construction. The mast has a tubular cross-section. There is a welding gap between the plate and the ringstiffener. The distance between ringstiffener and this welding gap should be two millimetres, but it was ten millimetres in fact. The cables vibrated in transversal direction. Therefore, a stress-concentration occurred, which was responsible for a fatigue crack. So the crack was caused by brittle fracture.

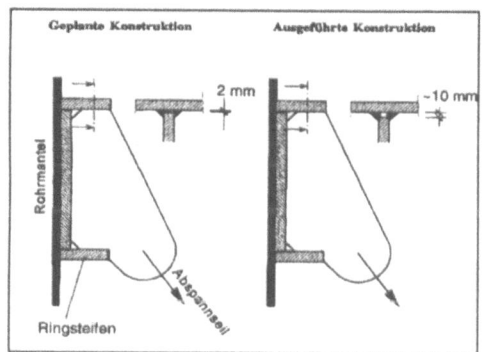

Figure 20. Planned and performed construction.

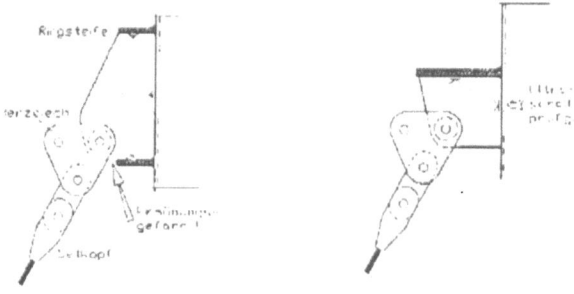

Figure 21. Connection constructions

What can we learn from that collapse? The plates should be carried out in such a way that fatigue can be excluded. The left figure of figure 21 shows an example of a connection construction with an unfavourable transfer of shearing forces. The disadvantage of this solution lies in the great difference of the stiffnesses between the plate and the ring stiffener. Thereby the danger of fatigue arises. The right figure presents a connection construction with a semi-rigid plate and welds that can be checked regularly. Here the danger of fatigue is less than in the first connection detail. However, the obstruction in the transversal direction is still relatively large.

A better construction for the connec-tion between cable and mast is the construction with a cardan-joint. This joint prevents longitudinal and trans-versal stresses in the transverse plane to the cable-shank plane.

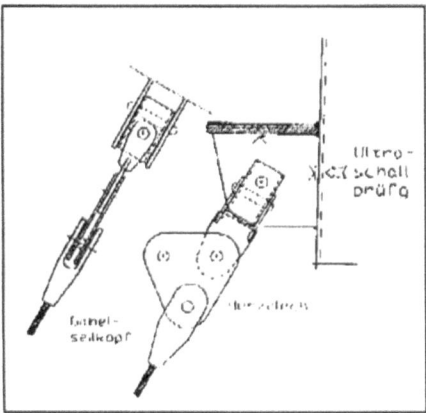

Figure 22. Connection construction with cardanjoint.

4.2 MAST IN BUDAPEST [2]

The mast in Budapest, that is described here, was built in 1933. It shows the typical features of a mast built in that time. It has a height of 314 metres, which was, back then, the world record. The cross-section of the mast shows a rectangle with a width of 14.65 metres. At each side of the cross-section, there are two cables – altogether eight cables - starting from a level of about 140 metres. The cables were made of steel wires. Every cable is divided into three or four separate cable parts between the isolators. The isolators are made of cast iron and porcelain.

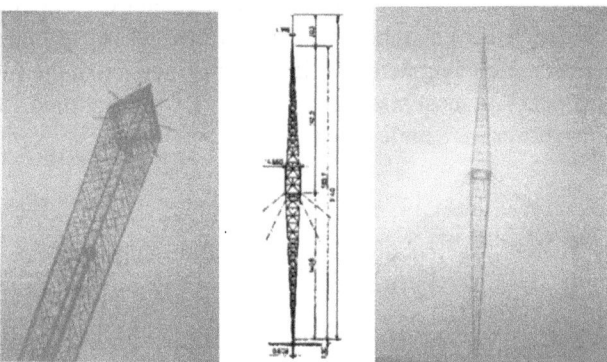

Figure 23. Mast in Budapest.

Figure 24 shows the details of the connection between foundation and cable. The empty holes are meant for temporary cables (during the time of mounting).

Figure 24. Connection: foundation-cable.

Figure 25. Renovation of the cables.

The mast has a long and interesting history. In the beginning, the mast was a radio transmitter, it was a self-radiating mast. In 1944, the mast had been destructed. Two years later, it was reconstructed. A complete restoration of the mast followed in 1968. Until 1975, the mast was again used as a transmitter. In the 80ies, the Department of Steel Structures of the Technical

University of Budapest conducted an investigation and gave an expert opinion about the status of the construction. It was decided that the mast has to be treated as a technical monument.
After that, a reconstruction including a renovation, the justification of the cables and corrosion protection was carried out. The mast was rehabilitated for the last time 13 years ago. In the meanwhile, signs of corrosion were noticed once again.

5 VIBRATIONS OF A CABLE [3]

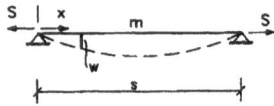

s - length of the cable
f - frequency of the vibration [Hz]
m - distributed mass [kNs²/m²]

Figure 26. Calculation of the cable force.

The important question in the field of masts that are mounted by cable guys is how to calculate the tension force S in a cable. The situation in figure 26 can be taken as a starting point. The cable shows a sine-shaped deflection w. This deflection depends on the derivative of the time t and on the derivation of the distance x.

Differential equation of the cable under vibration:

$$m\frac{\partial^2 w}{\partial t^2} - S\frac{\partial^2 w}{\partial x^2} = 0$$

Assumption:

$$a = \sqrt{\frac{S}{m}}$$

It follows:

$$\frac{\partial^2 w}{\partial t^2} - a^2\frac{\partial^2 w}{\partial x^2} = 0$$

with the solution:

$$w = C \cdot \sin\omega\, t \cdot \sin\frac{\pi x}{s}$$

ω = eigen angular frequency

$$\omega = \frac{a\pi}{s} = \frac{\pi}{s}\cdot\sqrt{\frac{S}{m}}$$

$$\boxed{f = \frac{\omega}{2\pi} = \frac{\pi}{s\cdot 2\pi}\cdot\sqrt{\frac{S}{m}} \Rightarrow S = 4\cdot s^2\cdot f^2\cdot m}$$

For practical purposes, the vibrations due to artificial excitation have to be counted.

6 DETAILS [4]

A critical area is the transition between the cable and the guy anchor. It has to be carried out very carefully. The advantages of an elastomere-sleeve are

- the decrease in difference of the stiffnesses
- the decrease in sensibility against corrosion due to a better prevention against moisture and
- the higher attenuation of the cables.

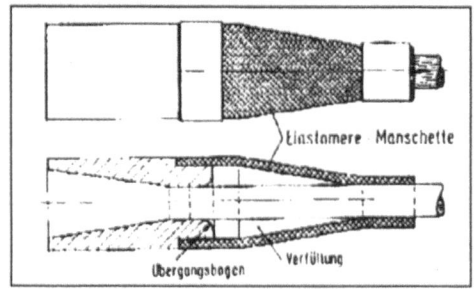

Figure 27. Elastomere-sleve.

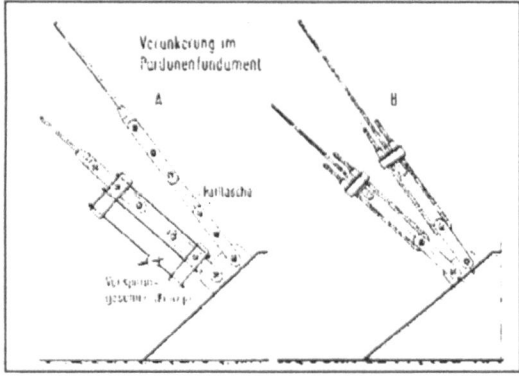

Figure 28. Detail foundation-cable.

There are two possible ways of performing the transition between the cables and the foundation. One can either use plates with a hole, so called eye-bars, or one can use thread-bars. The first solution with eye-bars is the more typical one. With regard to the pre- and poststressing of the cables, however, the solution with thread-bars is the better one. But the stressing of the cables is not the decisive argument here. Masts are wind-loaded, but they are calculated for predominantly static loading. This is not right. If repeated loading occurs (mainly with exposed structures), the stress-concentration has to be especially taken into consideration. Therefore, the solution with thread-bars is inferior to the solution with eye-bars.

7 FURTHER EXAMPLES

7.1 MISSOURI, USA [4]

In 1988, a TV lattice mast, which was 610 metres high, collapsed in Missouri, USA. Three persons died. What had happened? The mast consisted of three solid steel-posts. Gusset plates were welded onto the posts. Bracing rods with endplates were screwed to the gusset plates on one side. After a relatively short period of time, cracks could be observed at the transition of the bracing rods and the endplates. Altogether, 25 cracks were observed. The de-cision, however, was to substitute all 1,130 bracing rods. When 800 bracing rods had been removed at a level of 146 metres, the collapse appeared. As a result from this collapse, stress-concentrations with buildings which are susceptible to vibrations and loaded by wind forces should be avoided.

Figure 29. Collapse of the 610m-Mast

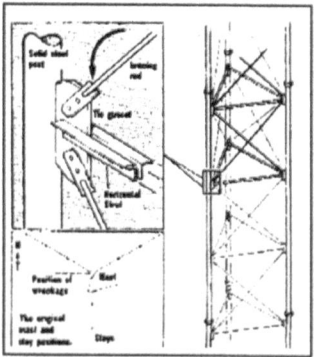

Figure 30. Detail of the connection.

7.2 MUNICH, GERMANY [4]

The figure 31 shows a planned and controlled blasting of a 256-metre lattice mast close to Munich. One of the cables was cut at a defined point and so the shank of the mast collapsed at a predetermined breaking point. This example teaches us that the collapse of only one cable can lead to the collapse of the complete building.

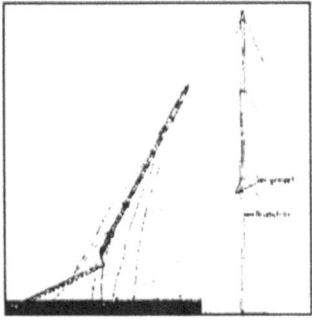

Figure 31. Planned collapse.

7.3 MANZANARES, SPAIN [5]

The next figure shows the anabatic wind power station Manzanares in Spain, which was built in 1982. It is an area of 30,000 square metres covered by a transparent coating. The chimney – a guyed mast – is situated in the middle of this area.

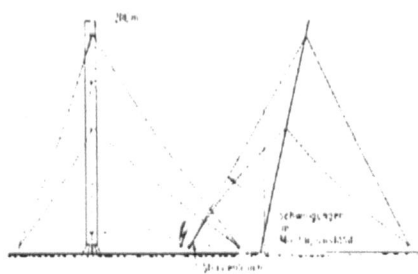

Figure 32. Wind power station Manzanares.

Warm air gets up into the chimney where a propeller is installed and a generator is producing electricity. During the procedure of assembling, the structure was supported by only two cables. The moving of the shank of the mast led to vibrations in the cables. A dangerous situation occurred during a thunderstorm. Due to the vibrations, the lower cable broke down. A plastic hinge developed in the supporting structure. This example shows that details are of great importance.

8 Conclusion

For the reconstruction and rehabilitation of masts and towers, the consideration and calculation of statics are absolutely necessary. Even the smallest (and possibly inconspicious) details can be important. The collapse of only one structural element can lead to the collapse of the whole construction. Especially with buildings which are subjected to stresses of wind-loading, stress-concentrations should be avoided in order to prevent fatigue of the material. The design of the whole construction on the one hand and of the details on the other should absolutely correspond to the planned structure.

9 REFERENCES

[1]   Skov, K., Pasternak, H.: Antennenturm Habichtsberg – ein Beispiel für Bauwerkser-haltung. Bauingenieur 76 (2001) 92-94

[2]   Massányi, K.: Die Stahlkonstruktion des 314m hohen Funkturms des neuen Großsenders Lakihegy (Übersetzung eines 1934 gehaltenen Vortrags in deutsch). Periodica Polytechnica, Vol. 28, No 1-4, Budapest 1984

[3]   Palkowski, S.: Statik der Seilkonstruktionen. Heidelberg, New York, Springer 1990

[4]   Petersen, C. : Abgespannte Maste – historische Entwicklung, Stand und Perspektiven. Vorträge zum Festkolloquium Prof. Scheer, TU Braunschweig, Institut für Stahlbau 1992

[5]   Schlaich, J: Das Aufwindkraftwerk. Deutsche Verlagsanstalt. Stuttgart 1994

[6]   Pasternak, H.: Zum Einsturz des 646m-Mastes bei Warschau. Bauingenieur 68 (1993), S.226

Thanks from the autor to A. Schwarzlos.

CHAPTER 4

STRENGTHENING TECHNIQUES FOR BUILDINGS

A. Mandara
Seconda Università di Napoli, Napoli, Italy

1 Introduction

Structural and functional restoration of existing buildings represent today an outstanding activity within the field of constructions. Such a problem arises as a consequence of the great increase of demand in urban centres for buildings fitted with up-to-dated features, but also of the need to preserve valuable constructions from ageing, natural hazards and, last but not least, human misuse.

This chapter deals with the techniques currently adopted for strengthening buildings in urban areas. In this view, its main purpose is to concisely outline the present state-of-the-art in this field, as resulting from the wide experience gathered in Italy in the last decades. Because of the incredible value of its building heritage, which comprises the largest amount of monumental works in the world, strengthening and refurbishment practices have been, in fact, particularly developed in this Country. Also, the presence of many seismic areas over Italian territory has involved a significant step-up of the knowledge in the field of earthquake protection techniques, with a direct fall-out in the practice of structural rehabilitation of buildings.

In this framework, a general description is given herein of the most important intervention techniques on foundation structures, masonry works – including arches and vaults – floor structures and r.c. members. Attention is focused in special way on the discussion of design principles and constructional steps of cases under consideration, with a particular emphasis to provisions to be adopted in seismic areas.

2 Reinforcement of foundation structures

2.1 General

The structural upgrading of foundations is usually necessary for two reasons:

- damage or inadequacy of foundation structures;
- lack of bearing or excess of settling in foundation soil.

In the first case current provisions for repairing and strengthening of masonry and reinforced concrete structures can be applied, which is why this aspect is dealt with in the following. The second case is concerned with a lack of foundation soil, which can be involved either by inherent poor soil properties, or by inadequate design. This may happen quite often in old masonry or r.c. buildings, which in most cases have been built without specific account for geotechnical problems. Most of times, damage of foundations arise as a consequence of soil settlings, that is of displacements which may occur across the whole lifetime of the building. Such displacements, either absolute or differential, can be due to (Figure 1):

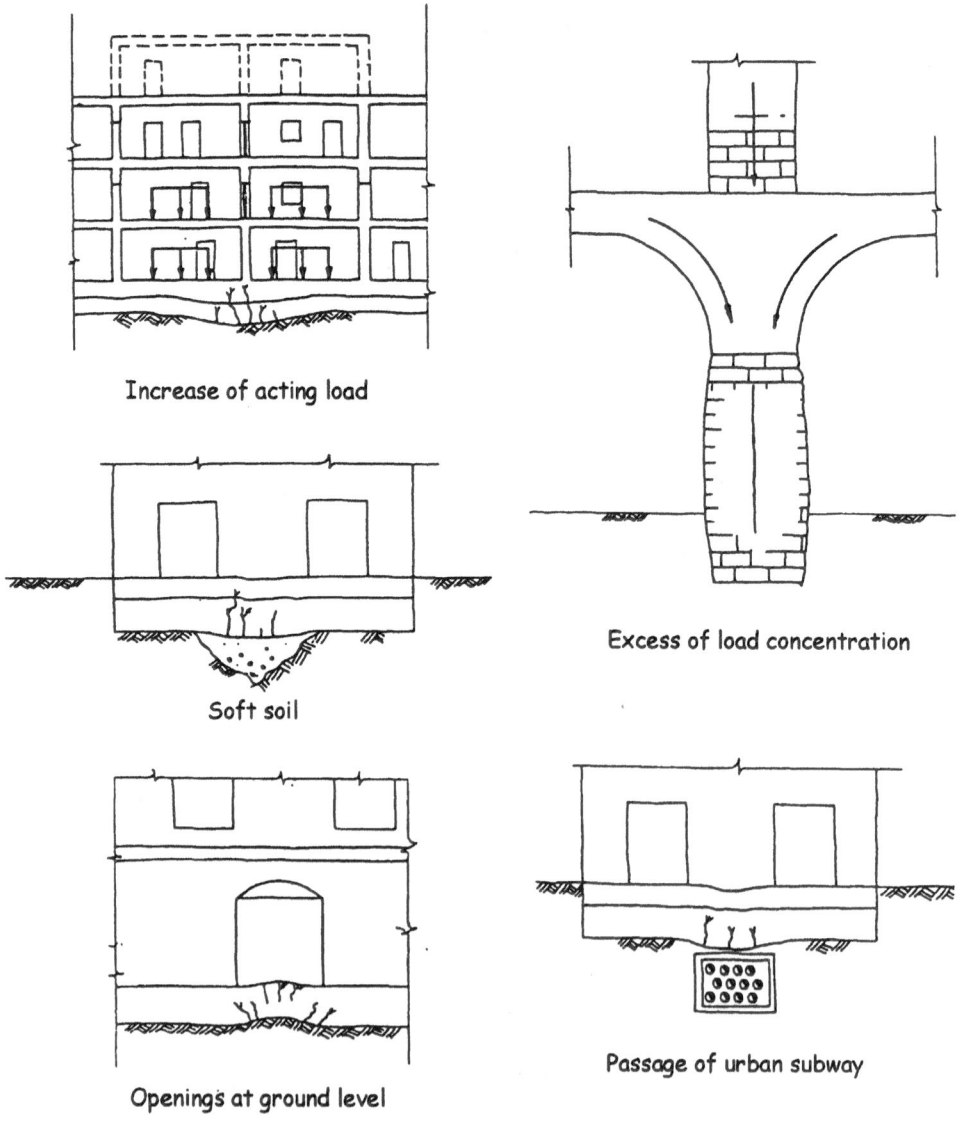

Increase of acting load

Soft soil

Excess of load concentration

Openings at ground level

Passage of urban subway

Figure 1. Causes of foundation damage.

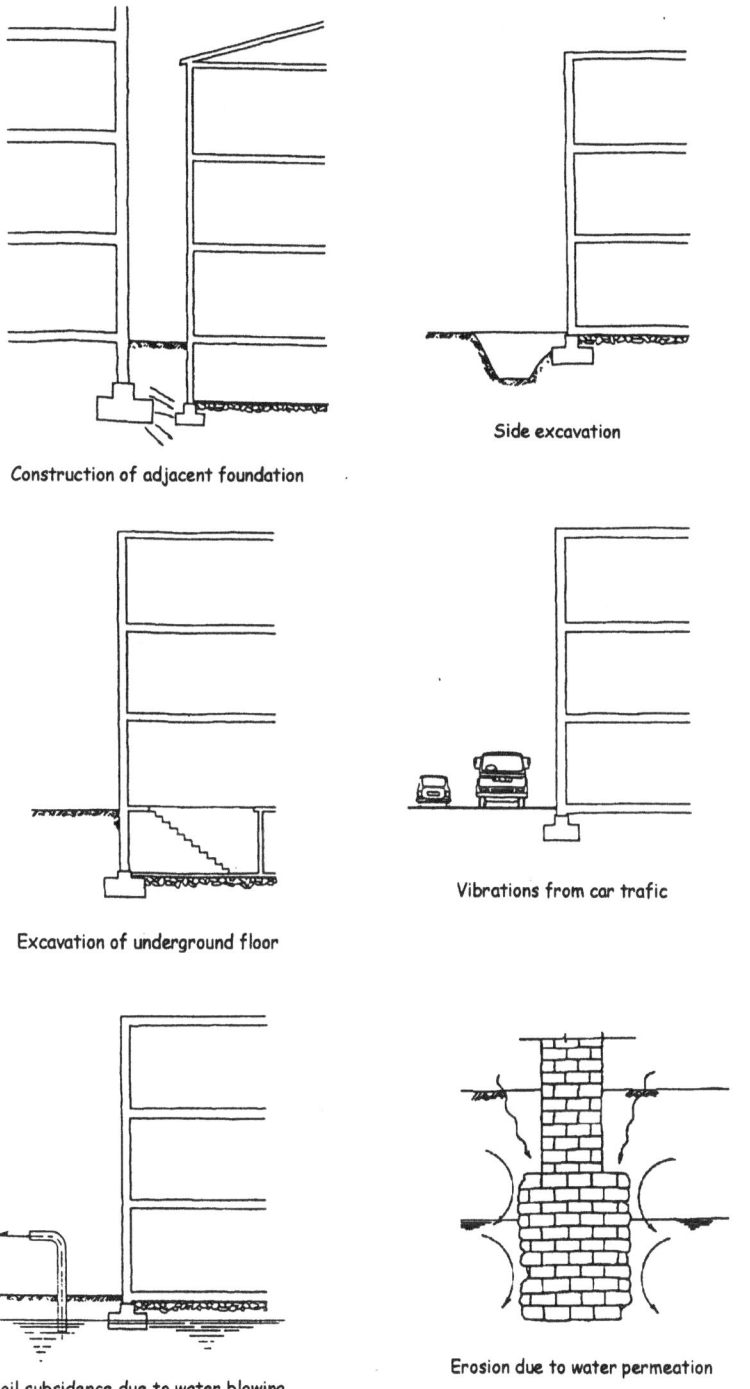

Construction of adjacent foundation

Side excavation

Excavation of underground floor

Vibrations from car trafic

Soil subsidence due to water blowing

Erosion due to water permeation

Figure 1 (ctd). Causes of foundation damage.

- insufficient depth of foundation bed, which prevents soil portions lying aside to stabilize the foundation itself;
- excess of acting load, which can be also involved by a transformation of the building, for example as a result of the construction of additional storeys, or due to the modification of its structural layout;
- variation of subsoil water level, which can produce soil consolidation and, hence, increase of displacement independently of applied load;
- excavation aside the existing foundation, for example made during the construction of a adjacent building, or due to improper digging operations made into the building underground floor;
- seismic actions, which can involve a strong increase of both horizontal and vertical load acting on foundation, depending on the building configuration.

It should be also pointed out, as outlined in the following, that a damage in the foundation soil generally involves a given amount of damage to the superstructure, which may result in some repairing and/or strengthening operations to be necessary as well. In other cases, strengthening interventions of foundations are applied not owing to existing damage but they may be required if new, more demanding working conditions are intended for the structure. In all these cases, the consolidation interventions on foundation must be carefully evaluated and executed in order to reduce the state of stress in the soil to an acceptable level. At the same time, the sequence of execution steps must be optimised in such a way to reduce the disturbance to the existing structure to a minimum, so to avoid excess of deformation in the overhanging superstructure.

As being stated, the most common operations for the strengthening of foundations can be grouped as follows:

- underpinning (by means of additional masonry or r.c. elements, or piles);
- base enlargement;
- improvement of soil properties.

Interventions referred at first two points aim at either reducing the state of stress on the soil or transferring the actions to stiffer, more resistant soil layers. The third option, instead, is applied in order to improve the mechanical features of the foundation soil itself, so as to achieve a higher value of the foundation collapse load. In the following, attention is mostly paid to the strengthening intervention on foundation structures. Some general information will be also given on underpinning by means of small diameter piles. The case of soil improvement, apart from a very few general considerations, goes beyond the goals of this description.

2.2 Underpinning by additional masonry or r.c. base

The concept underlying underpinning is the insertion of additional structural elements between the existing structure and the foundation soil. The new structural configuration, therefore, relies on two elements arranged in series, so as to carry the same amount of load as in the existing structure. The additional members, typically consisting of a new masonry or r.c. base, are designed in such a way to deepen the bed of existing foundation down to more resistant soil layers. In this way the stabilizing contribute of a thicker soil layer can be properly exploited, with a corresponding increase of foundation limit load. If needed, the horizontal cross sectional area of additional elements can be also enlarged as respect to existing foundation, in order to reduce the magnitude of applied stresses. Forces transmitted to the soil usually follow the same path as in the existing structure, except in those case where underpinning is applied in order to allow building subways for hydraulic or electric works.

From a general point of view, underpinning can be applied to both masonry and r.c. buildings, even though a closer interpretation of its basic concept is achieved in the case of masonry constructions. In r.c. buildings, and mostly in case of plinth foundation, underpinning is usually obtained by means of an additional r.c. beam or plate or, as an alternative, by piling. In both cases, the new elements can be arranged in such a way to transform the existing foundation from discontinuous to continuous.

Considering masonry buildings, underpinning can be applied according to the techniques shown in Figures 2a,b. As additional elements both masonry and r.c. basements can be used, sized in such a way to limit the stresses on the soil to the required values. If necessary, a multi-layered new foundation bed can be arranged. As an alternative, when additional rooms are needed in the building underground floor, underpinning can be also made by means of arches founded on r.c. plinths, so to reduce the amount of added material.

The most delicate aspect of underpinning is that it must be generally applied by excavating downside the building for a relatively long section, which can involve settling into the superstructure. The maximum allowable free span along masonry walls depends on both material and structure mechanical properties, as well as on the damage conditions of the superstructure. Sometimes, the design plan requires the building to be completely cut alongside its foundation base. As a consequence, a settling of the superstructure generally occurs, with a corresponding amount of damage. This can be reduced, even though seldom eliminated completely, by planning the sequence of construction phases carefully, e.g. by eliminating small portions of soil downside the masonry walls one at once, in such a way to rely upon an arch effect in the masonry. In any case, the use of appropriate propping systems, as the one shown in Figure 2c, is advised. In the same way, the use of enforcing devices, such as wedges and/or expansive mortars, or even flat hydraulic jacks is always recommended in these operations. Under certain circumstances, such enforcing procedures must be applied two or three times, depending on both soil and superstructure features, in order to achieve the best performance in terms of final displacements. To this purpose, the common practice is to use hard wood wedges, to be replaced after some days by larger wedges, so as to increase the degree of applied co-action. Once settling of overhanging masonry is completely developed, the wedges are removed and substituted by a brick layer, completely grouted with expansive mortar.

In any case, once the displacements due to the execution have occurred, underpinning will prove highly effective in reducing the additional ground settling involved by live loads acting in the building. This is due to the fact that, thanks to the series arrangement, the whole applied load is transferred to soil through the new foundation bed. In the end, when all settlings in the superstructure are stabilised, the new structural configuration of the foundation will be much more effective in reducing the displacements involved by applied live loads. For this reason, underpinning is particularly useful when soil and/or foundation structures exhibit damage due to overload or to inadequate design.

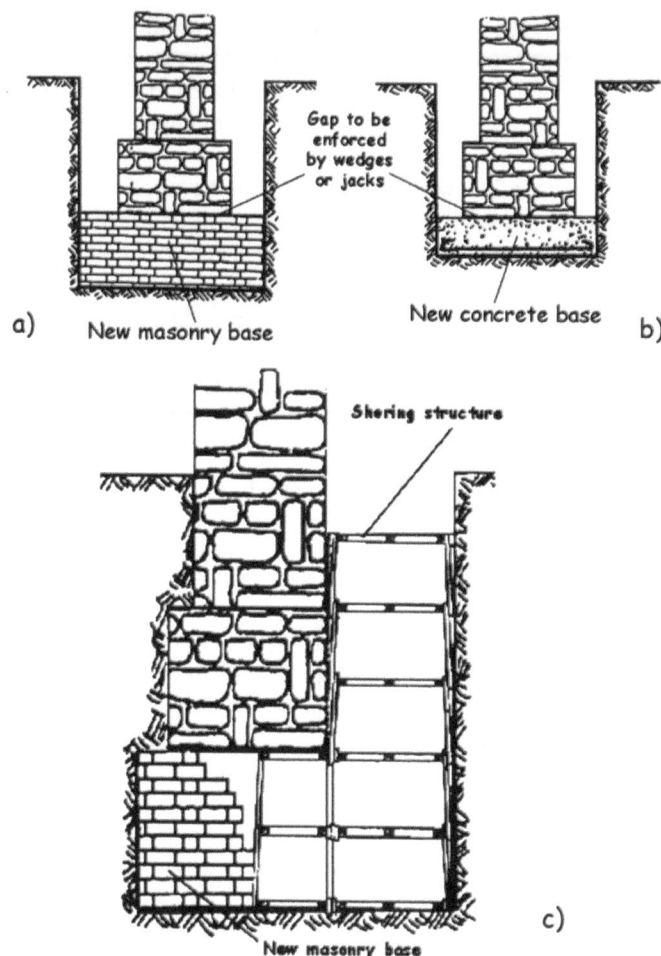

Figure 2. Examples of foundation underpinning in masonry buildings (a,b) and provisional shoring of a masonry wall during excavation (c).

2.3 Base enlargement

Compared with underpinning, base enlargement is based on a different mechanical principle, namely that to add new resisting elements working in parallel to the existing ones. As deep excavation downside the existing structure can be avoided, base enlargement may represent a cheaper alternative to the higher cost of underpinning, and in particular to the use of piles. Some possible options are shown in Figure 3. The most usual solutions are based on additional r.c. elements placed alongside the existing foundation (Figures 3a,b), in such a way to enlarge the contact area on the soil. Steel cold formed elements, sometimes tied against the base masonry in order to impress a given amount of lateral confinement, can been used as well, also as disposable formwork (Figure 4). In this case great attention has to be paid in order to prevent corrosion of steel elements by avoiding direct contact with soil.

As one can easily observe, the intervention effectiveness depends on two factors: on one hand both strength and stiffness of the system connecting additional and existing elements, on the other hand the actual degree of collaboration which can be ensured between them. As elements are arranged in parallel, they will share the acting load according to their stiffness and ultimate load bearing capacity. Therefore, it is very important that such connecting system is as much rigid and resistant as possible. At the same time, the use of controlled-shrinkage concrete is strongly recommended in order to ensure the highest degree of collaboration between materials, in particular in case of base enlargement of r.c. plinths or beams. To this purpose, epoxy resins can be also used at the interface between materials.

As new elements are cast aside the existing base, they will be enforced only by loads applied after their construction. Thus, in order to make these elements effective, additional displacements are needed, which can cause possible damage in the superstructure. Such risk can be reduced by trying two solutions. The first one is to eliminate all live loads from the building and, if possible, also a part of dead loads (e.g. furniture, partition walls or even floor structures, if their eventual substitution is planned), before laying down the base enlargement. In this way, all loads applied later will insist on the enlarged foundation acting as a whole, that is on both existing and new elements. The second way is to adopt active methods in order to apply a given preload to additional elements, for example by means of hydraulic flat jacks placed as shown in Figure 3c. In this case the value of the force acting on the new elements can be adjusted so as to meet the specified design requirements. In addition, preload can also be applied at different construction stages, depending on the planned load application procedure. In particular, some existing damage in the superstructure can be also partially or totally recovered by means of such active devices.

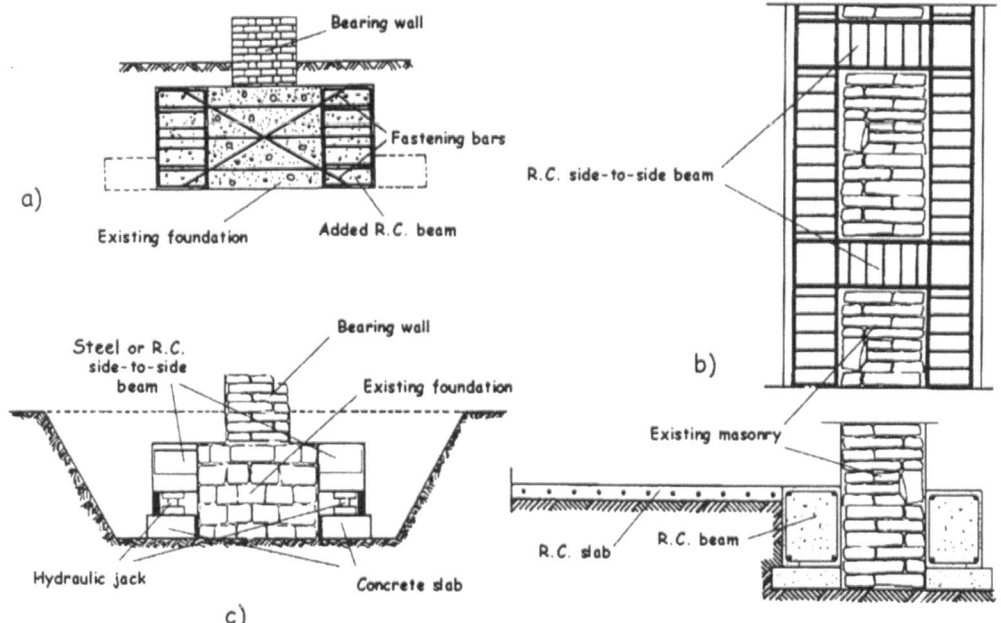

Figure 3. Cases of passive (a, b) and active (c) base enlargement.

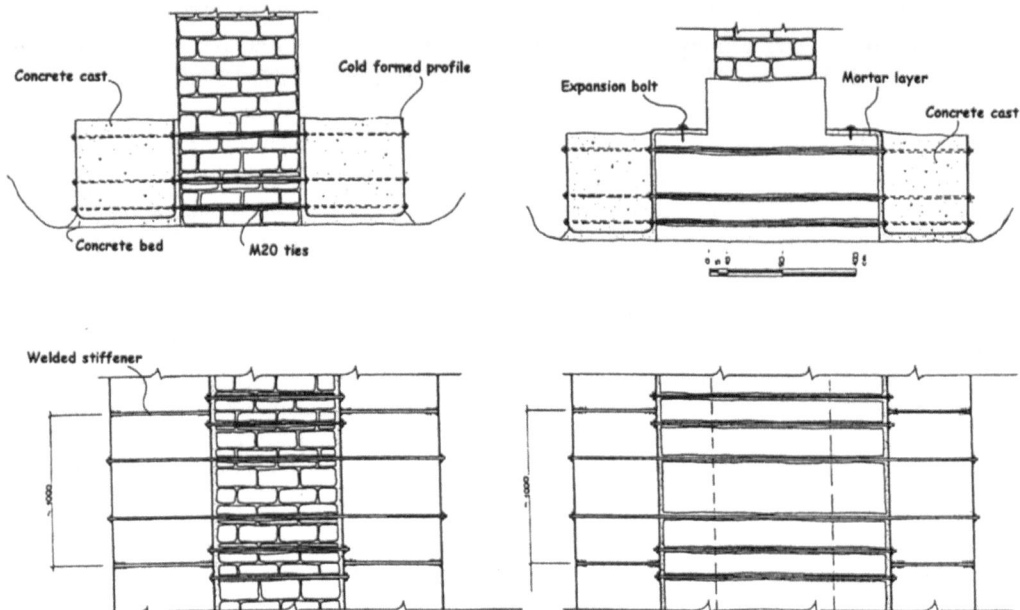

Figure 4. Example of base enlargement by means of tied lateral concrete beams cast into disposable cold formed steel profiles.

Compared with underpinning, the base enlargement is less effective, but does not involve significant additional damage to the superstructure during the construction of the new elements. A possible damage can occur when the portion of soil aside the existing foundation is removed. Nevertheless, this problem can be minimised if the building has been partially unloaded and/or if suitable propping has been adopted. For the same reason, a preliminary strengthening of both existing foundation and superstructure can be useful before excavating alongside the base walls.

For its inherent features, base enlargement is mostly recommended when a significant increase of active load is planned, independently of any possible damage existing in the foundation soil. This may occur, for example, when a new use is foreseen for the building, or even when additional storeys are to be added to the existing construction. In this view, base enlargement can be also obtained by means of a stiff concrete bed cast beneath the whole building area (Figure 5), so as to achieve a continuous foundation. In this case, provided that enough stiffness and resistance is ensured to the connection system, a very effective reduction of soil stresses can be achieved.

2.4 Underpinning by piles

As a more advanced and up-to-dated alternative to underpinning and base enlargement by means of masonry or concrete elements, the use of foundation piles can be considered. Sometimes there is not enough space for digging aside or downside the existing foundations, or there can be the risk that underpinning or simply removal of surface soil could involve a great damage in the superstructure. In these cases, piling underside the building can represent a suitable solution for satisfying the intended design requirements, in particular in case of strongly damaged buildings or when the rate of displacement is steadily increasing. Similarly, the use of piles can be convenient in case of complex soil stratification, as well as when the resistance of superficial soil layer is very low. To this purpose, many kinds of pile can be used, depending on the conditions to be met. Generally, low diameter bored piles are preferred, in order to reduce the magnitude of vibrations transferred to the superstructure. Modern light-weight drilling machines are today available, able to be easily introduced into small rooms for the execution of holes for piles with diameter up to 250-300 mm. This kind of pile, also called "micropile", can be considered as a modern version of that very old technique based on the use of wooden piles for the construction of small houses upon very weak soils. It has been largely developed in the last years thanks to the availability of high pressure soil grouting techniques which involve an improved bearing capacity by lateral friction. Similarly the use of high strength steel tubular members has allowed a great increase of individual pile ultimate bearing, which can reach 600-700kN for a tube with a diameter lower than 15cm. For such features, this technique is very appropriate to historical and monumental buildings, too. Possible micropiling solutions are shown in Figures 6a-d.

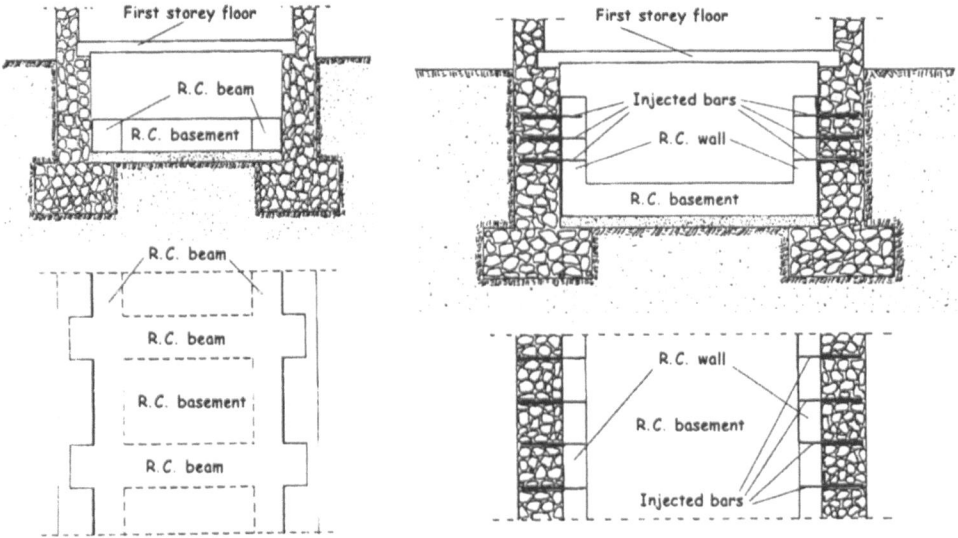

Figure 5. Base enlargement by new reinforced concrete basement.

Piles are also adopted as they allow to overcome many of the drawbacks related to conventional underpinning and base enlargement techniques. For example, deep foundations can be underpinned without expensive and dangerous excavations, which can involve severe damage to superstructure. In addition, piles are very effective when an increase of the foundation resistance against horizontal actions is required, in which case it is recommended that a part of piles is sloped to vertical ($\alpha = 10^{\circ} \div 30^{\circ}$, see Figures 6c,d). In this way, piles work partially in axial load and partially in shear.

Also, piles prove to be suited for strengthening the foundation of single structural elements, namely columns or males, e.g. in those case where a rather significant transformation of the global structural scheme has involved a strong increase of load acting on foundation. Together with required strengthening interventions on these elements, in such conditions the use of piles can significantly reduce the amount of additional settling caused by overload. A possible solution for the strengthening of a masonry column with its foundation is shown in Figure 7, in which the column has been reinforced by means of a r.c. jacketing.

Piles are also very effective for increasing both soil stiffness and ultimate collapse resistance. In addition, if properly executed, for example as shown in Figure 8, they do not involve any disturb to existing structures. In some cases, piles can be also used in order to stabilise or even recover existing settlings. This may happen, for example, in case of tall constructions (e.g. bell towers, free-standing columns, etc.), where prestressed piles can be used in such a way to control differential settlings upstream and downstream the construction. Such a solution was used in the restoration of a bell tower in Burano, near Venice, in order to contrast a continuously increasing slope of the tower, and was also proposed for a partial correction of out-of-plumb displacements in the Leaning Tower of Pisa.

2.5 Jet grouting

Jet grouting is one of the latest developments in the field of techniques for improving soil mechanical properties (Pinto et al., 2001). It is based on the execution of boreholes in the soil, which are subsequently grouted by mortar injected at very high pressure (higher than 30Mpa). Unlike common grouting, jet grouting technique is based on the disintegration of the soil as a consequence of the very high pressure mortar injection. To make this, very small diameter nozzles, placed at the end of the drilling perforation rod, are used. The grouting itself takes place after the complete borehole drilling, starting from its bottom. As a result, either planar or cylindrical grouted elements are created around each borehole for a depth variable depending on design requirements. The distance between boreholes should be determined in such a way that the injected bulbs interlock between each other without leaving untreated soil portions. Depending on soil properties, grouting diameters up to 80-100cm can be reached.

Contrary to micropiling, for its inherent features jet grouting is quite appropriate to soils characterised by high deformability and porosity, when it is necessary to create a more stable basement to the building or when retaining walls for protecting excavation works are needed. In such conditions it is possible to take advantage of the mobilization of a great amount of lateral friction in grouted soil elements, with a significant end bearing capacity as well. In addition, when used in weak soils, jet grouting greatly gains in stiffness and water tightness as respect to micropiles. On the other hand, jet grouting involves a great mortar consumption, as well as a difficult control of execution in those situations characterised by subsoil cavities or lack of homogeneity.

In conclusion, it can be said that advanced underpinning techniques, such as micropiling and jet grouting, can yield a much higher performance than conventional strengthening methodologies for foundation structures. In addition, a negligible disturbance is brought to existing structures, and this can play a very important role when operating on damaged or historical buildings. In most case, this performance outbalances the higher cost and the need for specialised workmanship involved by these techniques.

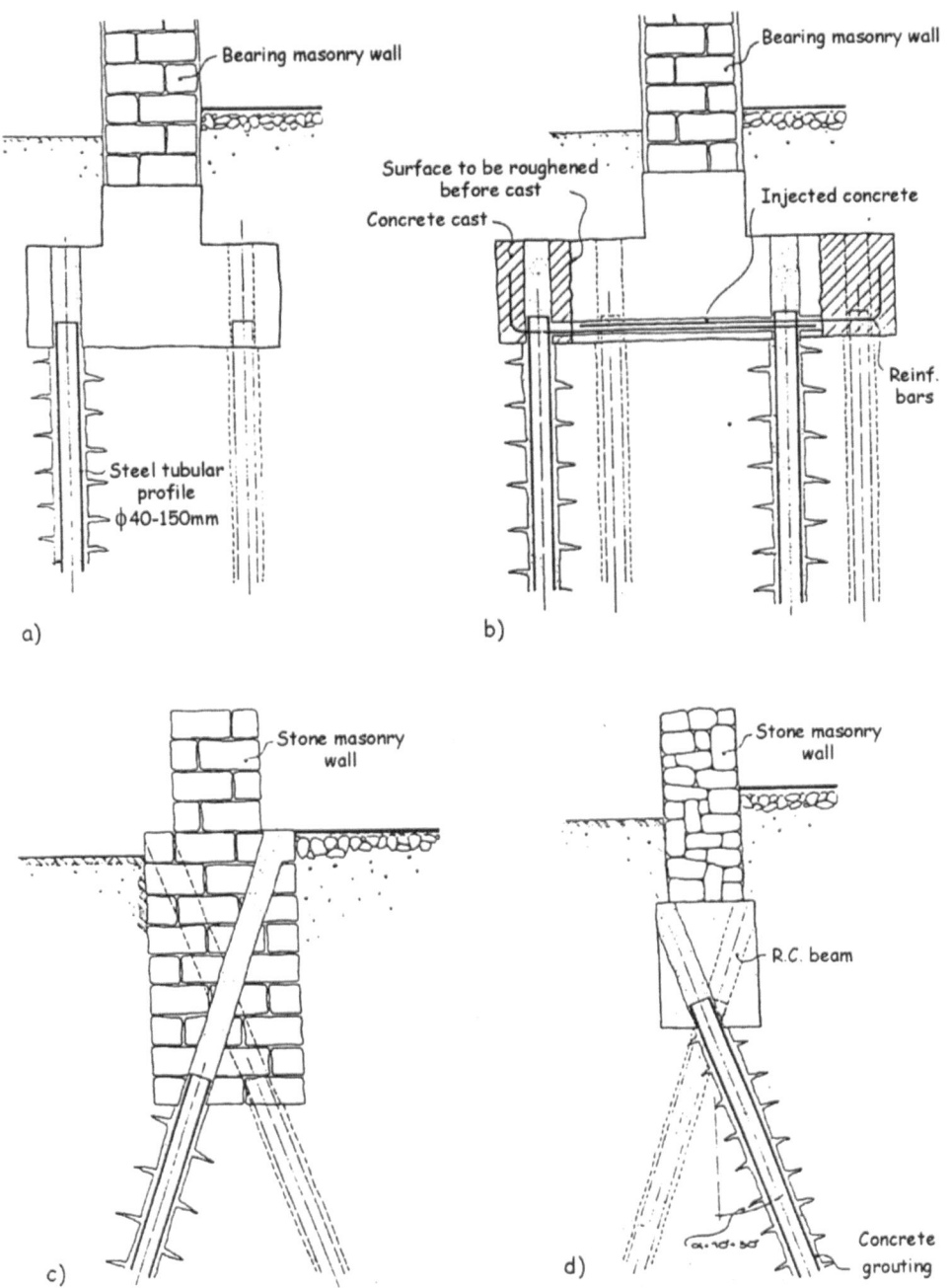

Figure 6. Underpinning by piles of an existing foundation:
a) Intervention on simple masonry or r.c. base; b) Underpinning by piles with base enlargement;
c) Sloped piles bored into a masonry base; d) Sloped piles bored into a r.c. base.

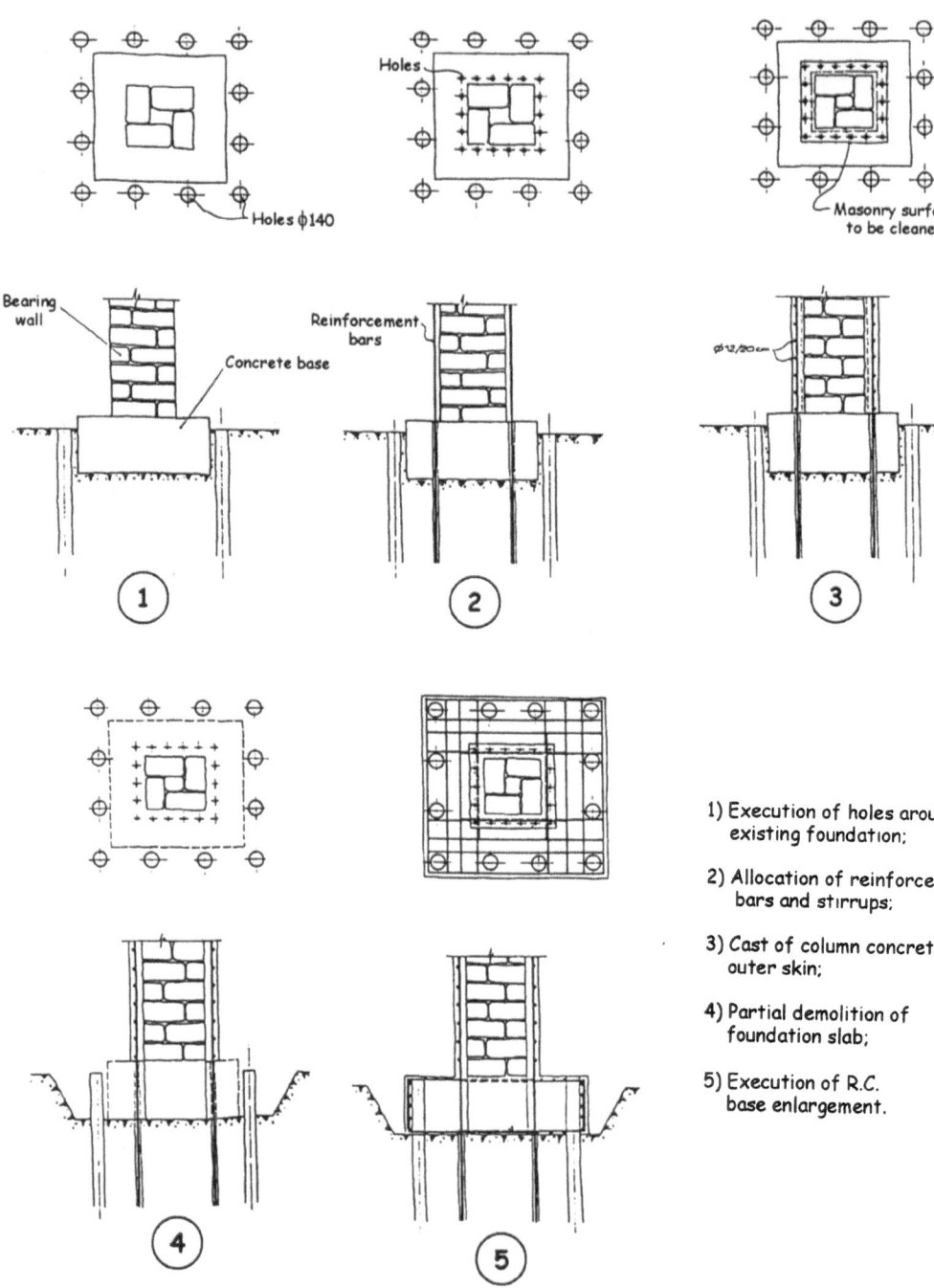

Figure 7. Underpinning by piles as a consequence of column strengthening with corresponding base enlargement and indication of relevant execution steps.

Step 1:
Execution of holes

Step 2:
Allocation of steel
tubular members

Step 3:
Injection of concrete
under pressure

Step 4:
Pile concrete filling

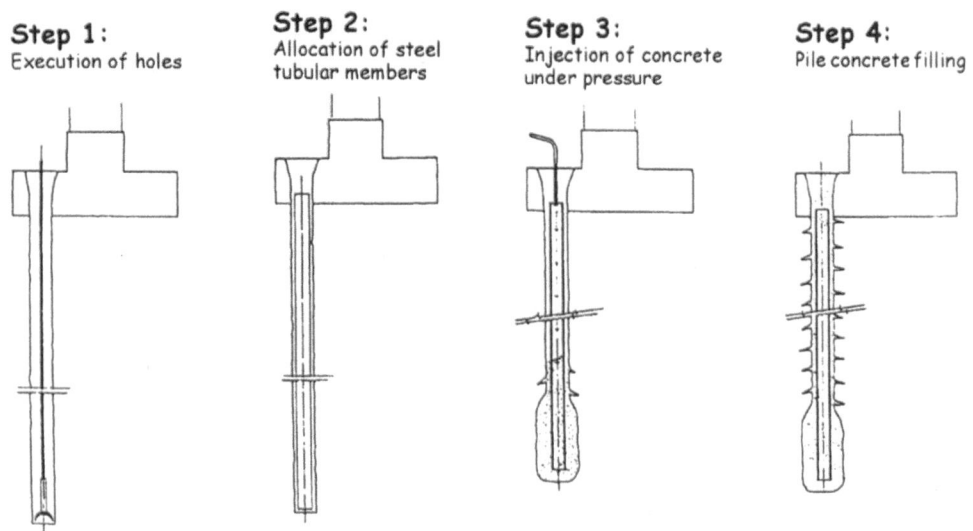

Figure 8. Typical execution steps of a bored steel tubular pile into an existing masonry base.

3 Consolidation of masonry structural elements

3.1 General

Repairing and strengthening of masonry members can be carried out according to the following basic principles:

- passive consolidation;
- active consolidation.

Passive consolidation aims at improving the load carrying capacity of existing members by means of additional provisions conceived so as to increase either cross sectional areas or mechanical properties of materials, but without any modification of the stress pattern due to internal co-action. The full strength of masonry is exploited only through additional displacements, so that the effectiveness of such interventions may involve a given amount of structural damage. Thus, particular attention has to be paid to reduce this damage, by means of additional enforcing elements, or studying the sequence of construction steps much carefully.

The most widely adopted practices are:

- local repairing of joints, cracks and damaged parts;
- masonry strengthening by pressure grouting (injection of concrete mortar);
- masonry reinforcement by steel concreted bars;
- coupling or substitution of existing masonry elements with r.c. or steel members;
- suspension of existing elements to overhanging structures.

Active consolidation is based on the introduction of a co-action into masonry, so that a multi-axial state of stress can take place. This involves a significant increase of the compressive strength of masonry, with a corresponding improvement of both ultimate strength and ductility. Contrary to passive consolidation, this practice is able to develop its action without additional displacements, which is why it can contrast or even reduce the damaging causes. This is made possible by suitable enforcing procedures, which have to be carefully studied in order to provide the best effectiveness. The most common active systems are:

- – steel tie-beams placed alongside bearing walls;
- – steel tie-beams at the base of thrusting structures;
- – prestressed steel bars into masonry elements subjected to axial load;
- – confining of compressed load-bearing masonry elements.

3.2 Local repairing of joints, cracks and damaged parts

When applied to masonry joints the practice of local intervention it is also named "rejointing", and is aimed at restoring or increasing the mutual connection between masonry blocks in order to enhance both compressive and shear strength. Rejointing can be superficial, when applied so as to achieve a simple sailing of masonry outer surface, or deep, in which case it can produce a significant increase of global masonry resistance. In the latter case, it can be supplemented by a pressure grouting of the masonry wall core (see next paragraph) which is able to transform the whole masonry mass in a monolithic body.

In case of locally damaged masonry, local repairing technique takes the name of "strengthening by substitution" or, after the italian term, "cuci e scuci" (Figure 9). It is known since the ancient age as the most suited to the conservative restoration of historical masonry. It is essentially based on the elimination of cracked and damaged parts, as well as on their substitution with new elements (blocks and mortar) which have to be, as a rule, quite similar to the existing ones. As this practice allows to obtain a masonry texture quite close to the original one, it is widely followed in the rehabilitation of historical buildings. In this case, the main drawback is represented by the difficulty to find workmanship sufficiently skilled with old techniques and materials really similar to those used at the first construction stage. As an alternative, when original materials are not available, different kinds of stones or bricks can be used as well, provided that their mechanical properties are not very different from those of existing masonry.

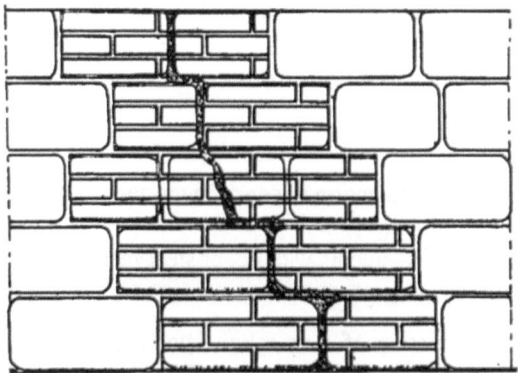 

Figure 9. Examples of masonry "cuci e scuci".

From the practical point of view, the structural effectiveness of this intervention mostly relies upon the bonding effect at the interface between in-situ masonry and added elements. If this contact loosens, e.g. due to mortar shrinkage, the operation loses a great part of its value, as too much additional displacements may be needed for getting the required degree of structural collaboration. This may involve damage to the superstructure. For this reason, it is recommended to adopt low-shrinkage mortars, as well as very low-thickness layers for embedding new brick or stone elements. At the same time, it may be useful to enforce suitable wedge elements before mortar setting in order to reduce the risk of loading the masonry without the collaboration of the new parts.

During the execution, like for any passive intervention, it is convenient to shore the superstructure by means of proper wooden or steel props, in order to avoid local or global crushing of the worked masonry. Sometimes, when acting with local repairing of masonry new cracks can occur in other parts of the building due to masonry settlings; a step-by-step procedure is often necessary to limit the damage possibly occurred elsewhere.

As a modern option for repairing local cracks and damage, filling by resins can be also used. These products, available in a large quantity of types with a wide range of mechanical properties, offer high strength features ($\sigma_c \sim 100$ N/mm^2), high adhesion to masonry surface, ease of modelling, very quick setting, capability of filling microscopic cracks, excellent corrosion resistance, etc. In addition, they have no shrinkage thanks to no thinner evaporation. Resins are used in extremely small thickness, which makes them suited to the surface treatments of brick and stone blocks, as well as to the restoration of existing cracks. In general their application is more appropriate to masonries with high mechanical features. They require surfaces to be treated with strong thinners in order to be completely clean and dry before the application of resins. This involves the use of highly specialised workmanship in both handling and application of such materials.

As bad points, they still have high cost, together with other drawbacks which can be summarised as follows: high fluage, particularly in pure resins; poor fire resistance (inflammability) with thermal softening at $70 \div 80°C$; brittle behaviour at failure; high degradation due to chemical ageing with loss of adhesion; strong decrease of strength when combined with inert material; low value of elastic modulus (approximately 1/10 of concrete elastic modulus E_c). Furthermore, like many interventions based on the use of injected products, the application of resins is not reversible, in the sense that it cannot be removed from the worked masonry, and this can represent a questionable aspect about their use in monumental building. Such considerations have recently oriented the use of epoxy-resins to the restoration of specific details, including microscopically cracked stone elements or rotten ends of wooden beams, where their features can be fully exploited with acceptable costs. Eventually, epoxy-resins can be used as additive elements for the preparation of special mortars provided with special features in terms of viscosity, shrinkage and water tightness. In this view they can validly supplement traditional cement-based products.

3.3 Masonry strengthening by pressure grouting

A step ahead as respect to local repairing is the reinforcement of masonry by mortar injection. This practice is based on the full mixing between masonry blocks and injected material, giving place to a more resistant system. Global properties of masonry in terms of both stiffness and strength are greatly increased as a consequence of this intervention. As mortar is injected under pressure ($p_{inj} = 10 \div 60$ N/cm^2), a careful analysis is needed before starting the intervention in order to avoid outer masonry leaves to bulge out the wall due to excess of internal pressure. This preliminary investigation, to be usually carried out by means of a trial-and-error procedure on small masonry samples, allows to define, together with the injection pressure, also the distance between holes (generally $20 \div 40$ cm but not greater than the wall thickness) and the water-to-concrete ratio ($1 \div 2$ depending on void volume in the masonry).

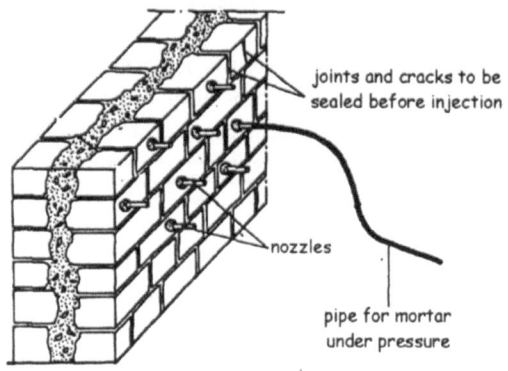

Figure 10. Masonry grouting by injected mortar

The executions proceeds via the following steps (Figure 10):

- elimination of both damaged plaster and masonry fragments from the cracked sites;
- execution of holes ($30 \div 50$ mm) with boring machine along the cracked masonry joints;
- insertion of pipes ($\ell = 10$-15cm, d = 3/4') into the holes;
- sealing of masonry joints and pipe holes in order to prevent the injected mortar to jump out of the wall;
- injection of water at low pressure into one of the pipes in order to saturate the existing masonry;
- injection of pressurised concrete mortar starting from the lower holes ($150 \div 200$ kg per cubic meter of masonry).

Perforation is usually made on one side for wall thickness up to 60 cm and on both sides, if practicable, for greater thickness. Care has to be paid in controlling the mortar absorption by the masonry. An insufficient filling, in fact, would cause voids to remain after the intervention with consequent weak points inside the masonry. An excess of injected mortar, on the contrary, could cause undesired lateral expansion of the wall. For these reasons, the use of low-shrinkage fillers is recommended in this application.

This practice is very effective in the strengthening of existing masonry, in particular in those cases where the internal core of the masonry wall is made of cohesionless materials, as usually happens in the so called "multi-leaf" masonry. At the same time, this procedure requires to be applied to masonry with good inherent features, in order to prevent the wall from mortar expulsion or even from side spalling of large outer portions of masonry. For this reason it is recommended that injection is made only after all joints have been properly repaired and sealed. It is thereby less effective in brick masonries with very close joints, where the injection pressure would be too much high. In this case grouting could be applied by hand for protecting the external skins of the wall from the entrance of moisture and aggressive agents, according to typical procedures adopted for rejointing. For the best effectiveness of the intervention, the injected mortar should possess adequate physical and mechanical properties, namely high fluidity, good adhesion and water tightness, low shrinkage and bleeding, together with capability to fill small or even microscopic cracks.

3.4 Masonry reinforcement by steel concreted bars

As a further solution, masonry can be also strengthened by means of steel bars embedded into bored holes and sealed with injected mortar. This practice, which represents a development of the one described above, could be also named as "Strengthening by reinforced perforations" or "Strengthening by grouted bars". It aims at a stronger connection between masonry layers, by increasing in the meantime its shear strength (e.g. against seismic actions) as well as its ductility at failure. For this reason such technique can be applied for repairing existing cracks (Figure 11a) or for strengthening openings and plat-bands (Figure 11b). Nevertheless, it is particularly effective for increasing the strength of masonry crossings and angles, in particular when disconnection has occurred as a consequence of earthquake or simply because of lack of mutual link (Figure 12). The effect of reinforced perforations is based on the lateral confinement of masonry arising owing to steel bars as a consequence of internal friction or due to end plates. As well known, these parts are the most stressed in case of seismic action and are extremely important for the achievement of the so-called "box-like" behaviour, which allows the optimal distribution of horizontal actions among the resisting walls of the building. When properly applied, this provision can be very useful for the improvement of the seismic response of a masonry building. In particular, it can be used to enhance the level of global ductility of the construction under seismic loads.

The above technology is suited to masonry walls with rather high mechanical features, otherwise a preliminary consolidation is recommended. It is also used for connecting new reinforcing elements to existing structures, in such a way to obtain a fully collaborating system. This is the case, for example, of steel or r.c. frames inserted into wall openings, and also of additional columns built alongside a masonry wall in order to increase its load bearing capacity.

In some cases one or two r.c. walls are built alongside the existing masonry walls so as to give place to a sandwich system. By means of this technique, also called "concrete jacketing" (Figure 13), a lateral confinement is obtained, which deeply improves both service and ultimate behaviour of masonry. This can be useful for the reinforcement of a wall opening, when the masonry needs to resist tension stresses, as well as for repairing side-to-side wall cracks. When using r.c. walls fastened with concreted bars, one may conservatively assume that the acting load is supported by r.c. walls only. In this case both existing masonry and internal ties prevent r.c. walls from buckling. The equivalent resisting scheme can be interpreted as a trussed beam subjected to design seismic forces or to notional horizontal actions (e.g. $1 \div 5\%$ of dead load) in non seismic areas. Use of rather liquid expansive concrete and 45° sloped (both horizontally and vertically), little diameter bars is recommended, fastened outside the walls. Concrete walls can be also obtained by means of a $4 \div 5$ cm thick "spritz-beton" leaf.

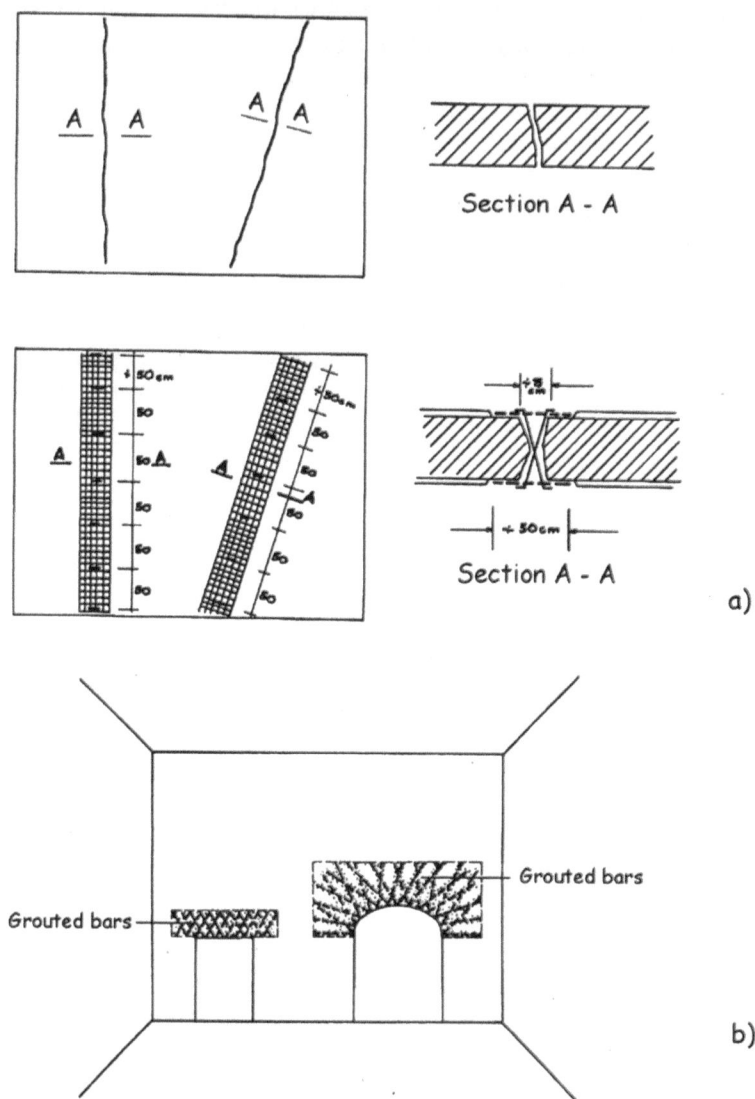

Figure 11. Repairing of localized cracks (a) and wall openings (b) by means of steel concreted bars.

Reinforced perforations have been widely used in Italy after the most recent earthquakes (Friuli 1976, Campania-Basilicata 1980, Umbria-Marche 1997) (Mazzolani and Mandara, 1989a,b). At that time, their application was encouraged by national and regional seismic regulations purposely issued after earthquakes. An *ad-hoc* reduction of the seismic behavioural factor β was also introduced in the italian seismic code for keeping into account the increase in global ductility involved by such technique. In spite of this, their effectiveness has been recently rather questioned, pointing out some aspects which proved to be not completely satisfying in the long term. Among these, there is the risk of corrosion for the steel bars, which is particular important at the light of the non reversible character of the intervention. In addition, their application does improve mechanical features of the masonry, but this happens only in the region involved by the strengthening. Elsewhere the resistance is that of the untreated masonry and failure can easily occur at the interface between reinforced and unreinforced areas. As a result, possible cracks can arise in zones other than those which would be expected without applying this provision. On the other hand, overall consolidation of building by concreted bars is not possible, due to their great cost. Eventually, their application to historical constructions has been often criticized because of the non-reversible feature, but also for a certain invasive character which could be inappropriate to artistic buildings. For these reasons, their use has recently become less common in the rehabilitation practice, where less intrusive and disturbing techniques are preferred as a rule.

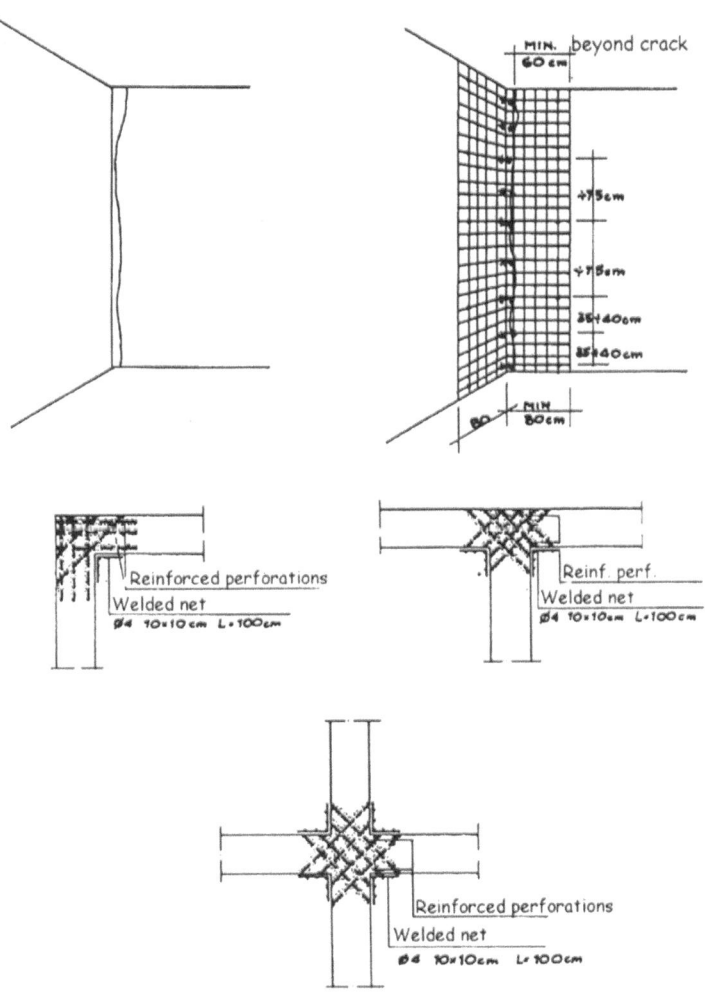

Figure 12. Application of concreted steel bars for the strengthening of wall crossings.

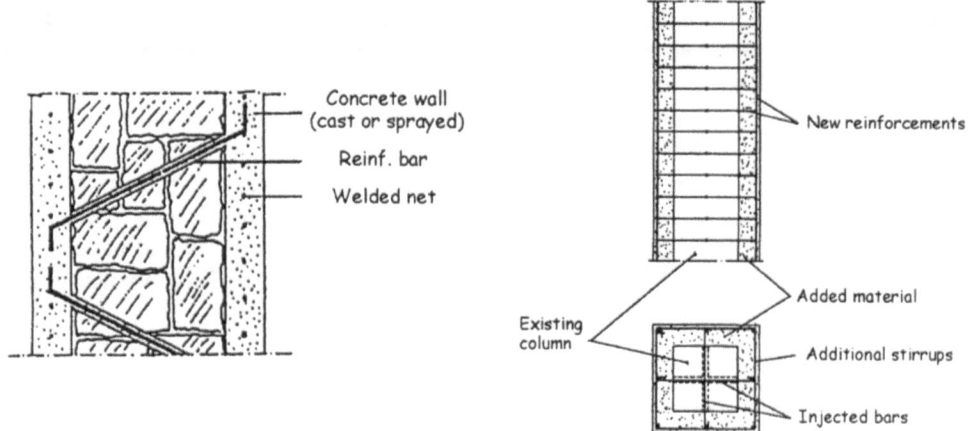

Figure 13. Reinforced concrete jacketing for the strengthening of masonry elements.

3.5 Coupling or substitution of existing masonry elements with r.c. or steel members

Among passive interventions the so called coupling interventions are worth being mentioned. They are based on the parallel arrangement of new and old structural elements, e.g. steel or r.c. columns placed aside an existing masonry wall (Figure 14). Coupling can be adopted for improving the resistance against both vertical and horizontal actions, including those cases when a portion or the whole of the wall is removed due to new functional requirements, in which case suitable provisions must be adopted in order to substitute for the eliminated parts. According to the basic principle of coupling interventions, the distribution of acting forces between existing and added resisting elements depends upon their mutual stiffness ratio. From the point of view of global structural behaviour, this may be not of great concern when the strengthening of simple elements against vertical load is concerned, but it can play a very important role in case of interventions made to reinforce openings or even when new elements are added in order to provide an higher level of safety against seismic actions. In this case the introduction of new elements can so deeply affect the stiffness distribution within the construction, that a global change of the structural layout is usually required, including foundation structures. This happens, for example, when additional braces are inserted in order to resist horizontal loads, or when large portion of masonry are taken off and substituted by appropriate supporting elements.

Coupling interventions can be either local or global, depending on whether their effect is limited to an area close to the intervention itself, or involves the overall behaviour of the building. Reinforcement of wall opening represent typical cases of local interventions. It can be made with steel (Figure 15) or r.c. elements (Figure 16), arranged as a simple plat-band or as a closed frame. The effectiveness of the intervention strongly depends on the connection between new and old elements. When r.c. elements are used, this does not represent a problem, as the concrete cast can fill up all surface voids of masonry. However, a proper use of concreted bars is recommended in order to create a strong connection between members. The frame elements can be sized in such a way to recover the loss of wall stiffness and resistance involved by the opening.

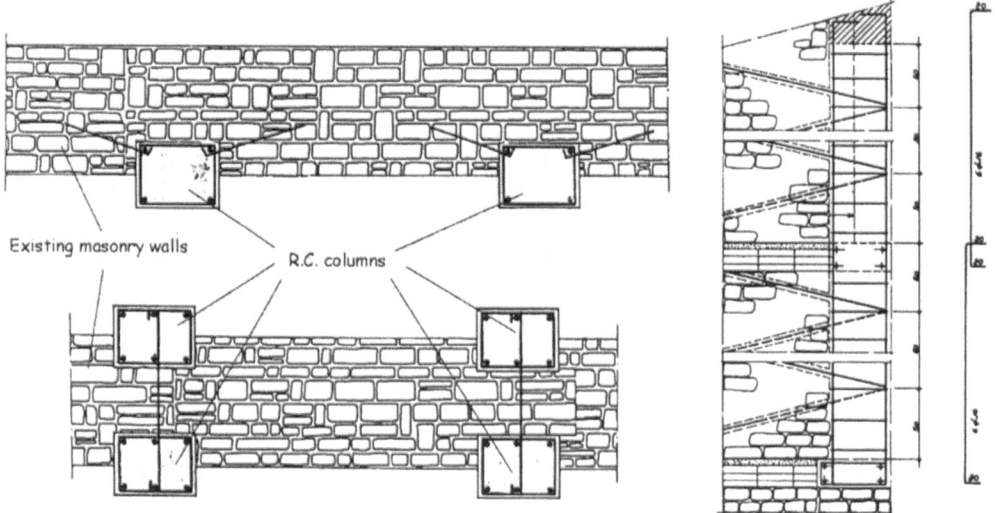

Figure 14. Coupling of masonry walls with r.c. columns.

If steel beams are used, a suitable wedging system is required as a rule, in order to enhance the effectiveness of the intervention. In addition, the vertical stiffness of steel plat-bands has to be carefully checked, as an excess of displacement could involve damage in the upper masonry structure. Steel elements, however, offer easy of transportation and erection, lightness and immediate availability of load bearing capacity, which can turn very useful in such operations. The connection to surrounding masonry can be easily made by means of steel ties grouted into bored holes. In the same way, bolted ties are recommended to fasten steel beams to each other through the masonry wall.

The above principle can be also used for strengthening masonry walls due to the complete elimination of a wall at a lower storey. In this case the problem is much more delicate as the redistribution of stresses in the upper walls can easily involve damage spread everywhere in the structure. In addition, the strong modification of the stress pattern generally requires the foundation of the building to be substantially revised. Some possible schemes are shown in Figures 17 and 18, referring to the use of steel-based solutions. Similar provisions can be used in case r.c. elements are adopted.

Concerning interventions shown in Figures 17 and 18, it is important to point out as the whole of the load formerly transmitted by the eliminated wall is now concentrated on the side walls, as well as on their foundation, both of which must undergo an appropriate strengthening treatment. Also, the problems related to the deformation of steel elements and to the application of strongly concentrated loads have to be solved. A possible solution can be that to strengthen the upper walls by means of a r.c. or a steel skin, so as to increase both stiffness and ultimate resistance. Figure 18 shows such an application, in which a double steel plated web has been used for strengthening the wall. Note the connection system between webs and to the masonry, usually made with steel bolted ties. In both solution in Figures 17 and 18, if necessary, vertical side walls can be reinforced by means of additional steel or r.c. columns, e.g. according to the scheme shown in Figure 14.

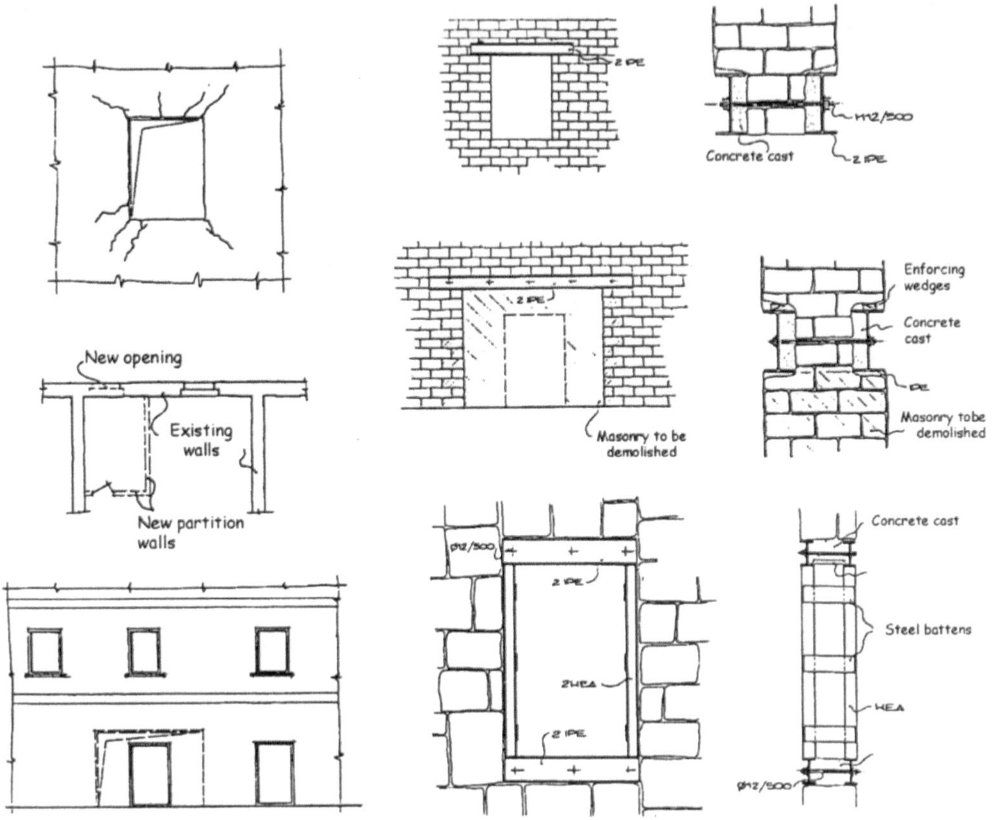

Figure 15. Use of steel profiles for the strengthening or enlargement of a masonry wall opening.

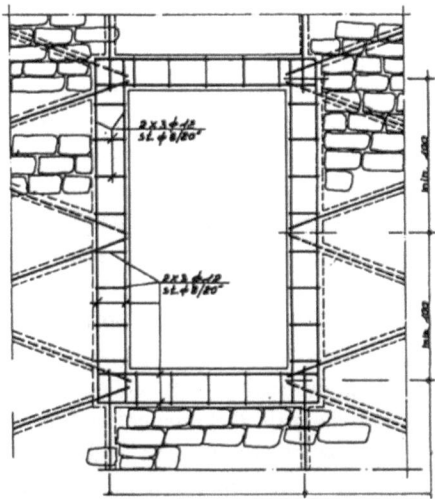

Figure 16. Use of r.c. closed frame for the strengthening of a masonry wall opening.

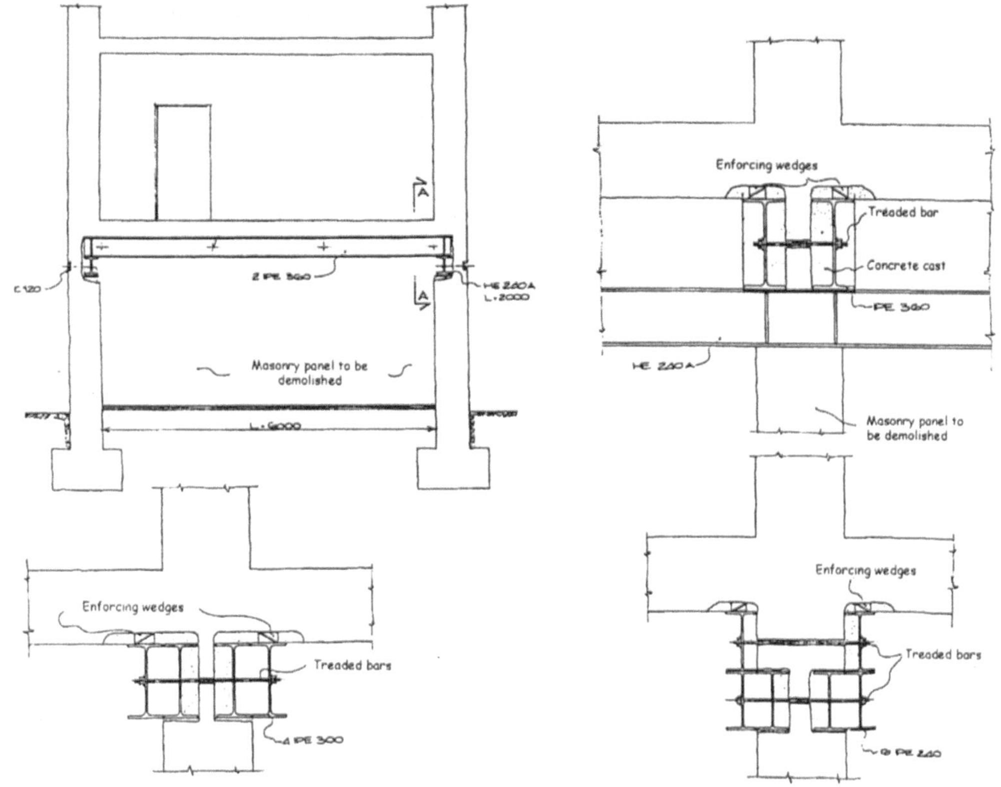

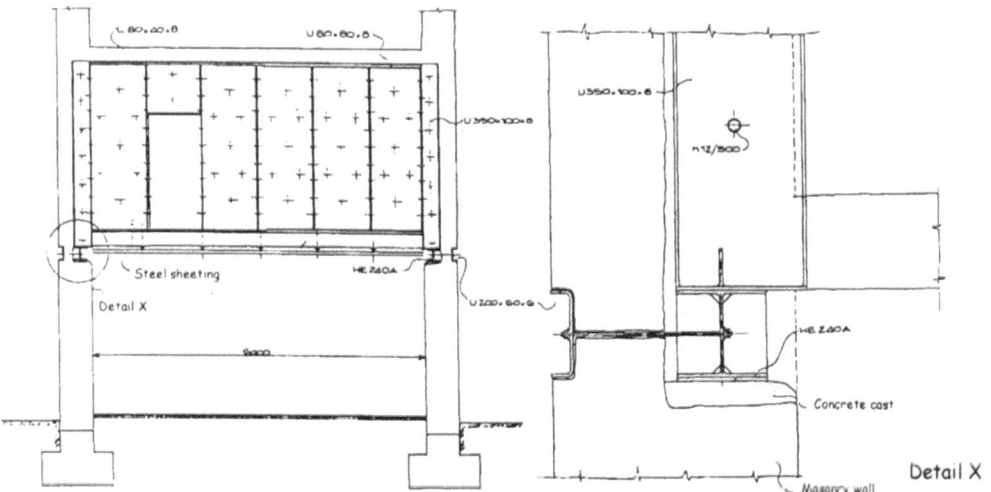

Figure 17. Use of multiple steel beams as supporting elements for superstructure after the elimination of a masonry wall and relevant constructional details.

Figure 18. Same as in Figure 17, but using a double sided tied steel plated panel.

The most demanding case of coupling between additional and existing structures is the insertion of braces for increasing the resistance against horizontal loads. Such braces are conceived in such a way to create stiff cantilever structures, able to withstand a share of global horizontal action. This is a typical retrofitting intervention adopted in seismic areas to upgrade the safety level of buildings in case of earthquake. Braces can be made of both steelwork and r.c., depending on the intended design requirements. Again, steel elements provide great effectiveness with a negligible increase of weight, especially when the concentric St. Andrew's cross configuration is used (Figure 19).

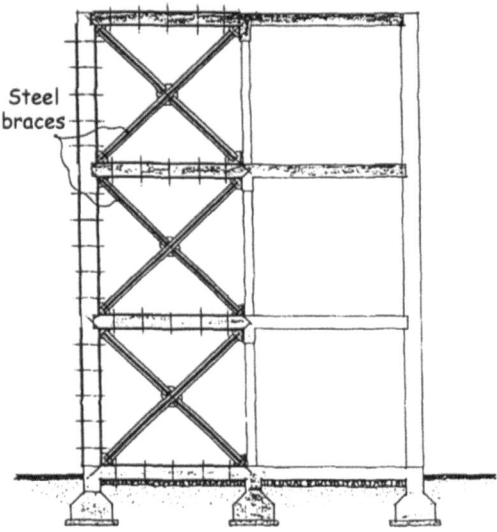

Figure 19. Insertion of steel bracing in a masonry structure for increasing resistance to horizontal actions.

As well known, the reticulated configuration is characterised by great in-plane stiffness with a minimum material demand, as structural resistance is exploited in the most rational way. Nevertheless, the resulting stiffness of the braced structures may result excessive, so to affect the global behaviour of the building to an unacceptable extent. In particular, most of global horizontal forces would converge to braces and, hence, to their foundation, involving a great modification in the seismic response of the building. This would also create serious problems in the transmission of internal forces between bracings and both floor and foundation structures. For this reason it can be sometimes convenient to adopt less rigid systems, like those shown in Figure 20, in order to fit the global brace stiffness to the building features. In particular, eccentric bracing systems offer stiffness properties which can be easily adjusted in order to match the actual horizontal building stiffness. Also, they provide enhanced ultimate ductility as respect to the classical concentric St Andrew's cross brace configuration. Such alternative systems have to be properly designed, so as to achieve the required global stiffness, but without producing a strong modification of the resisting scheme of the construction. In any case, the most delicate aspect of this kind of intervention is the connection with existing elements, due to the very high values of forces transmitted thereby. Figure 20 shows an example of such connection. It is therefore recommended to use as many braces as possible, complying with the internal distribution needs of the building, in such a way to reduce the extent of these forces to a minimum. For these reasons, an accurate structural analysis is recommended when the construction of reticulated structures is planned for stiffening buildings, in order to achieve the best performance with a reasonably reduced modification of the resisting scheme of the building itself. Such analysis should cover both elastic serviceability and ultimate inelastic behaviour.

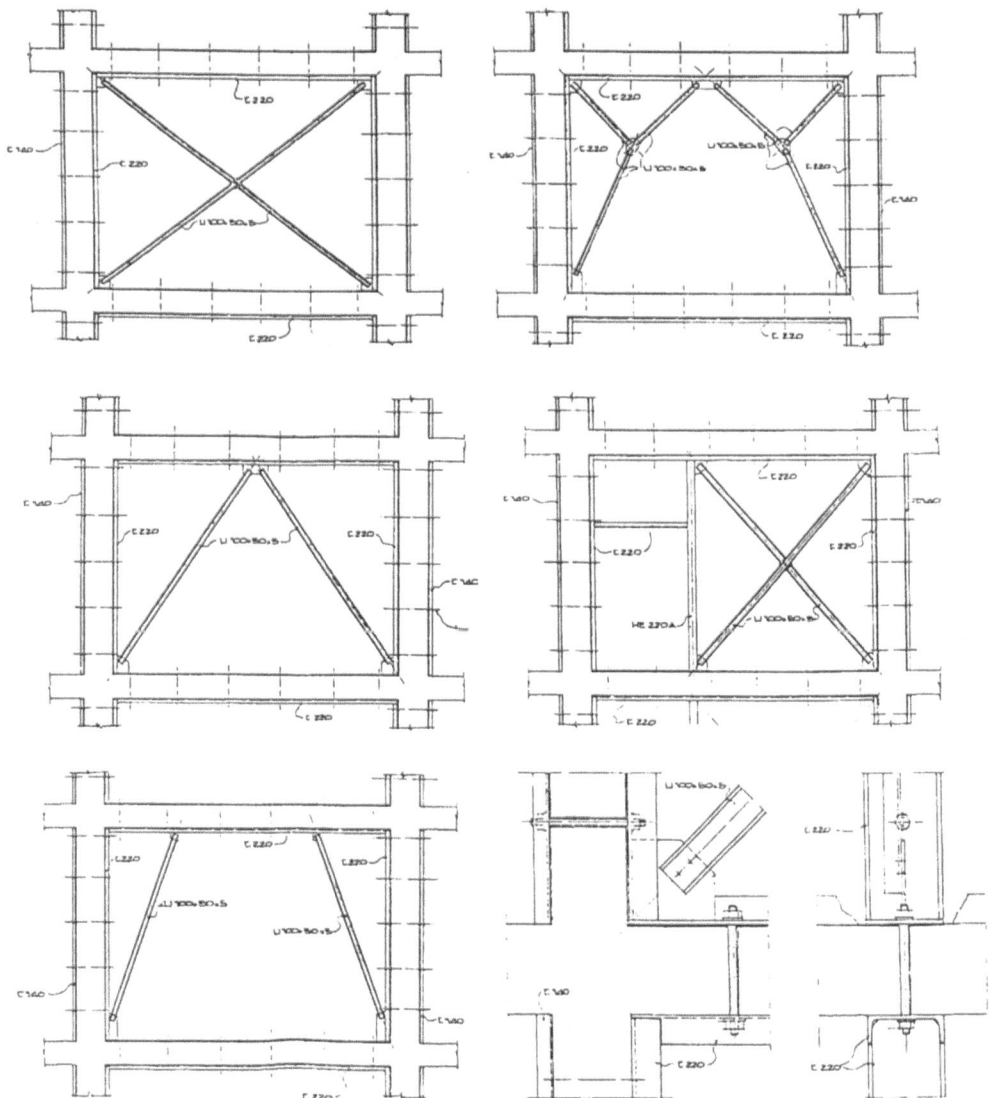

Figure 20. Possible schemes for steel bracing liable to be used in masonry buildings as an alternative to the St Andrew's configuration (top left) and connection details with masonry walls (bottom right).

3.6 Strengthening by means of steel tie-beams

It represents the oldest and most common system of active consolidation, whose effectiveness is widely recognised on the basis of a millenary experience. Tie-beams can be profitably used for many purposes, depending on the specified design requirements. In particular, they can be intended for linking together the walls of a masonry building in order to enhance its box-like behaviour (Figures 21 and 22), or for improving the connection between floors and vertical walls, or for contrasting the thrust of arches and vaults, or for increasing the compressive stress in the walls so as to improve their strength against shear actions. They can also be arranged vertically, so as to provide bearing walls with higher resistance to horizontal actions. Furthermore, in the view of active interventions, tie-beams can be also adopted for recovering some kinds of both local and global damage, namely local spalling of masonry walls, arch or vault displacements, wall out of plumbs and so on (Figures 23a,b). In some cases, the original position of a whole church façade, disconnected from the main bay of the building, can be restored by means of suitable tying elements, provided a suitable contrast element is found. In Figure 23c such contrast is given by the abside. In the same way, ties can be used for recovering cracks into stone elements working as plat-bands, or for reducing the deflection into roofing timber elements.

Use of both bars (corrugated or not) and profiles is possible for ties as well, depending on the load to be transmitted. The installation of ties is very simple, especially when put upward floors, and just a few provisions need to be adopted. In particular, the application of ties is recommended for masonries with good strength features, so that they can easily absorb the load concentration involved by end plates. To this purposes, such contrasting devices should be conceived in such a way to reduce surface stresses in the masonry, so as to limit any permanent deformation which would compromise the effectiveness of ties (Figure 24). Also, ties require to be fastened to stiff walls or, in general, to compact and heavy elements, in order to maximise their effect. Similarly, the application of a little prestressing to ties is advised, in order to ensure the required degree of structural collaboration with the masonry structures before too large displacements are attained. For the above reasons centering of tie-beams with respect to masonry resisting sections is very important and also great attention is to be paid to temperature at erection stage, so as not to impair the intervention effectiveness. In the end, when very long tie-beams are adopted a little upward tilt is suggested in order to offset the vertical deflection due to self-weight.

Ties are also very effective for eliminating the thrust of arches and vaults (Figure 25). This is a widely adopted practice, as it allows to reduce or even annul horizontal forces acting on the impost elements. The application of ties at the impost level of arches or vaults transforms these structural systems into a statically determined scheme, whose influence on the bearing structures if the building is limited to vertical action, only. Suitable tightening devices placed along the tie-rod make for the magnitude of impressed co-action to be easily controlled and modified if necessary.

In all the above referred cases the application of ties is able to significantly increase the global strength level of the construction. In addition, it also represents a fully reversible intervention, which is why the use of tying elements is recommended in the structural restoration of historical and monumental buildings. Last, but not least, tying systems are very cheap, whatever the technology adopted, in both material and construction cost.

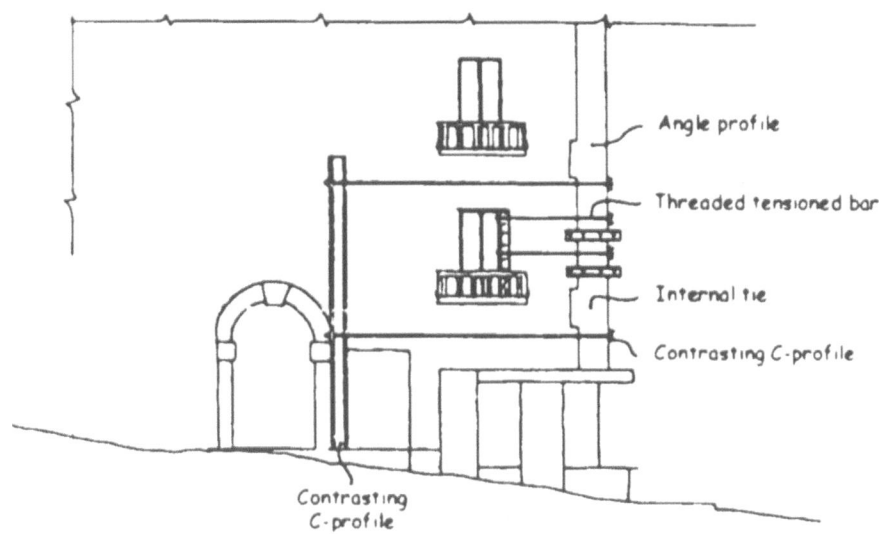

Figure 21. Use of steel ties for retaining wall angles in a masonry building

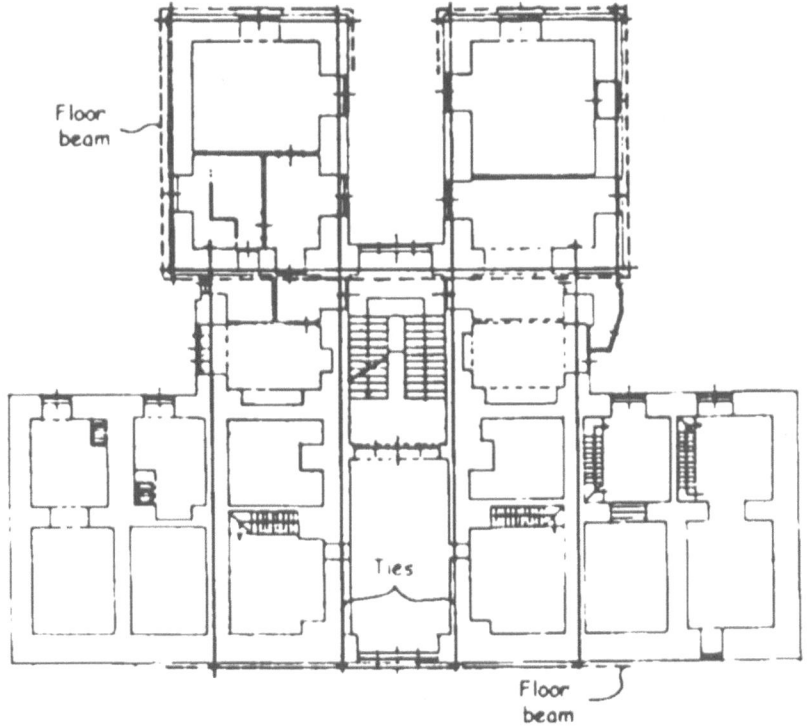

Figure 22. Scheme for a possible location of steel tie rods in the plan of a masonry building for the improvement of its box-like behaviour

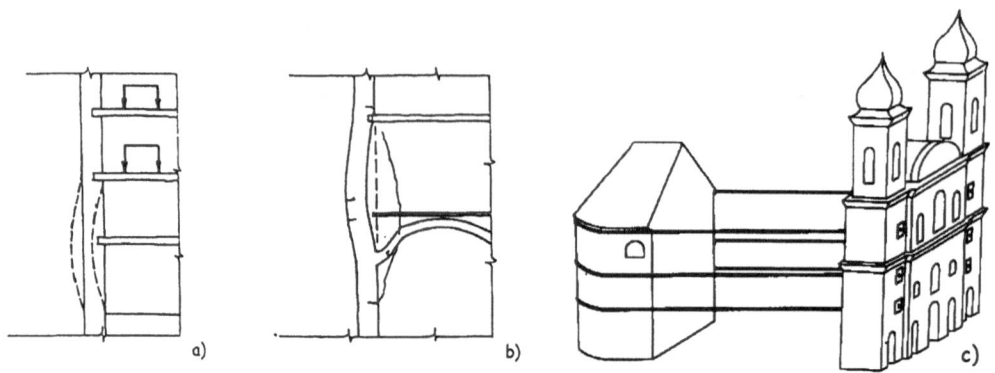

Figure 23. Examples of damage recoverable by means of steel ties.

In the very last years, special devices based on the use of Shape Memory Alloys have been used in combination with steel tying elements for the seismic protection of existing constructions. Shape Memory Alloys (SMA), mostly Ni-Ti or Cu-Al-Zn alloys, may be regarded as "smart" materials, as both their yield stress and modulus of elasticity strongly increase as long as temperature increases within the so-called transformation temperature range, due to a solid martensite-austenite phase change. Such range is limited by M_f and A_f, that is by the temperatures where only full martensitic or full austenitic structures can exist, respectively. The above transformation can be induced by either mechanical stress or temperature change, resulting in the capability to recover, spontaneously or by heating, initial large strains due to load. In the former case, the behaviour is defined *superelastic* (Figure 26a) and leads to the complete strain annulling after unloading. Due to different loading and unloading paths, a given amount of energy is dissipated over the cycle. In the latter case, complete strain recovery may occur by heating material above A_f, which is why this behaviour is called *memory effect* (Figure 26b). If the element is constrained, internal stresses arise, which can be very high owing to the great material stiffness in the austenitic phase. In structural restoration problems, this memory effect can be exploited for impressing a state of co-active stress to structural elements, such as in the case of confining interventions of structural members working in compression. SMA elements can be arranged at a temperature lower than M_f, where pure martensite exists and material can be easily modelled up to the required strain, and then heated above A_f, in order to apply the required degree of co-action. In addition, the superelastic effect, because of the energy dissipated in a full loading cycle, allows the construction of seismic protection devices (Dolce et al., 2000, Pegon et al., 2000), as those used for securing the tympanum of the Basilica of St Francis in Assisi (Italy) (Croci et al., 2000) (Figure 26c). Prestressed vertical steel tie-rods fitted with SMA end devices have been used in the Bell tower of S. Giorgio in Trignano (Italy), in order to increase both flexural strength and energy dissipation capability of the tower (Indirli, 2000). The basic properties of SMAs in comparison with constructional mild steel and other structural materials are shown in Table 1 (Mazzolani and Mandara, 2000), where average values of unit weight γ, elastic modulus E, conventional yield strength $f_{0.2}$ and ultimate strength f_t, ultimate tensile strain ε_t and linear thermal expansion coefficient α are reported.

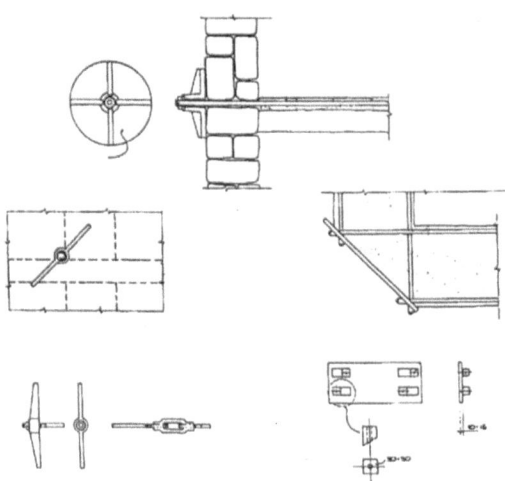

Figure 24. Details of tie-to-wall anchorage and fastening.

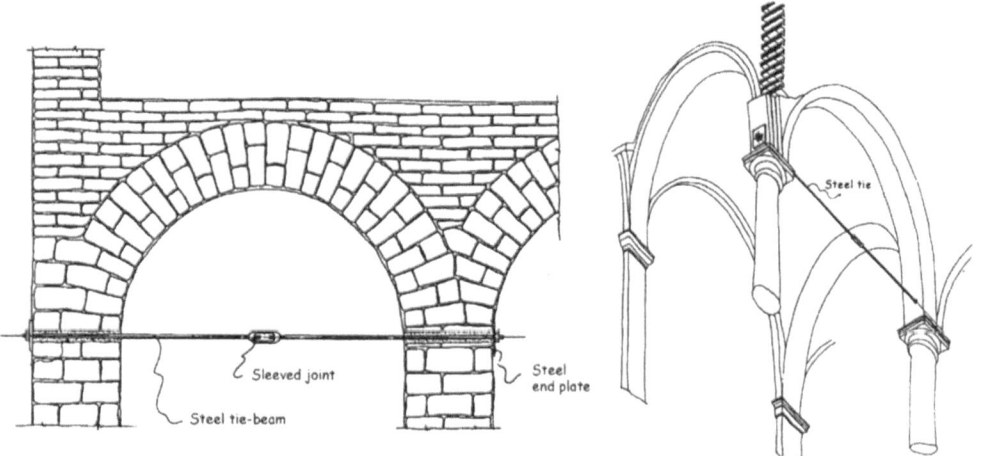

Figure 25. Use of steel ties for eliminating the trust in arches and vaults.

Table1. Synopsis of mechanical features of Shape Memory Alloys compared with other metal materials.

MATERIAL	γ (g/cm^3)	E (kN/mm^2)	$f_{0\,2}$ (N/mm^2)	f_t (N/mm^2)	$\varepsilon_t \times 100$ (A_s)	$\alpha \times 10^6$ (C^{o-1})
SMA Ni-Ti (Nitinol)	≈ 6.5	28 ÷ 75(*)	100 ÷ 560(*)	750÷960(*)	15.5	6.6 ÷11(*)
Mild steel	7.85	206	235 ÷ 365	360 ÷ 510	10 ÷ 28	12 ÷ 15
Stainless steel	≈ 7.8	≈196	200 ÷ 650	400 ÷ 1000	10 ÷ 40	17 ÷ 19
Aluminium alloys	≈ 2.7	65 ÷ 73	20 : 360	50 : 410	2 ÷ 30	24 ÷25
Titanium alloys	≈ 4.5	≈ 106	200 ÷ 1000	300 ÷ 1100	8 ÷ 30	6 ÷ 7

(*) Values referred to martensite and austenite, respectively.

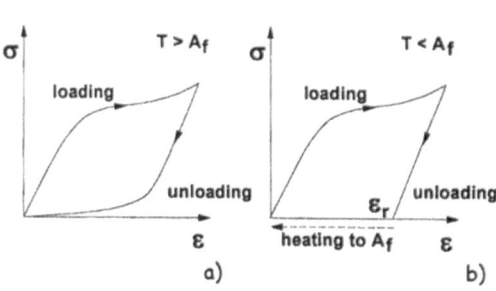 

Figure 26. Basic behaviour principles of SMA (a,b) and SMA energy dissipation devices installed on the roof of the Basilica of S. Francesco in Assisi – Italy (Courtesy FIP) (c).

3.7 Confining of vertical load-bearing elements

The confinement of load-bearing masonry elements working in compression, such as columns, males and walls, is very effective when a higher structural performance is required to these members. The beneficial effect of confinement is slightly evident in terms of elastic stiffness, more significant in terms of maximum load bearing capacity and even greater as far as ultimate ductility is concerned. This is due to the tri-axial state of stress occurring in the element, which strongly reduces the onset of cracks parallel to the direction of applied load. As a consequence, no local instability phenomena occur inside the member and this results in both strength and ductility at failure being greatly enhanced.

Typical needs for confinement interventions arise when single elements subject to axial load have to be strengthened due to an increase of acting load or as a consequence of crushing cracks. Both such situations can occur, for example, when the building destination is changed and higher loads are applied, or when modifications of the structural layout is changed, for example as happens when a wide opening is made through a masonry wall (Figure 27).

Confinement can be applied in several ways, depending on practical situations and performance requirements (Mandara, 1992). For example, concerning vertical masonry elements, it can be both radial and transverse, depending on the shape of resisting cross-section. Radial confinement is based on the encircling effect of external ties only, in which the resultant of tensile stresses produces an uniform compressive stress on the member surface. This technique is very effective in case of circular or smoothly edged polygonal sections, where no or negligible stress concentrations arise at cusp points of the element. In addition, it does not produce any modification to the existing column, which is why it is fully and immediately reversible. In practice, the radial confinement is obtained by means of steel plated elements encircling the whole cross section of the element (Figure 28a). In ancient applications, the installation of such rings were made at high temperature (300 ÷ 400 °C), so that the thermal shrinkage occurring with cooling could involve the required degree of co-action. The use of fork joints made for an easy fastening of rings to each other. More recently, the hot-arranged configuration has been substituted by bolted rings, which do offer the possibility to adjust the confinement force according to the loading process on the element and, thus, are less sensitive to daily and seasonal thermal changes. In all cases of radial confinement, it is strongly recommended that vertical plated elements are placed downside the encircling rings, so as to reduce the stress concentration due to confinement. Also, this provision allows the distance between rings to be increased.

Radial confinement can be also applied to polygonal sections, when no perforations can be made through the element. Nevertheless, in these cases the use of encircling elements would result in an excess of stress concentration at section edges and, as a consequence, a strong increase of shear and tensile stress would arise in the central regions of the member sides, with a possible spall out of masonry. In these cases it is necessary to arrange suitable steel angle profiles at the sections edges, so as to reduce stress distribution peaks (Figure 28b) or, as an alternative, it is possible to use passing ties at section angles in order to prevent lateral expulsion of masonry (Figure 28c). The execution steps of a typical confining intervention are shown in Figure 29.

Similar in principle to steel encircling elements, wrapping sheets or strips of FRP (Fibre Reinforced Polymer) material can be used (Triantafillou, 1998). This technique has been recently developed for the strengthening of both masonry and r.c. elements, in particular with the use of carbon fibre (CFRP) in a plastic matrix. Such materials have very low weight, great chemical stability and high mechanical properties, in particular as far as ultimate tensile strength is concerned. For this reason, their use as confining elements is particular convenient, as a very thin, fully bonded layer can be applied to members. Even though the inherent behaviour of FRP is generally rather brittle at failure, their confining effect on compressed members can produce a significant increase of both global strength and ductility.

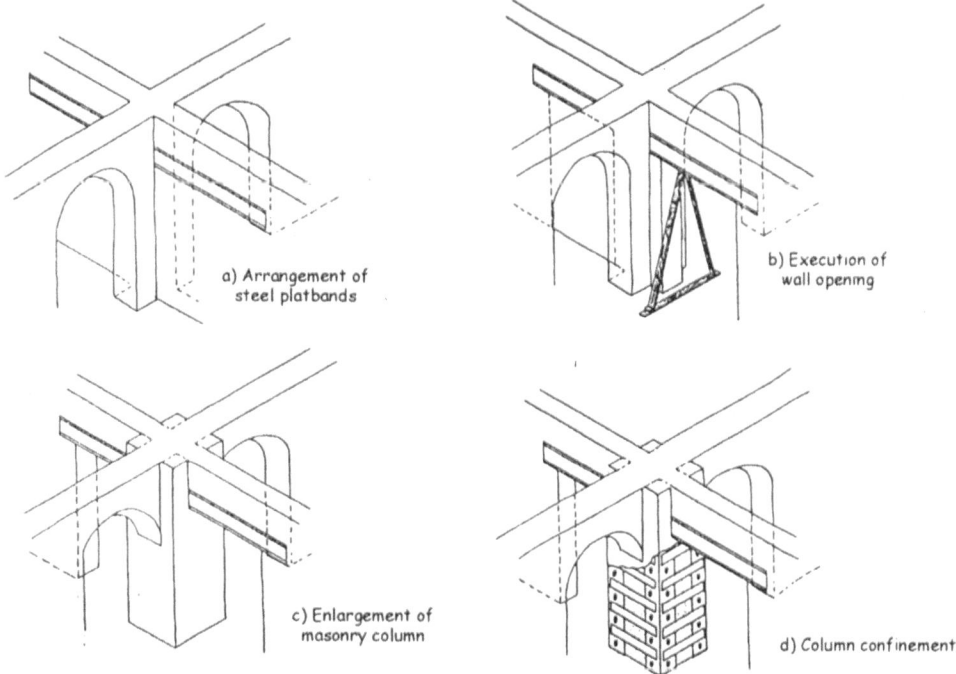

a) Arrangement of steel platbands

b) Execution of wall opening

c) Enlargement of masonry column

d) Column confinement

Figure 27. Typical situation demanding for confinement of vertical bearing elements.

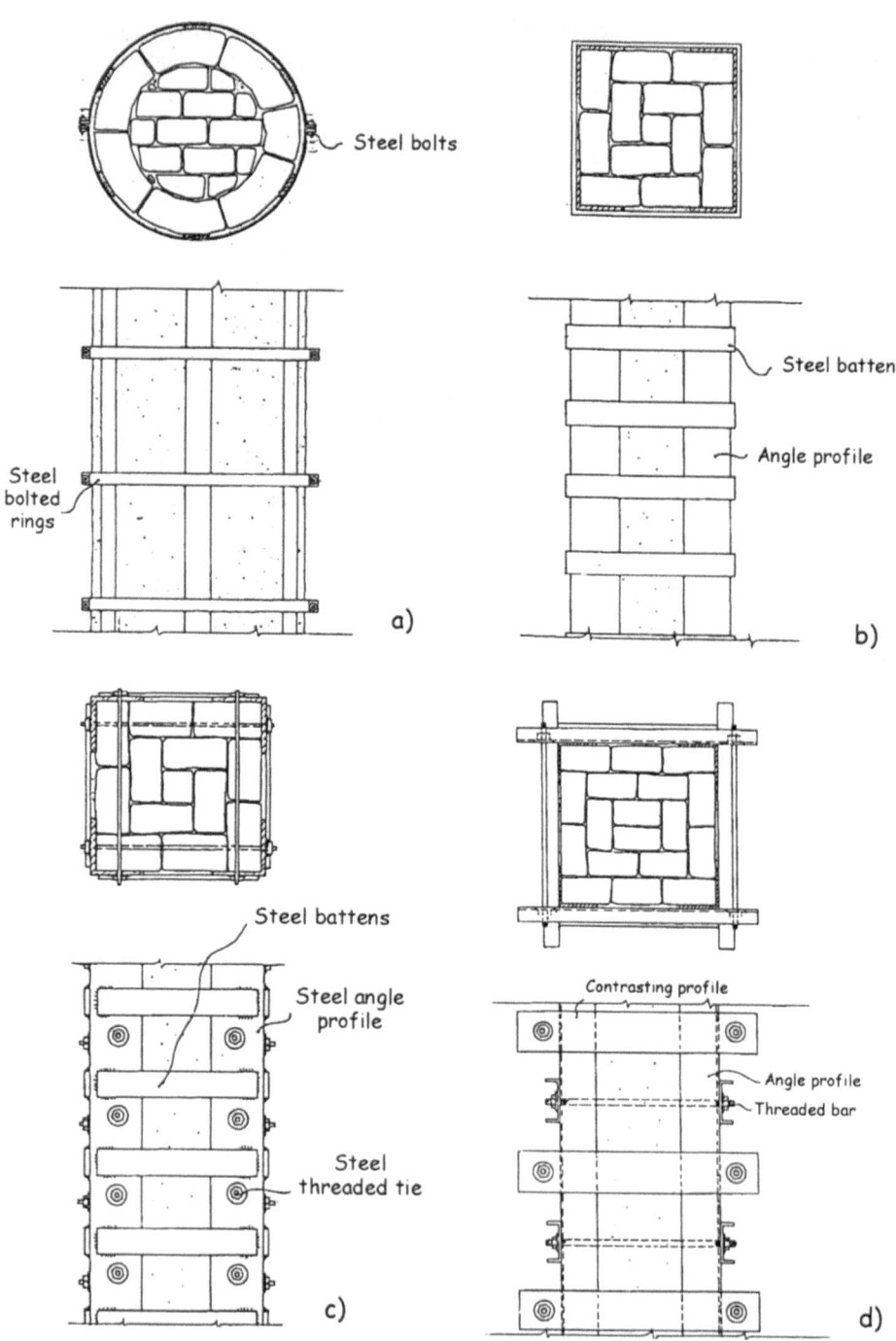

Figure 28. Confinement of masonry columns by steel encircling and/or passing elements.

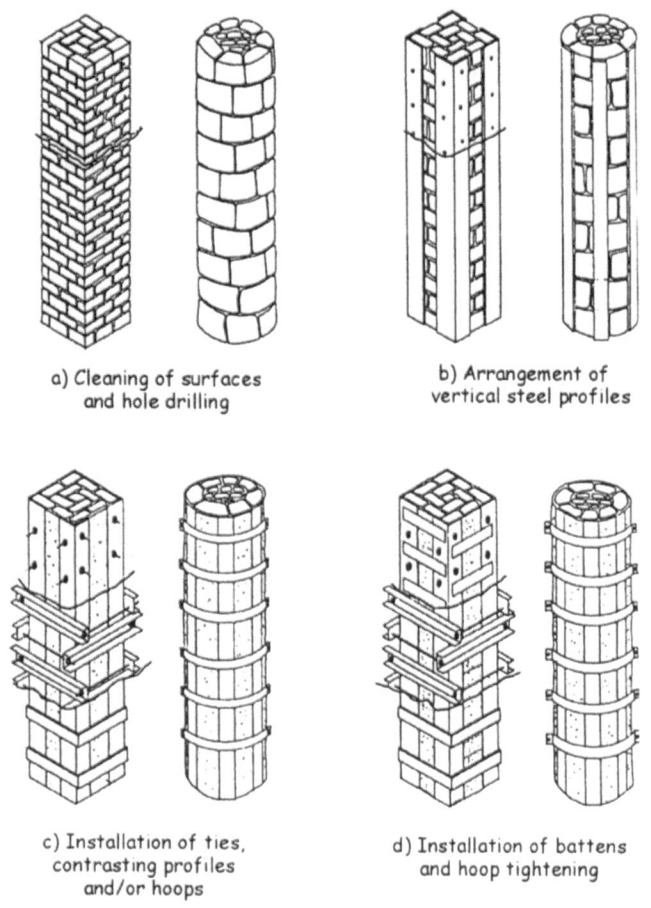

a) Cleaning of surfaces
and hole drilling

b) Arrangement of
vertical steel profiles

c) Installation of ties,
contrasting profiles
and/or hoops

d) Installation of battens
and hoop tightening

Figure 29. Execution steps of a confining intervention on a masonry column.

This technique is relatively easy and cheap as far as application procedure is concerned, even in complex situations. Nevertheless, it requires a careful preparation, that is full plastering of the surface underlying the FRP film and section edge smoothing, in order to avoid damage to fibres. This operation could be rather difficult in old or irregular masonry, as the member has to be completely coated before applying the confining film. In addition, the intervention is in practice hardly reversible, which can bother the designer in case of application to monumental constructions.

When any stress concentrations at edges is to be avoided, transverse confinement can be used. Basically conceived for masonry walls as well as for polygonal section columns, transverse confinement usually requires members to be drilled across the whole thickness, so that confinement can be applied in one or two perpendicular directions by means of passing ties. Such ties have to be conveniently terminated with plates or angles in order to limit surface stress concentrations to values compatible with masonry strength. Tying elements can be also external (Figure 28d), in which case very rigid elements have to be placed between tendons in order to achieve a rather uniform lateral pressure on member. Such kinds of intervention, which does not require any perforation, are most of times intended as temporary strengthening systems, for example for arresting the propagation of cracks due to excess of axial load, before definitive refurbishment works take place.

Transverse confinement can be also applied to reduce the lateral deflections of a masonry wall about to fail due to local instability of external leaves. This condition often occurs in old constructions, whose walls are characterised by the typical double sided stone masonry filled with cohesionless material (multi-leaf masonry). In such cases internal disconnection may occur with a consequent reduction of instability limit load of the wall. Steel ties inserted at convenient points and suitably enforced can reduce the out-of-plane deflection of the wall, thus increasing its overall load bearing capacity. Of course, radial and transverse confinement can be also combined with each other, depending on structural requirements. For example (Figure 28c), some additional transverse ties can be adopted in order to prevent masonry elements from jumping out due to stress concentration at edges. Similarly, transverse confinement can be adopted for the strengthening of highly stressed elements, such as foundation walls (Figure 30), e.g. to allow the construction of urban subway (Figure 31).

The problem of transverse confinement of masonry walls with internal steel tying elements and end plates has been recently faced by Mandara and Mazzolani, 1998. The study has been focused upon the definition of a theoretical model for predicting the effect of lateral uniform confinement on masonry walls subjected to compressive load (Figure 32a), by accounting for the inelastic behaviour of both masonry and steel. The method has been calibrated on the basis of a F.E.M. numerical simulation, whose reliability has been checked in turn by means of a direct comparison with experimental data available in literature (Ballio and Calvi, 1993). Compared to existing models, mostly concerned with concrete, in the procedure developed the number of parameters to be fitted empirically is drastically reduced. With a suitable choice of these factors, the proposed model has proven to interpret either experimental or numerical results with a satisfying degree of accuracy. The comparison between numerical and analytical models is shown in Figure 32b, where curves referring to strength enhancement factor due to confinement k, confinement stress σ_c and Poisson modulus ν are also plotted for a 800×800 mm masonry wall with thickness $t = 500$mm. The cases of confinement by means of 600×600 mm and 800×800 mm steel rigid plates tied by steel bars with a global 2000mm^2 cross-section are referred to. Factors α, a and b, whose values are shown in Figure 32b, too, are the only parameters which characterise the theoretical model proposed.

In most cases of confinement bolted steel elements are generally used, as they allow the prestressing force and, hence, the degree of co-action, to be easily applied and controlled over time. In any case, and in particular when a co-action is relied upon, attention should be paid to the effect of differential thermal changes which can occur between the masonry and the reinforcing elements, due to the different thermal expansion ratios. A poor consideration of this may lead to loss of effectiveness of encircling or confinement elements and even to the cracking of stone blocks. For this reason, in some cases it may be convenient to use alternative materials such as, for example, titanium alloys. They offer, in fact, a very low linear thermal expansion coefficient ($6 \div 8 \times 10^{-6} C^{o-1}$), which is very similar to that of volcanic or metamorphic rocks, such as granite and marble. This allows titanium elements to be used in redundant or prestressed systems with no risk to impair the effectiveness of intervention due to thermal changes. Similarly, no additional co-active state of stress would be involved. For this reason, titanium reinforcement elements have been used, not only for confinement purposes, in the restoration of monuments such as the Parthenon in Athens and the Colonna Antonina in Rome, where titanium alloy stirrups have been inserted and hidden into existing stone blocks (Giuffrè & Martines, 1989). They proved to be far more effective than conventional steel elements used before, which had involved many cracks due to corrosion and excess of thermal dilatation. In the same way, some concrete bridges in Japan have been repaired with titanium rods. Values of the linear thermal expansion coefficient for several metal materials are given in Table 1 reported above.

Whenever possible, internal confining elements have to be grouted along their full length, in order to avoid possible corrosion problems which would result in the onset of expansive rust. Similarly, external elements have to be embedded in high strength mortar, in order to reduce both corrosion and stress concentration. When, for reversibility reasons, or when important applications are faced, no mortar is wanted to be used, it may be convenient to use stainless elements, which can be installed without risk of chemical corrosion, even under aggressive environmental conditions (Mazzolani and Mandara, 1999).

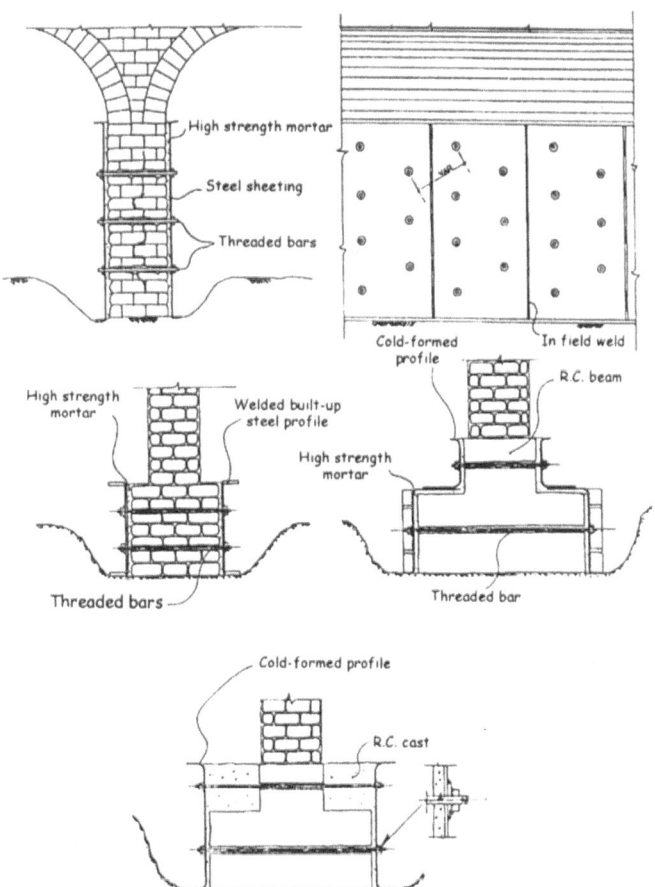

Figure 30. Lateral confinement of foundation structures by means of steel cold formed elements.

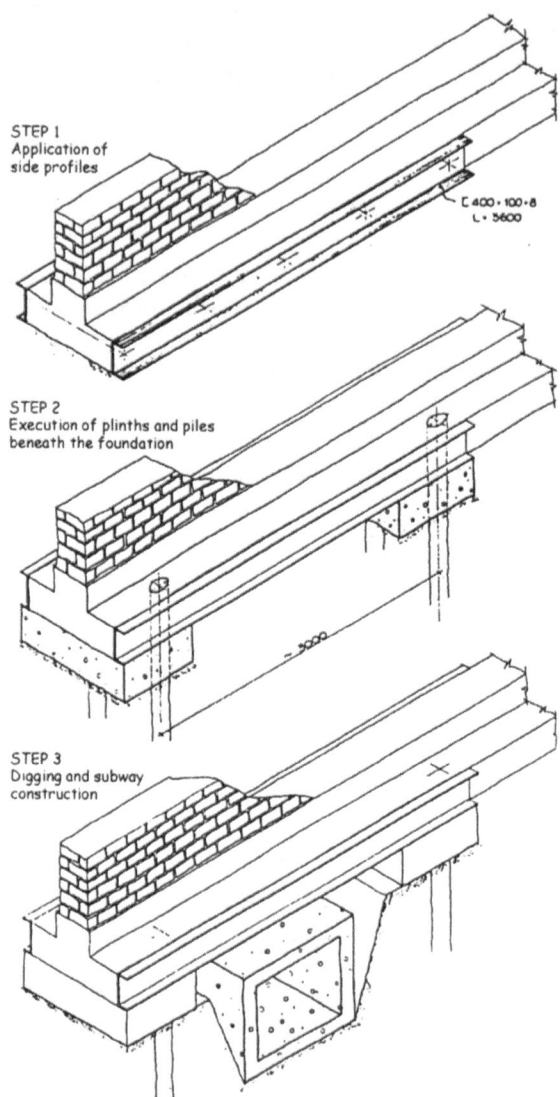

STEP 1
Application of
side profiles

⌐ 400·100·8
L·5600

STEP 2
Execution of plinths and piles
beneath the foundation

STEP 3
Digging and subway
construction

Figure 31. Lateral confinement of foundation structures by means of cold-formed steel profiles for allowing the execution of an urban subway.

3.8 Strengthening of arches and vaults

Arches and vaults are horizontal structures conceived to work in the same way as vertical masonry walls, that is predominantly in compression. As a consequence, most of strengthening provisions for masonry elements referred to above apply to arches and vaults as well. Nevertheless, refurbishment techniques have to comply with the problem of thrust, which is an inherent feature of such structures. Most of times, structural ineffectiveness or damage in such elements occur due to failure in resisting their thrust, which is in turn closely related to the amount of applied load. The effects of an inadequately resisted thrust are usually evident at the impost level, that is where the thrust reaches its maximum value. This can result in an excess of lateral displacement as well as in a downward deflection at key-stone. Depending on the loading conditions, the stress resultant along the arch contour can also exceed the resisting section, involving cracking and excess of deflection at reins (Figure 33). The possible intervention techniques can be from either downside or upside, depending on the structural morphology and on how the structure can be accessed. In any case they have to be tailored on the specific target to be reached, also considering the possible damage conditions. In addition, sometimes the designer has to cope with decorations or frescos on vault ceiling, in which case the intervention technique must be carefully studied in order to limit as much as possible the interference with existing valuable works.

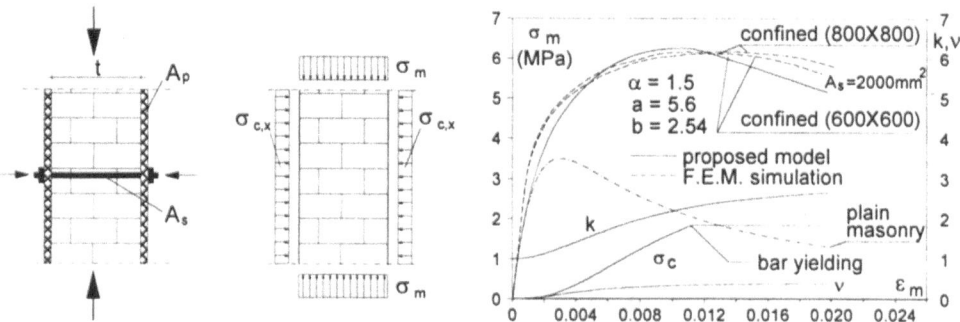

Figure 32. The model of confined masonry proposed by Mandara and Mazzolani, 1998 (a) and the comparison of analytical procedure with FEM results (b).

From the general point of view, two situations can be distinguished, namely the one when the structure only needs to be strengthened and the one when a certain damage exists and, as a consequence, repairing operations must be done. In the first case, provisions for the structural reinforcement are strongly depending on the load distribution on the arch or vault, as well as on the impost structures. In order to increase the effectiveness of the strengthening operation, in fact, a vault can not be reinforced without considering a suitable revision of the global load acting on both vault and supporting structure. For this reason, it is generally convenient to degut the vault upper filling in order to reduce the magnitude of the acting load. Either light filling material, or a new floor, as well as a so-called cellular filling, namely a gridwork of thin masonry walls erected on the top of the vault, can be adopted in lieu of existing heavy material (Figure 34a). As an alternative, when the impost structure is not able to withstand the increase of load, the vault can be also suspended to a new floor. In all cases, after the elimination of filling material, the vault upper side can be easily strengthened by means of a concrete slab reinforced with a steel welded net and conveniently fastened to the vault by means of stirrups grouted with high strength mortar (Figure 34b). When the required thickness for such concrete cap is not higher than 4 ÷ 5 cm, then the concrete cast can be made by "spritz beton". In general, the application of such a concrete cap results in the vault to be provided with a good out-of-plane bending stiffness, which reduces the magnitude of thrust acting on impost structures. In some cases, the concrete slab is connected at vault sides to continuous r.c. beams, conceived in such a way to transfer a share of the global thrust to stiffer structures placed on the front of the vault (Figure 34c).

When thrust is to be completely eliminated, then the simplest way to proceed is to insert side-to-side steel tie-rods. Such ties can be placed at both key-stone and springer level, depending on the requirements for internal clearance (Figure 25). Under certain circumstances, ties made with steel *poutrelles* can be arranged at springer level or slightly above it, so to be used as supports for overhead pipes and electric cables (Figure 34d). Ties can be also inserted along a different direction, if only a correction of the thrust line is required. In general, the most effective configuration is at springer, in which case the use of ties can reduce significantly the magnitude of transverse displacement. When such allocation is not possible, it is convenient to stiffen the springer regions with suitable provisions – e.g. grouted cross rods – so as to offset the geometric eccentricity between horizontal thrust and tie axis (Figure 34e).

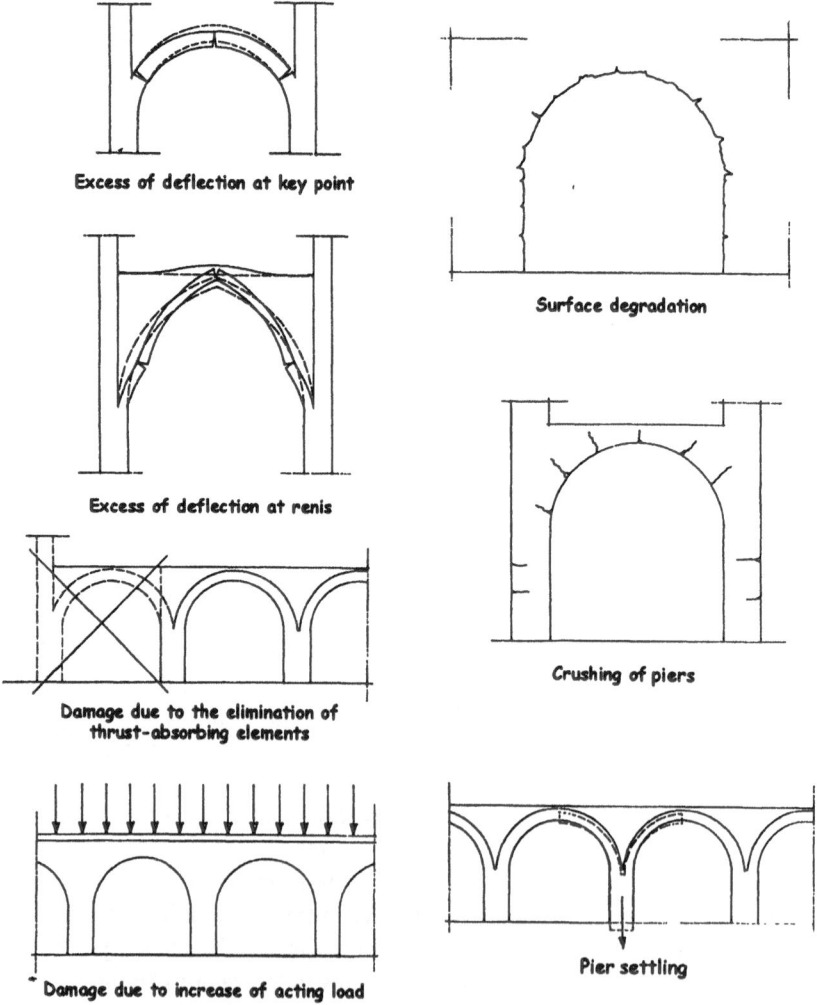

Excess of deflection at key point

Surface degradation

Excess of deflection at renis

Damage due to the elimination of thrust-absorbing elements

Crushing of piers

Damage due to increase of acting load

Pier settling

Figure 33. Possible damage conditions in arches and vaults.

When a damage condition exists, restoration of both cracks and undesired displacements should be accomplished. Small cracks at key-stone are rather frequent in both arches and vaults, and can be easily repaired with mortar or resin if their size does not exceed a few millimetres. In this case, repairing is mostly a superficial concern and it is generally possible operate from downside, even if the vault is decorated. On the contrary, larger cracks (2 ÷ 4 cm) can be restored by means of "cuci e scuci" or with injected concrete mortar (Figure 34f). In such a case, the use of steel welded can be appropriate when many cracks are present (Figure 34g). When cracks are even larger, then it means that also displacement at impost and/or key-stone have become unacceptably large, which involves the partial or total recovering of such displacements in order to restore the original geometry of the vault. In this case, the use of carefully placed steel ties can yield a quite satisfying result, in the spirit of the active strengthening. In any case, when possible, the use of wooden or steel wedges for enforcing the repaired crack is recommended in order to reduce the extent of impressed displacement. Alternatively, when no decorations are present, a concrete slab can be sprayed on the vault soffit, in which case an effective anchorage to upper vault body is strongly advised (Figure 34h). As a proper reinforcement system for such slab a flexible steel net can be used.

Figure 35 depicts two solutions for the suspension of a vault to an overhanging floor purposely built in order to reduce load on vault and, hence, thrust acting on springers. Figure 36 shows two steel-based strengthening solutions for arches, possibly integrated by a base tie. A steel tendon is also used for stiffen the rampant vault of a staircase (Figure 37).

As for the restoration of masonry members, strips made of FRP (Fibre Reinforced Polymer) have recently started to be used in the strengthening of arches and vaults, too (Valluzzi and Modena, 2001). The use of materials resisting in tension only, like FRP, can be useful, in fact, to prevent the onset of plastic hinges in the structure and, however, to increase both load bearing capacity at collapse and ultimate ductility. This may be very important, for example, in case of unsymmetrical load distribution, which, under certain circumstances, can result in the thrust line to go out the thickness of the arch or vault. The tensioned side of the hinge is effectively constrained by the bonded fibre strip, which can rely upon a large adhesion area over the arch or vault surface. Considering the possible collapse mechanisms of these structures, it is convenient to place FRP strips on both intrados and extrados in order to contrast both hogging and sagging moment. In any case, attention should be paid to the premature detachment of FRP strips, which could impair the attainment of a ductile collapse mechanism.

4   Strengthening of floor structures

4.1 Timber floors

Timber floors are generally typical of very old masonry buildings and, for this reason, they can frequently present important damage conditions. They are commonly made of a double structural order, in which the main system is obtained by 60 ÷ 80cm spaced oak or chestnut stocks and the secondary one is made of flat elements. The connection between structural elements of floor is made by means of stirrups or simply by nails. As a result, the floor deck is characterised by an insufficient degree of in-plane stiffness, which involves a poor performance under seismic actions. In addition, no particular provision is commonly adopted at the beam-to-wall support in order to connect horizontal and vertical structures to each other.

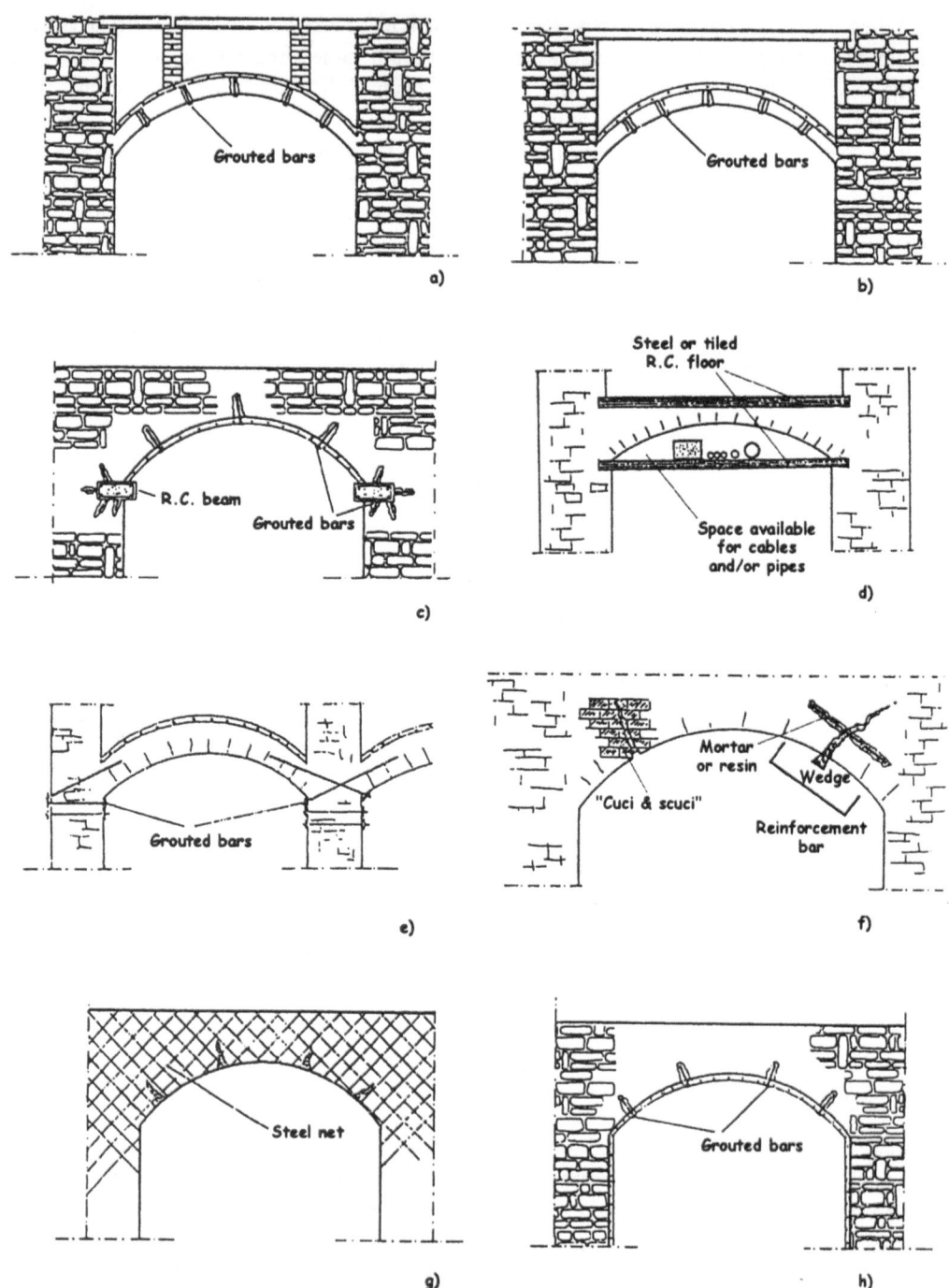

Figure 34. Strengthening techniques for masonry arches and vaults.

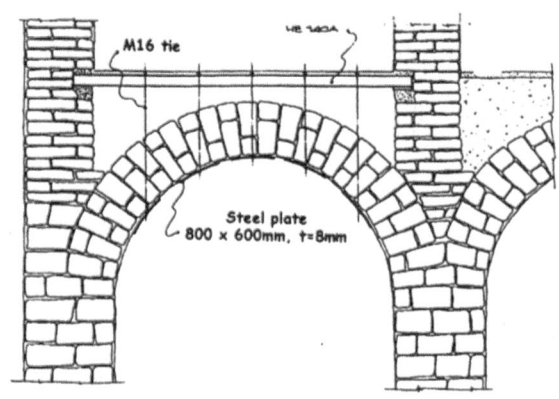

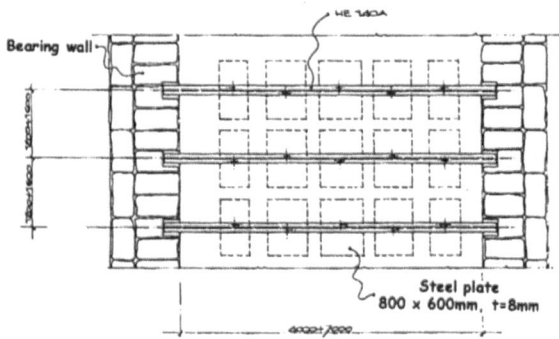

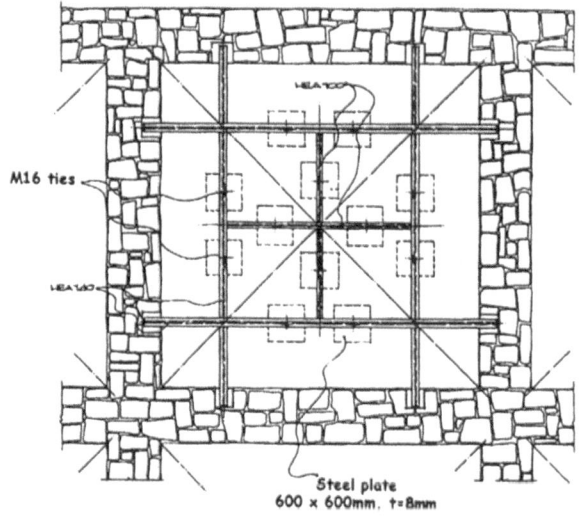

Figure 35. Suspension of masonry vaults to overhanging floors.

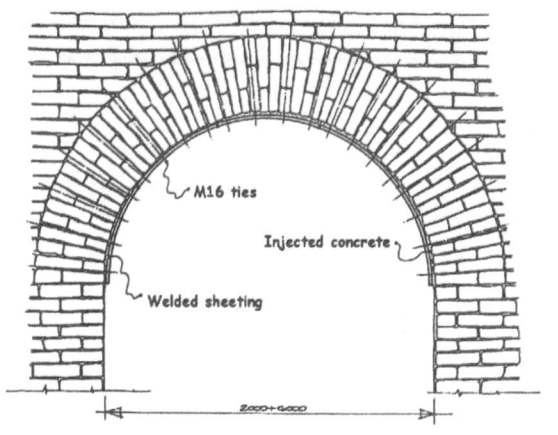

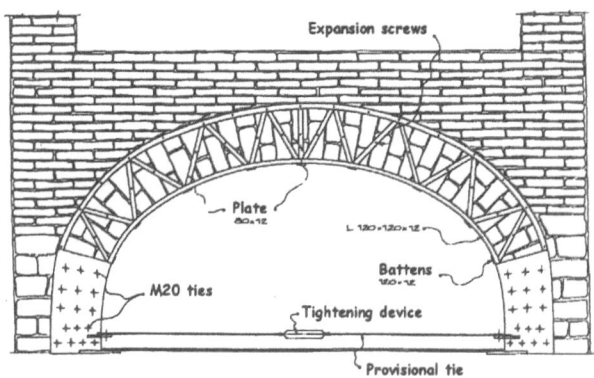

Figure 36. Strengthening of arches by means of steel elements or ties.

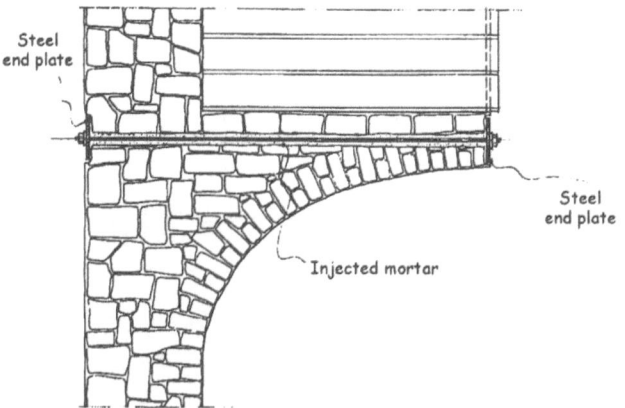

Figure 37. Steel ties for the strengthening of a vaulted rampant staircase.

Apart from cases of little structural concern, the timber deck is filled with a layer of incoherent material, adopted in order to have both a plane pavement and an additional thermo-acoustic insulation. This layer, which is completely ineffective from the structural point of view, is usually very heavy, and this may lead to an excess of floor deflection.

Damage conditions affecting timber floors are manifold and are strictly related to their inherent features. The most common of them can be summarised as follows:

- High creep deformations of both principal and complementary structure. This is an typical feature of wooden material, which is liable to yield large deformations under load. In structural applications, and in particular in floor decks, this behaviour involves additional provisions to be applied in order to offset the sag increasing over time. The most common of these provisions is to increase the floor filling layer, by adding new material and, hence, new load. As a results, the phenomenon is self-exciting and can produce unacceptable values of floor span deflection.
- High susceptibility to biological degradation. This is clearly a consequence of the organic character of wood, which results in a strong sensitivity to humidity, micro-organisms and atmospheric conditions. The biological attack is particularly evident at the beam ends, where the lack of air at supports encased into masonry can involve the onset of anaerobic, highly degrading reactions, usually leading wood to rotting. Furthermore, depending on environmental conditions, the effect of woodworms and bacteria can severely impair the structural integrity of timber elements.
- Easy degradation of end restraints due to vibrations and wood rotting. The initially poor degree of fixity at supports is worsened by both material bio-chemical degradation and mechanical shocks induced by applied dynamic load. The resulting effect is an increase of global floor deformability, followed by a corresponding reduction of user comfort at serviceability limit state. This problem stands in many cases due to the fact that sometimes, particularly in very old floors, engineers were used to design floors by assuming fixed end restraint conditions. This assumption is clearly far to be true, especially when end supports have been progressively damaged.
- Deterioration of the internal connection between floor structure and filler layer, which has gone vanishing over time, thus producing a further reduction of global vertical stiffness. As a results, wooden floors are frequently prone to exhibit very large values of elastic deformability, which are generally incompatible with serviceability requirements.
- In-plane cracks arising as a consequence of poor deck stiffness and resistance. This is usually a consequence of seismic actions, which are scarcely resisted by floor deck.
- Settling of bearing walls and consequent floor damage due to excess of concentrated load at beam supports, as a result of pounding effect from seismic actions.

When operating on timber floors, a preliminary decision should be taken on whether they can be repaired or need to be substituted. This decision is inevitably bothered by the fact that in many cases timber beams could have a great value from architectural point of view but, at the same time, they are in very bad conditions, so that they can be hardly restored and saved. If repairing existing timber elements is possible, interventions can be made at both local and global level. In particular, restoration operations can be executed either on wooden material itself and/or on structural elements, depending on specific needs.

Material can be locally restored by means of common techniques for wood surface treatment and protection, as well as with special products, such as for example epoxy resins, which are able to fill internal cracks and even to reconstruct severely damaged material portions. For these reasons, such products are frequently used for the restoration of beam ends, where degradation due to rotting and micro-organisms is sometimes very evident.

Timber structural members can be repaired and/or strengthened in several ways, depending on the performance to be achieved. Wooden beams can be locally repaired by means of steel hoops, when the extent of longitudinal cracks in some sections is so great to reduce the bending stiffness consistently. When a resistance upgrade of the single beam is required, and provided that the beam itself is in good conditions, it is possible to insert a lower running steel profile connected to the beam body by means of 45° sloped ties (Figure 38). In this way a corresponding increase of shear strength is ensured, too. The profile can be either a flat plate, a C-sectioned profile, or even a cold formed profile, as shown in Figure 38. Some alternative solutions, based on the use of additional steel profiles, are shown in figure 39.

Similarly, steel profiles can be applied upward, and connected to timber beams by means of suitable connectors, made of ties or nails (Figure 40). If such connectors are properly designed so as to absorb shear slip, then a real composite system is obtained, which greatly improves both strength and out-of-plane stiffness of the floor structure. In addition, this intervention does not involve but a negligible increase of overall floor height and also allows the floor ceilings not to be touched. This may result helpful when timber beams are worth being preserved.

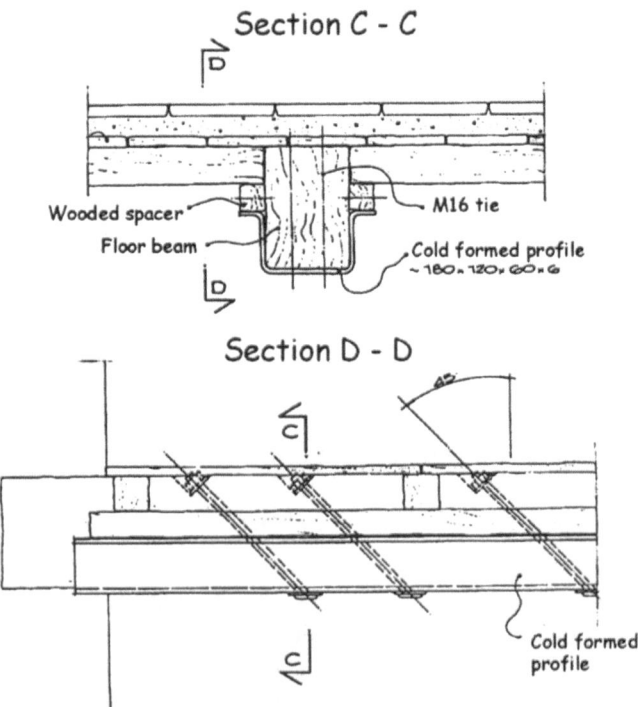

Figure 38. Strengthening of a timber beam by means of additional steel cold formed tied profiles.

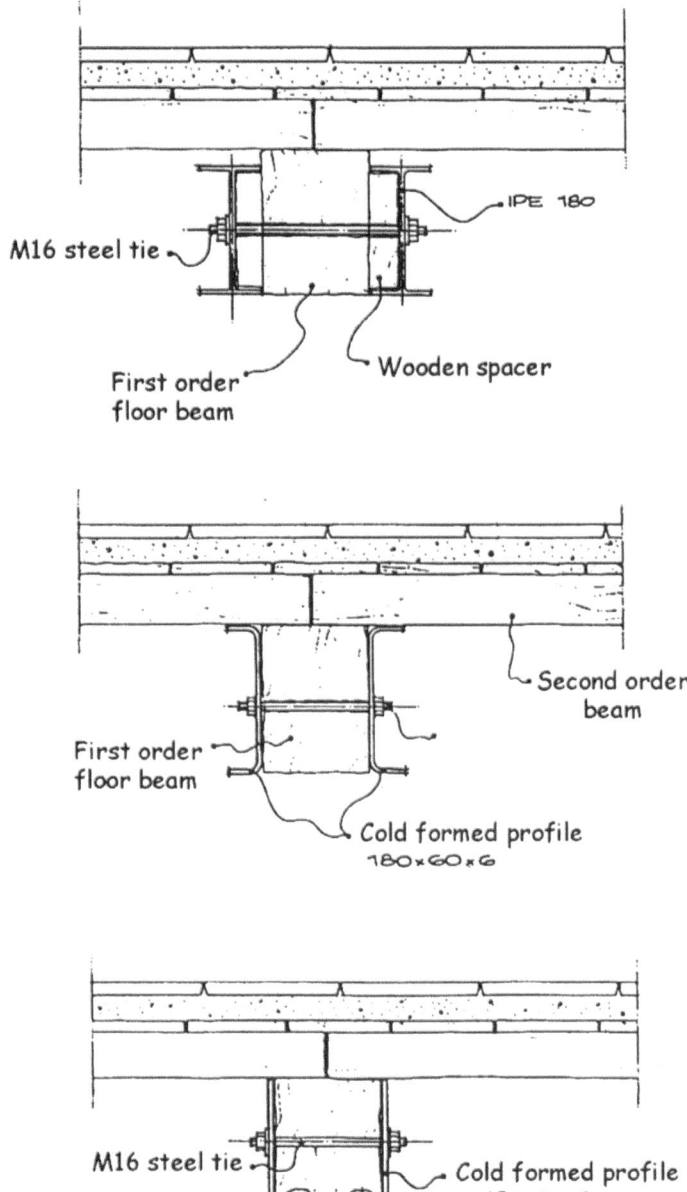

Figure 39. Steel-based solutions for the strengthening of timber beams.

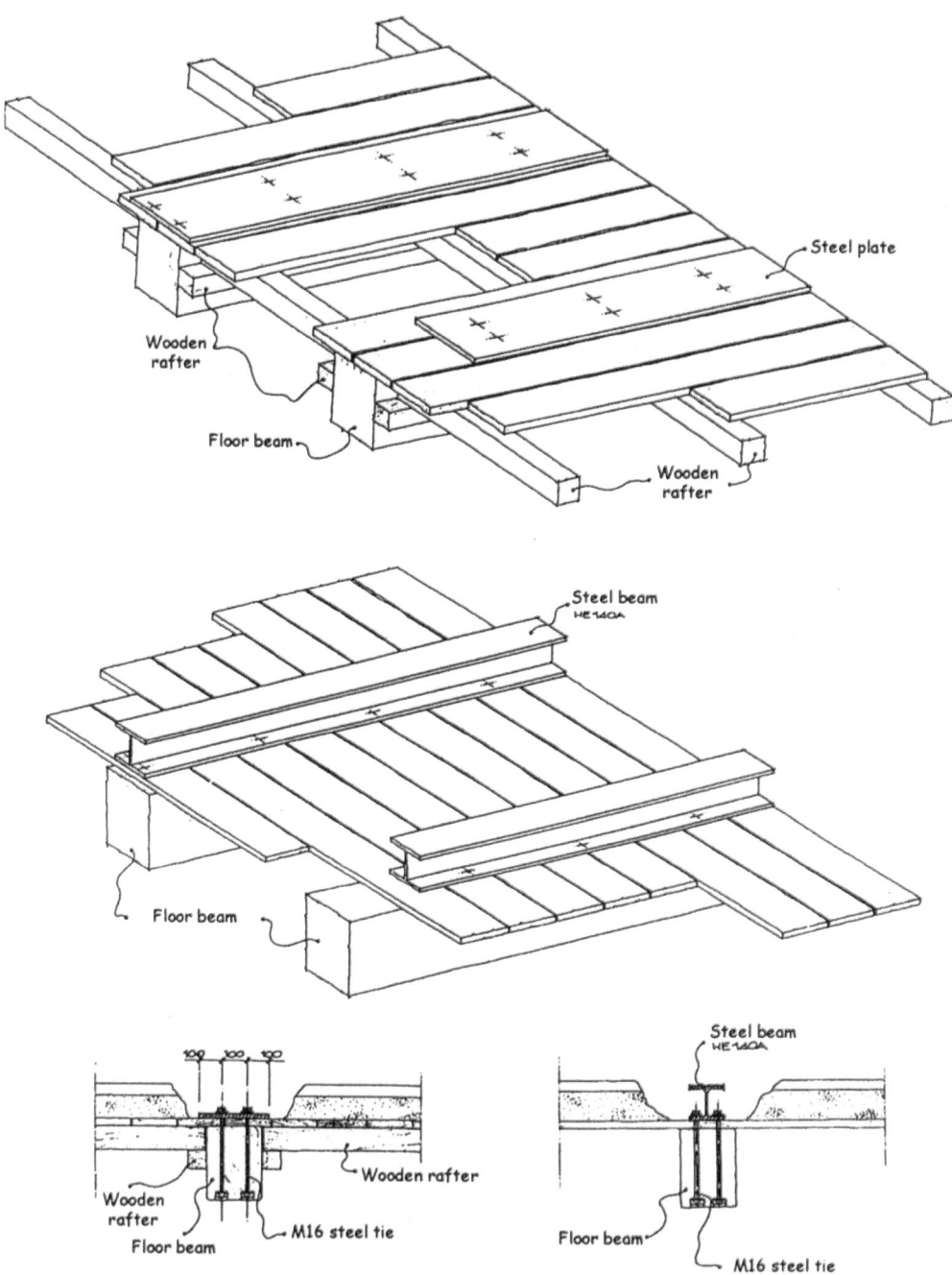

Figure 40. Upward strengthening of timber floor beams by means of steel elements (flat or profiles) tied through the timber beam with suitable connectors in order to obtain a composite system.

When a global increase of resistance is needed, for example to withstand higher live loads, then wooden floors can be reinforced by inserting one or more transverse beams in order to reduce the deck effective free span (Figure 41). This practice is very easy to built, cheap, and reversible, but does not involve any improvement of floor in-plane stiffness and, also, it can result in an excess of load on walls parallel to floor main order. For such reasons, this technique is generally used as a temporary strengthening system and, in any case, it is not advisable in seismic areas. As an alternative, and provided the general conditions are good, members can be individually strengthened with steel elements, so to have a parallel configuration or even a composite structural system, as already shown in Figures 39 and 40. It worth emphasising that these techniques do not improve in-plane stiffness of the floor, unless they are not integrated by *ad-hoc* operations, such as for example, additional timber boarding (Figure 42), concrete upward cast (Figure 43), steel sheeting (Figure 44), or diagonal steel ties (Figure 45). When applying such techniques, a rigid connection to side walls is recommended, in particular in seismic areas, in order to take advantage of the higher in-plane stiffness. This can be made, for the sake of example, as shown in Figure 45 .

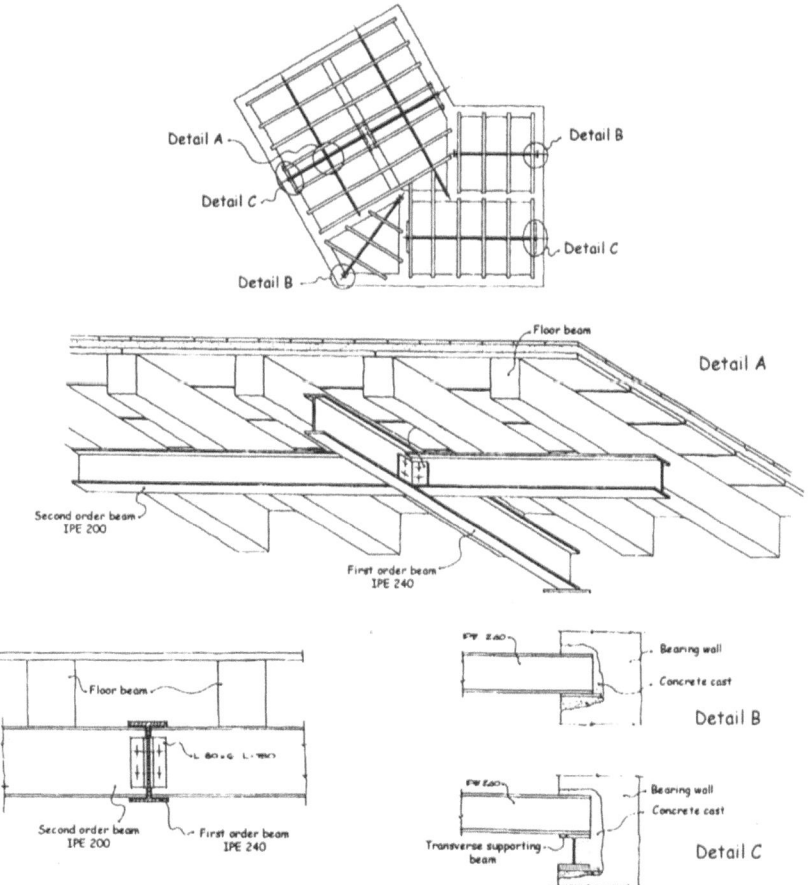

Figure 41. Reinforcement of timber floors by means of additional steel beam orders.

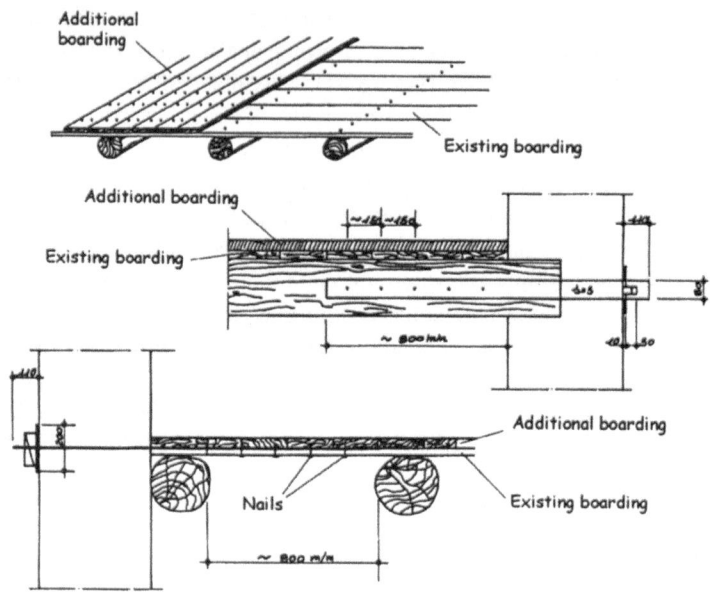

Figure 42. Stiffening of timber floors by means of an additional wooden boarding.

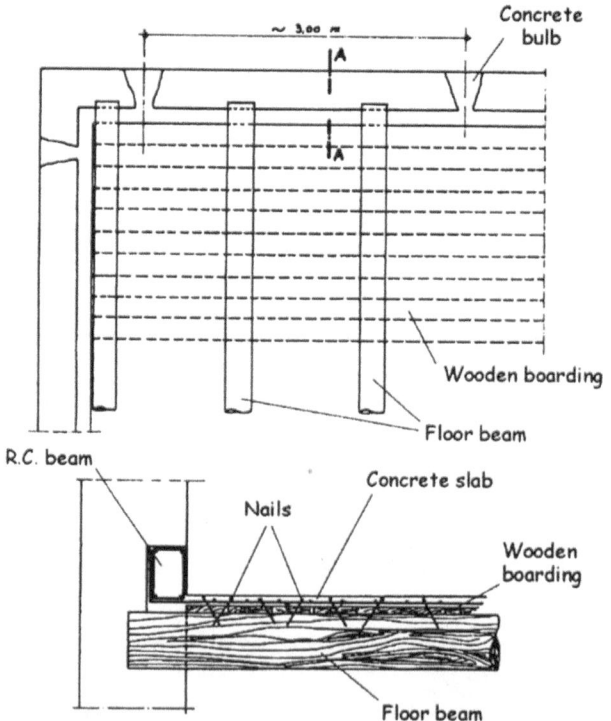

Figure 43. Stiffening of timber floors by means of an upper r.c. slab.

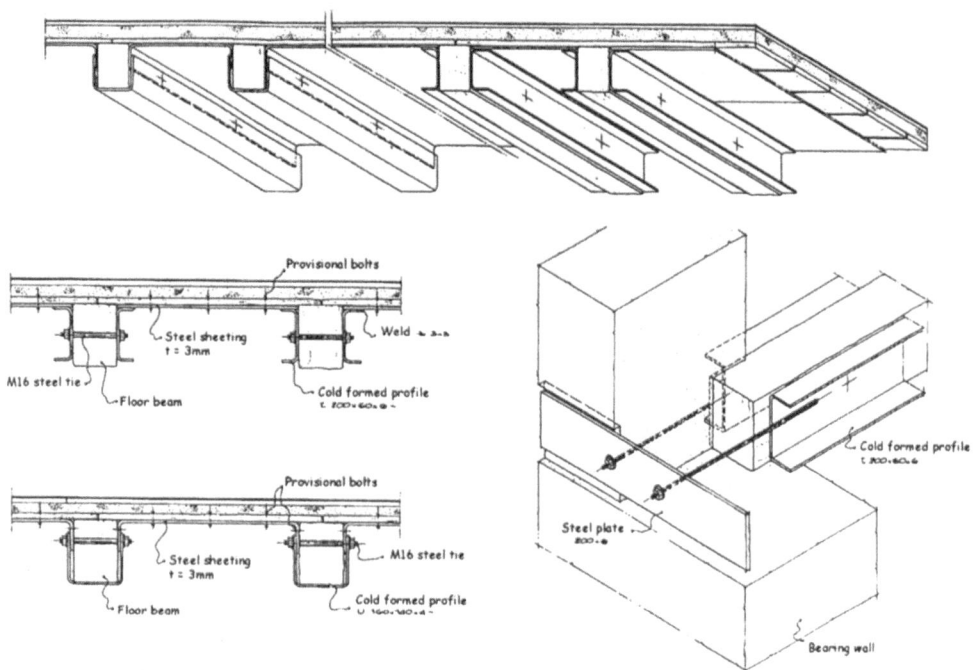

Figure 44. In-plane stiffening of timber floors by means of a intrados steel sheeting integrated by cold formed profiles for the strengthening of beams.

The above techniques, and in particular those relying on steel reinforcing elements, can be applied to other kinds of timber structures, such as for example roof trusses (Figure 46). Roof wooden structures share with floors most of the above mentioned problems and, therefore, can be treated, as a rule, with the same provisions. Figure 46 shows the use of cold-formed wrapping sheets, as well as that of lowered ties, both of them conceived in order to strengthen the upper roof truss against bending load.

In all strengthening operations carried out on timber floors, great attention is to be paid to the effectiveness of additional reinforcement elements, in order to ensure the maximum degree of structural efficiency, with the minimum level of displacement under load. This aspect is particularly important for this kind of floor, due to the high material deformability. Hence, a preliminary unload of timber members is recommended, followed by a proper tightening of existing and additional elements to each other. To this purpose, simple mechanical devices like wedges or jaws, or even hydraulic jacks, can be used (Figure 47). This will yield the maximum structural collaboration between members and, to some extent, will help to reduce permanent deformations of the floor. This provision, of course, can be applied to steel floors as well.

4.2 Steel floors

Steel floors started to be used in buildings at the beginnings of the 20th Century, as a more effective alternative to timber floors. They offer, in fact, enhanced structural performance in terms of both stiffness and resistance, greater durability and insensitiveness to environmental conditions. Notwithstanding, they share some behavioural aspects with timber floors, and exactly the poor in-plane stiffness, the concentrated forces at supports, the loss of connection with vertical walls and between beam and filling layers, and so on. As a consequence, apart from biological attack, damage conditions of steel floors can be quite similar to those of timber floors, namely:

- Degradation of steel beams due to corrosion;
- Degradation of end restraints due to both vibrations and repeated loads;
- Loss of internal connection between floor beams and filler layers;
- Excess of out-of-plane deformability;
- In-plane cracks due to lack of strength in the direction transverse to beams;

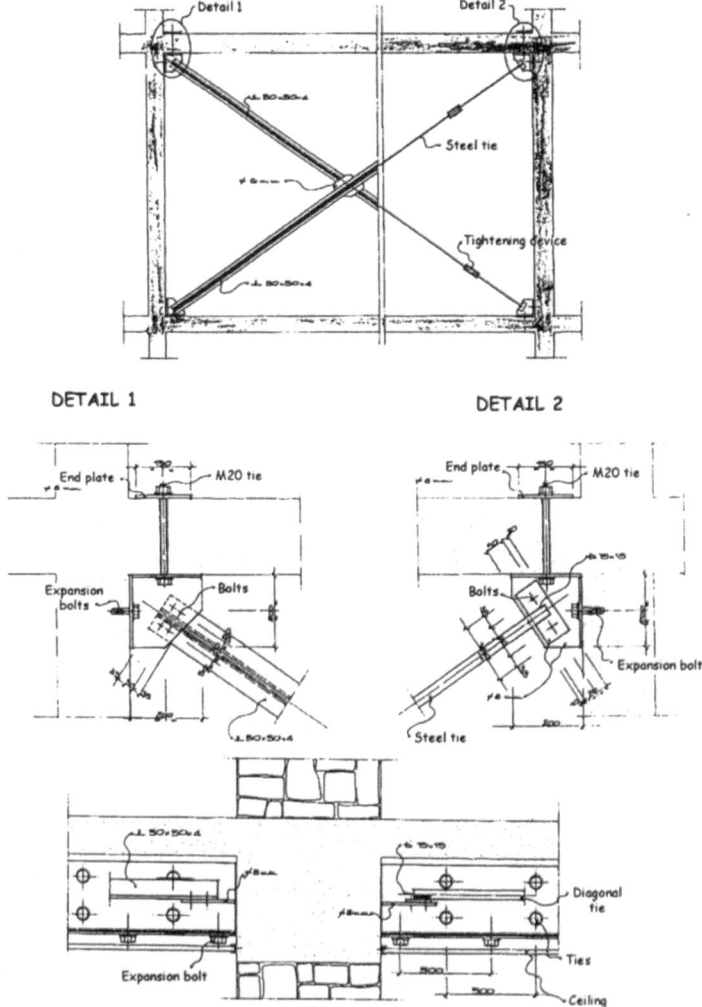

Figure 45. In-plane bracing of timber (or steel) floors and relevant details of brace-to-wall connection.

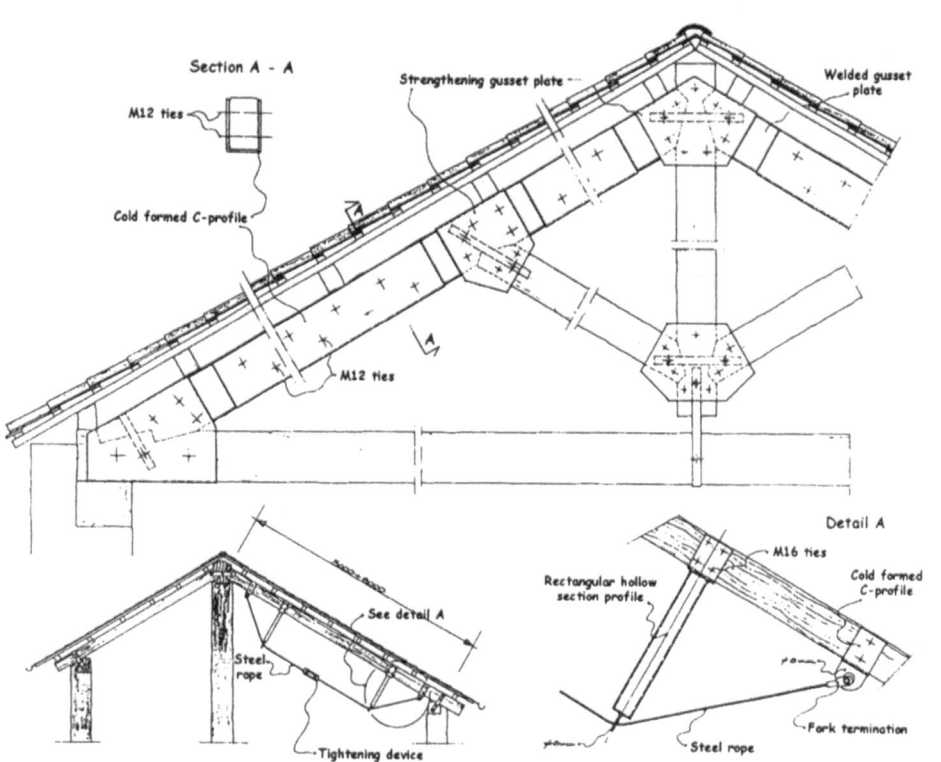

Figure 46. Common steel-based provisions adopted for the strengthening of timber roof structures.

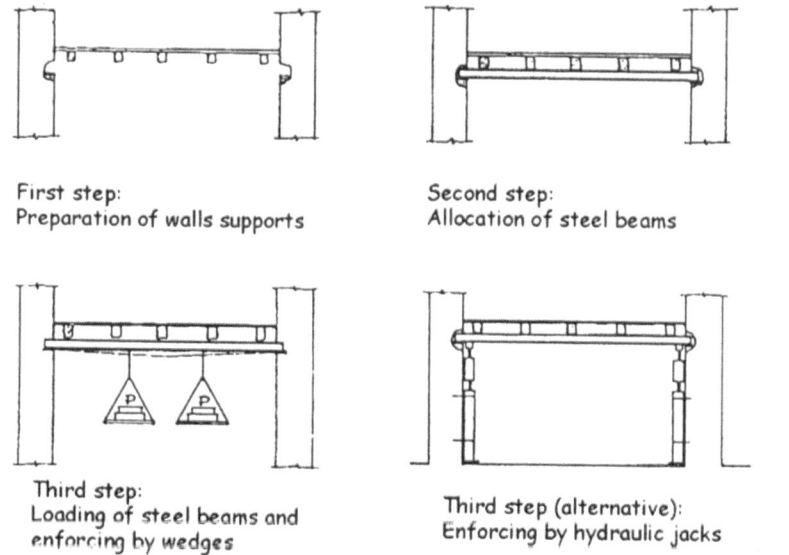

Figure 47. Enforcing procedure for additional strengthening elements against timber members.

Correspondingly, strengthening and reinforcement practices are rather similar to those adopted for timber floors, in the sense that operations at both local and global level are possible, depending on design requirements. The most important feature of steel floors is maybe that additional elements can be either welded, when they are themselves made of steel, or connected by means of welded or bolted connectors when they are made of concrete. This greatly simplifies the application of additional plates, profiles and also concrete slabs, so that a significant increase of both stiffness and strength can be easily obtained. Addition of welded elements can be made in several ways, depending on the constructional features of the floor. Some options of downward operations are shown in Figure 48. A tri-dimensional view of solutions E and F is illustrated in Figure 49.

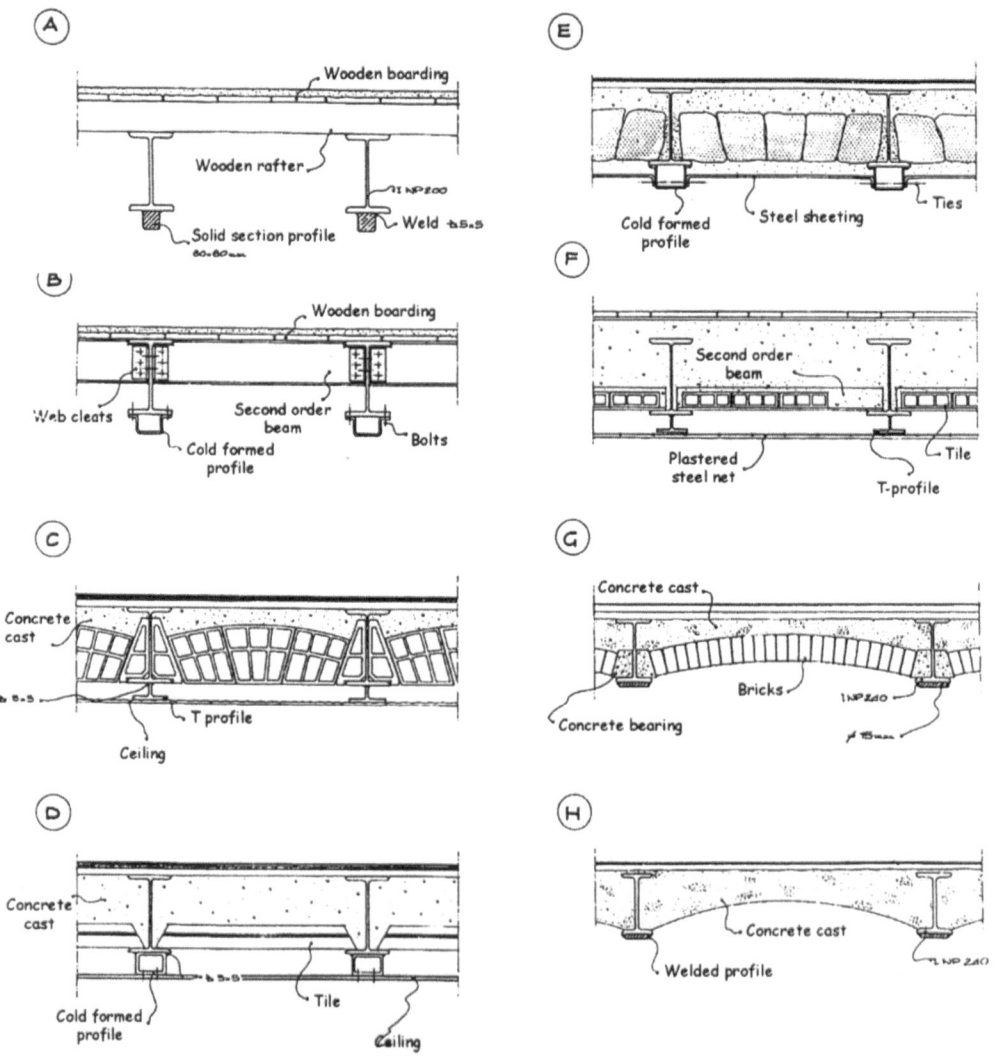

Figure 48. Downward strengthening options for steel floors based on welded added elements.

Alternatively, additional strengthening elements can be welded upward and, if necessary, properly connected or encased into an upper concrete slab in order to give way to a composite system (Figure 50). To this purpose, hollow tiled blocks can also be used as a lower concrete formwork. If the floor beams are not encased into concrete slab for at least one half of their height, then a suitable connection system should be provided. It can be obtained by means of stirrups, studs and also continuous profiles welded to the beam upper flange.

In all the above cases and, more in general, when re-use of beams is planned, the overturning of steel beams is recommended, so that the initial upward camber can offset the floor deflection under load. Also, similarly to timber floors, reinforcement of beam supports, as well as the adoption of special provisions for floor-to-wall connection are suggested, in order avoid pounding effects under seismic actions. Other strengthening interventions, with their relevant problems, are quite similar to those discussed for timber floors.

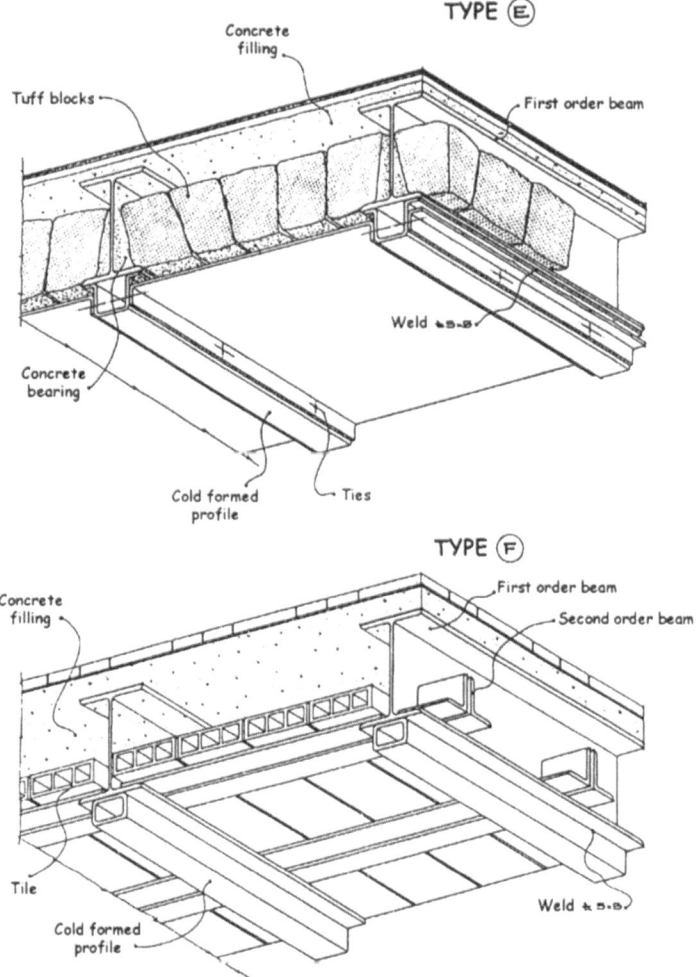

Figure 49. Tri-dimensional view of strengthening welded solutions E and F shown in Figure 48.

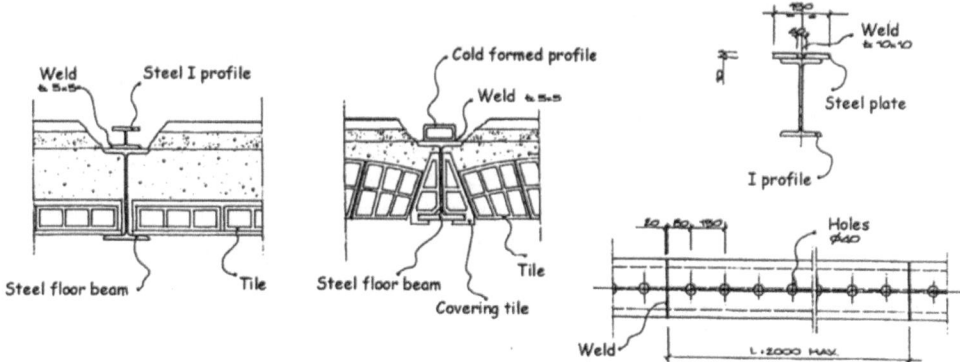

Figure 50. Upward strengthening solutions by means of welded profiles or connectors.

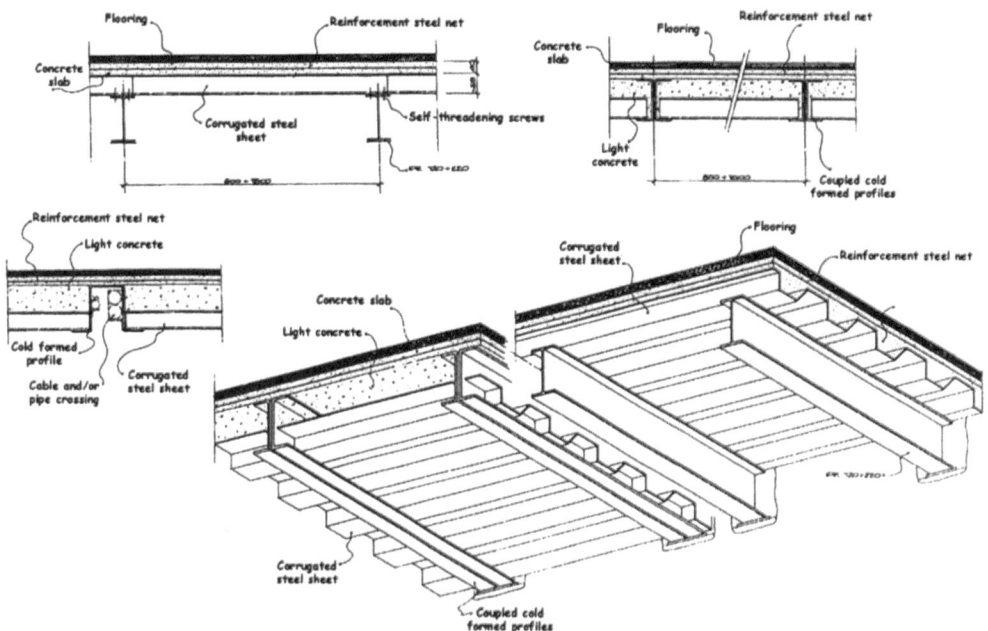

Figure 51. Some solutions for the construction of new floors into existing masonry buildings based on the use of steel cold formed corrugated sheet and profiles.

Steel elements are also widely used for the construction of new floors, when existing ones are to be completely eliminated. They provide, in fact, a better structural performance with a reduced weight. In these applications a profitable use can be made of innovative products such as cold formed profiles and corrugated sheets (Figure 51), which offer a great ease and flexibility of application, including provisions for an effective connection with side walls (Figure 52), or for the allocation of lighting systems (Figure 53). An important example of the so-called corrugated sheets of the 3rd generation (Ghersi et al., 2002), which is surely worth being mentioned herein, is represented by the TRP 200 profile, produced for the first time in

Sweden by Planjia (Figure 54). As respect to traditional corrugated sheets, this profile is higher (200 mm), larger (800 mm) and is ribbed in both transverse and longitudinal direction, resulting in a much higher stiffness compared with similar products (Landolfo and Mazzolani, 1989). This allows larger free spans (up to 5 ÷ 6m) to be covered without excess of deflection. The possibility to obtain a composite system thanks to *ad-hoc* connectors enhances the floor stiffness as well as the capability to cover spans as large as 7 ÷ 8m (Landolfo and Mazzolani, 1990). Also, profile dips can be usefully exploited to house light insulating material, resulting in a good performance from the point of view of thermo-acoustic comfort, too. Figures 55 to 59 and their relative captions illustrate some possible application examples of TRP 200, including details of both end support and floor-to-wall connection.

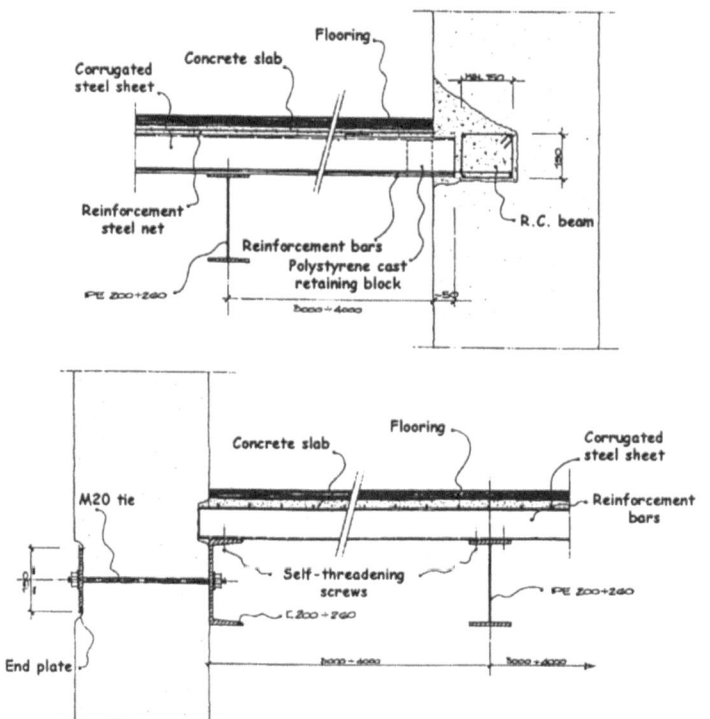

Figure 52. Provisions for floor-to-wall connection to be adopted in case of new steel floors.

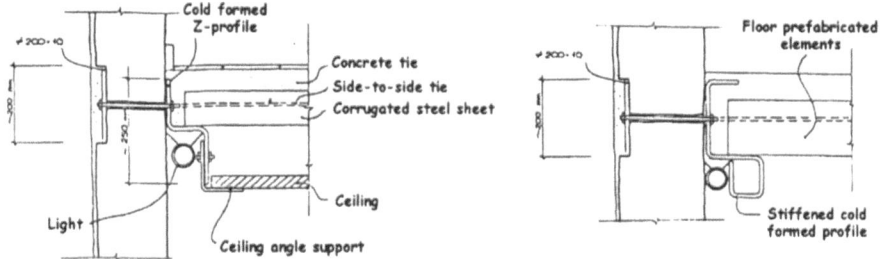

Figure 53. Use of steel cold formed profiles especially designed to facilitate connection with vertical walls and location of lighting system.

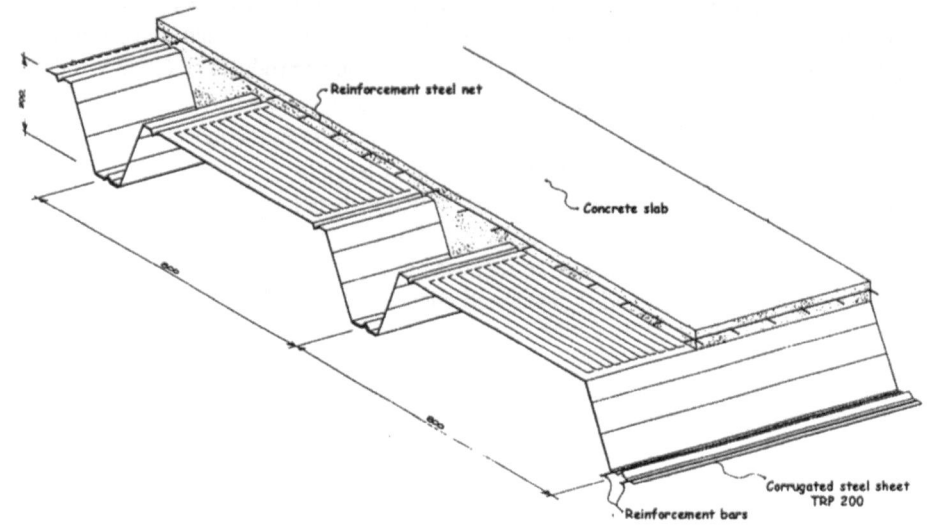

Figure 54. The steel corrugated sheet TRP 200.

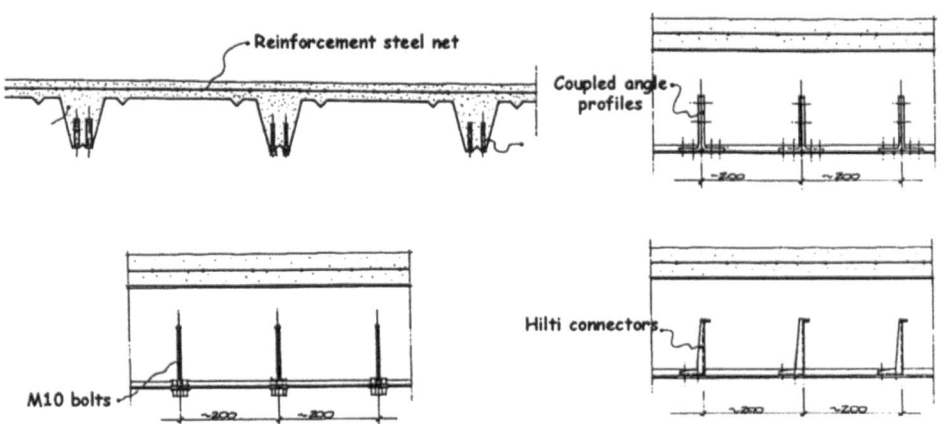

Figure 55. Connecting systems for obtaining a composite system from TRP 200.

4.3 Reinforced concrete floors

Reinforced concrete floors are quite different from both timber and steel floors, not only as far as the material is concerned, but also for some inherent structural features which make their behaviour completely different from that of floors made of discrete beams. R.c. floors, in fact, are in most cases obtained from a continuous cast which comprises an upper slab and side beam running along the floor perimeter. As a result, slab, side- and floor-beams behave as a monolithic whole and, hence, can guarantee very high in-plane and out-of-plane stiffness. For this reason, r.c. floor are suitable to behave as rigid floor diaphragm even in masonry buildings without the problems usually encountered with timber or steel floors. Due to the presence of the side beam, in fact, the connection with all vertical walls is uninterrupted, and this ensures an effective transmission of horizontal actions from weak to stiff walls. For this reason r.c. floors are used, when possible, as a substitute for old structures, in particular timber floor. As previously shown, in this case an effective connection with bearing walls can be obtained by means of suitable concrete bulbs cast into wall cut-outs purposely fitted.

Concerning damage conditions, it should be said that r.c. floors seldom suffer severe structural problems, apart from those conditions when highly aggressive environments can produce material degradation by chemical attack. Nevertheless such conditions rarely occur into buildings, which is why such cases are not covered in this chapter. Rather, in most cases a need for a structural upgrading of such floors raises when higher loads must be resisted, as happens, for example, when a more severe use is planned for the building. In this case it is possible to operate on the single floor beams in order to increase their load bearing capacity according to the techniques shown in Figure 60. To this purpose, steel elements can be used to supplement existing cross sections in either tension and/or compression. When used on both top and bottom of floor beams, such elements can be connected to each other by means of bolted ties working as common stirrups in r.c. members. If fully rigid connection between steel and concrete is ensured, then the design of such elements is quite simple, as they can be considered as a matter of fact as a sort of additional reinforcement. Nevertheless, caution should be paid when the connection between steel and concrete allow some slip which would impair the validity of linear beam bending theory.

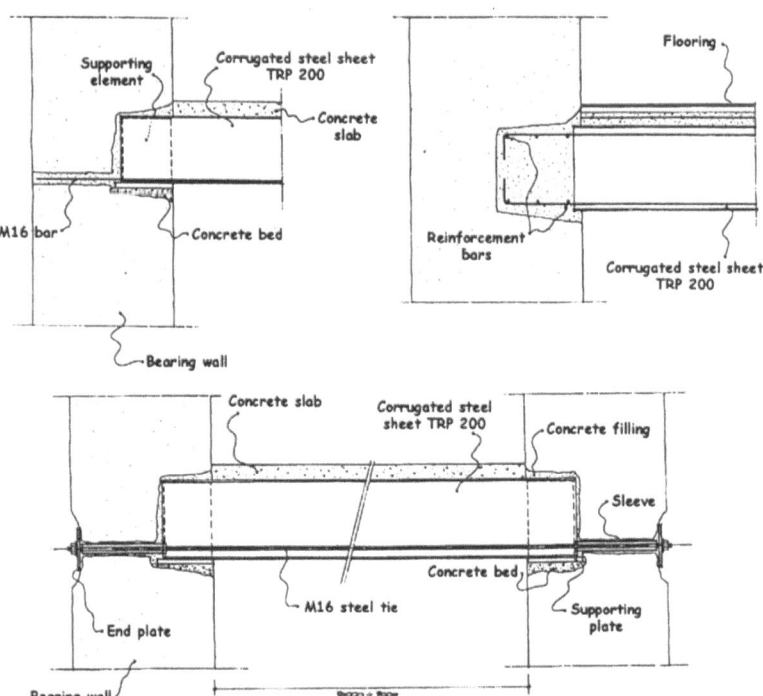

Figure 56. Connection of TRP 200 to bearing walls by means of partial encasing and end plates.

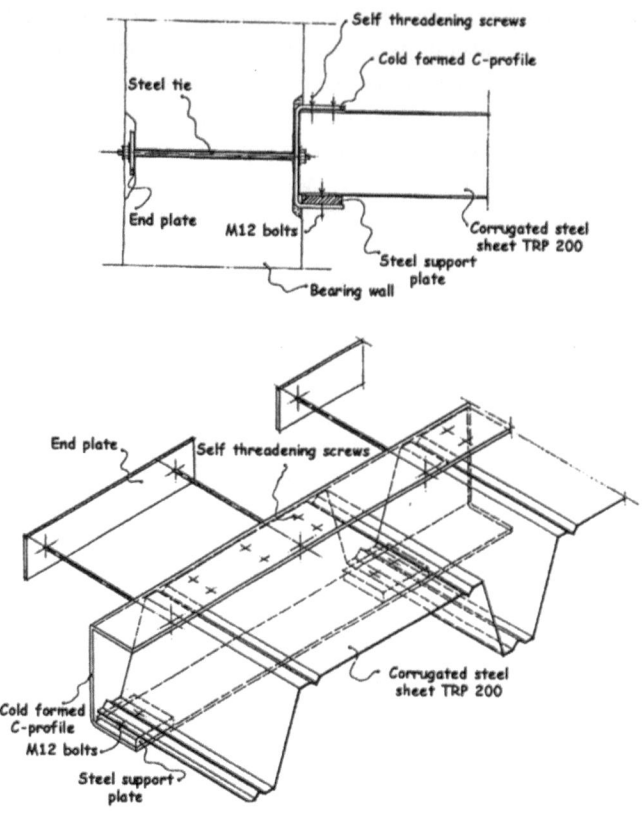

Figure 57. Connection of TRP 200 to bearing walls by means of a cold formed C-profile.

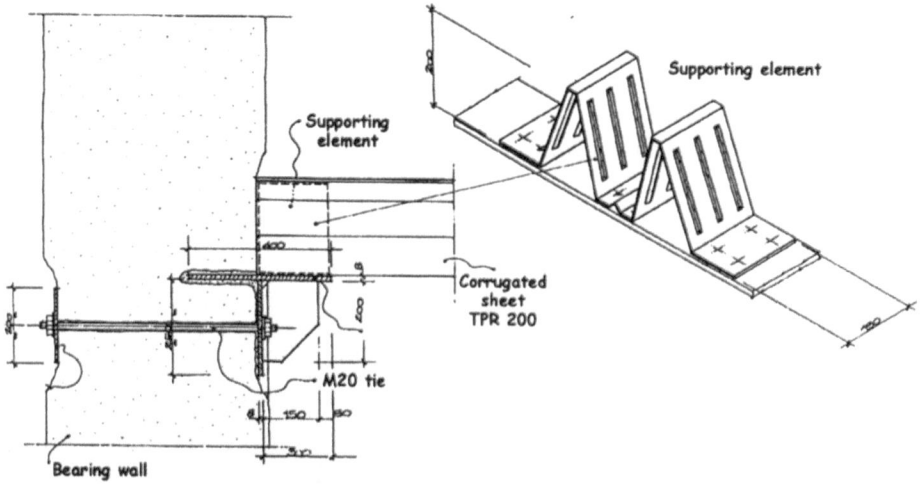

Figure 58. Use of a fitted supporting element for the connection of TRP 200 to bearing walls.

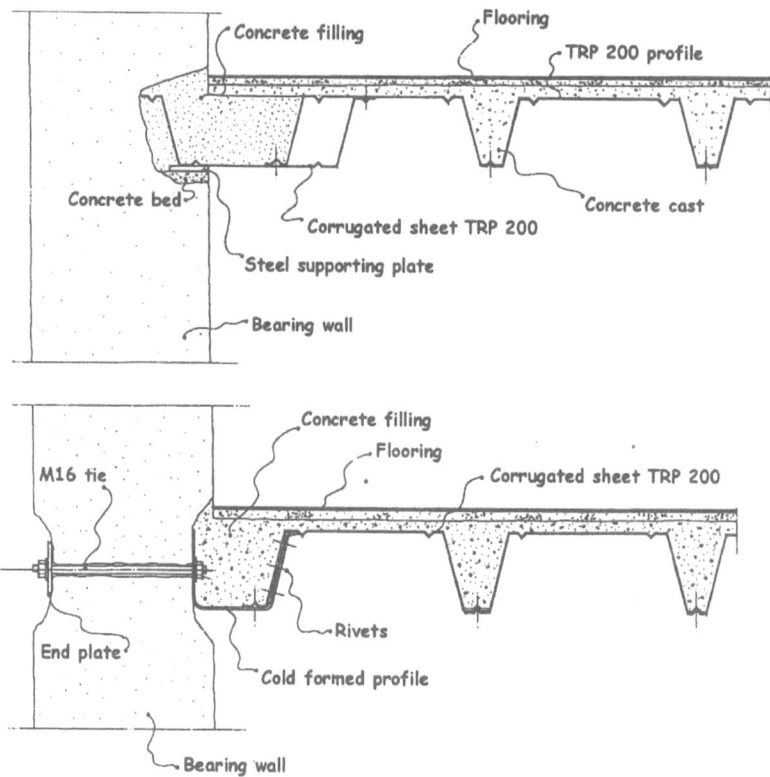

Figure 59. Options for connection of TRP 200 to side bearing walls.

In order to avoid slip, additional steel elements can be joined to the r.c. beam by means of expansion bolts or special fasteners. As an alternative, innovative products can be also used, such as for example epoxy resins or adhesive bonding mortar. The cross section of steel profiles can be either simply flat or variously shaped, depending on design requirements. In the same way as for timber floors, cold formed profiles can be adopted, in order to encircle the whole beam downside. In some cases, when an increase of load bearing capacity in bending is required only, simple tie rods can be allocated downside the floor beams with endpoints fastened to bolted contrasting elements. Such application has the advantage to be easily applied without interrupting the use of the floor. In addition, it can be adopted for a local reinforcement and is also completely reversible. In this application it is recommended that the whole amount of shear in the floor is born by existing concrete.

As for the strengthening of masonry members, an alternative to steel elements may be represented by CFRP strips epoxy-bonded to tensioned areas of floor beams. This practice allows the use of additional reinforcing elements without increase of cross section. Furthermore, it does not involve any problem of fibre-to-concrete adhesion. The design of such additional fibre reinforcement can be made according to r.c. bending theory. CFRP can also be used as a confining film in order to increase concrete ductility at failure (Spoelstra and Monti, 1999). Further information on the strengthening of r.c. members are given in the next section.

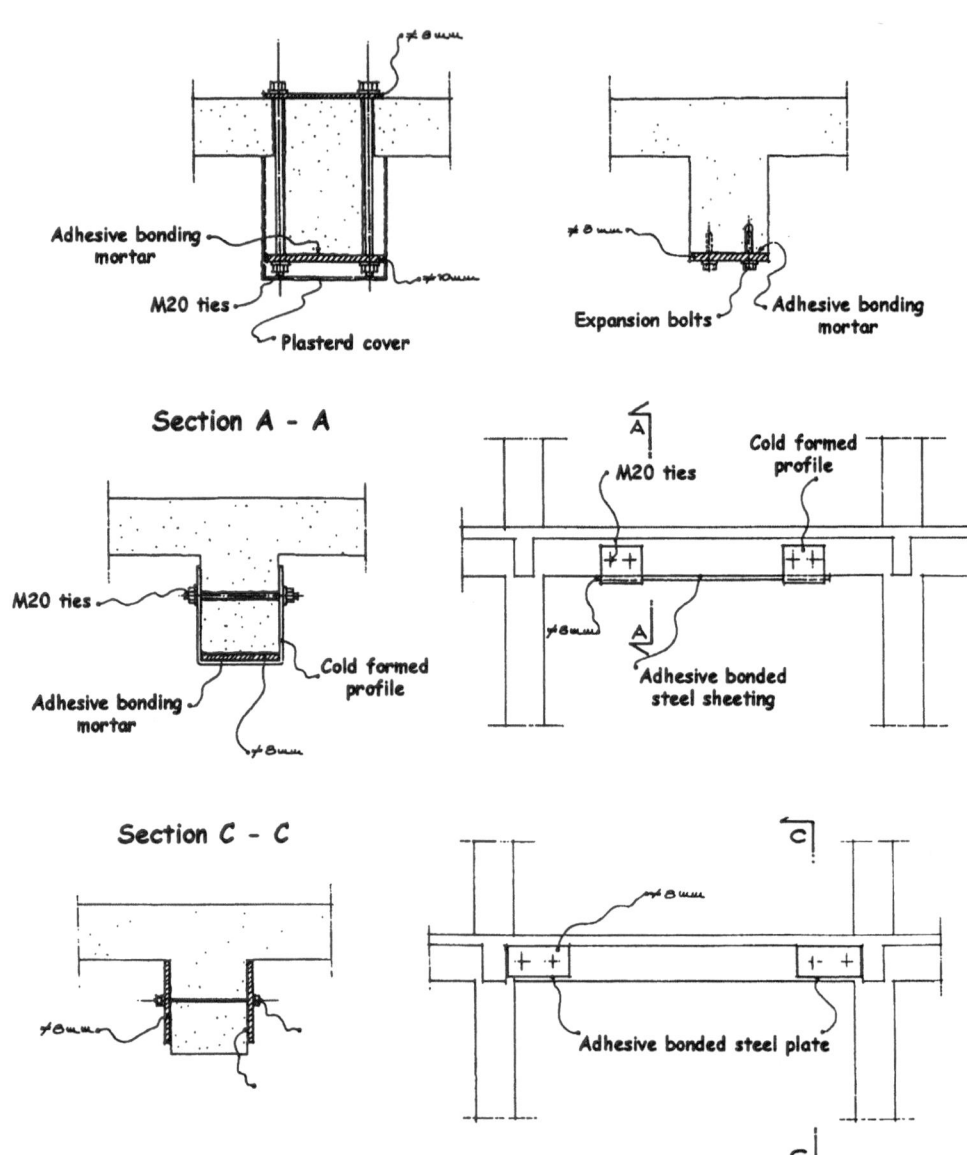

Figure 60. Common provisions for the strengthening of r.c. floors.

5   Strengthening of r.c. structures

5.1 General

Following information given on r.c. floors, the main provisions adopted for beams and columns are illustrated in this paragraph. The main problems related to the structural restoration of r.c. members can be summarised as follows:

- adherence and chemical compatibility of added materials with concrete;
- effectiveness of connection and fastening of additional elements to the existing ones;
- protection of the new reinforcements;

As being stated, materials adopted for strengthening r.c. structures should possess as a rule the following properties:

- elastic modulus of the same magnitude order as the material to be strengthened;
- adhesion higher than concrete internal one;
- low or even negative shrinkage;
- creep as low as possible;
- high durability.

Such features can be found, even tough not in the same amount, in materials like:

- steel;
- resins;
- special cement mortars;
- fibre reinforced polymers (FRP).

The main properties of such materials are well known and have been also highlighted in this chapter. As far as their application to r.c. structures is concerned, steel do have high strength, high stiffness, lightness, immediate availability of load bearing capacity, chemical stability with concrete and negligible creep. This allows very effective strengthening operations of both r.c. members and joints (Figure 61). Nevertheless, it is rather sensitive to corrosion, in particular when it is bonded outside the r.c. member; it also requires a careful preparation of concrete surface in order to achieve the best effectiveness. In addition, its elastic modulus and strength are too high compared with concrete, and this can cause problems in the stress transmission from steel to concrete, especially in those cases where r.c. and steel elements are arranged in parallel.

Properties of resins have been already introduced in the section devoted to masonry members. Most aspects discussed with regard to strengthening of masonry can be kept for r.c., too. In particular, resins can be profitably used for filling small cracks arising in tensioned areas. To this purpose, it should be considered that concrete cracking is, to some extent, a quite usual event in r.c. structures under serviceability loading conditions. Therefore, cracks are to be sealed only when are due to unforeseen actions (e.g. earthquake actions in non seismic areas) or when are incompatible with environmental conditions. Also, in applications on r.c. structures resins can be used for bonding additional concrete or steel elements to existing ones. In such applications, the good adhesion and chemical stability with both steel and concrete are fully exploited, providing a good effectiveness to the strengthening intervention.

Probably the most appropriate products for repairing or reinforcing r.c. structures are cement-based mortars, which are available in a very wide range of mechanical features. They offer full compatibility and high bond with both existing concrete and steel bars. Furthermore, contrary to other materials, e.g. steel, resins and FRP, they do not require highly specialised workmanship, as application techniques and constructional tolerances do not differ too much from conventional r.c. structures. As a results, they present ease of installation and, also, acceptable cost. The main drawbacks in the use of such products is represented by the shrinkage, which, if not properly controlled, can impair the effectiveness of intervention, involving additional cracking and detachment of added concrete skin. In order to avoid this, special products are available, which guarantee an adequate reduction of shrinkage. Anyway, as shrinkage is strongly influenced by environmental conditions (in particular temperature and relative humidity), the application of such materials should be carefully controlled and, if necessary, preliminarily executed on test specimens. Similarly, pH value at the material interface should be kept above 9, in order to avoid possible concrete carbonation.

Eventually, FRP wrapping can be applied in the same way as seen for masonry members. The use of this techniques is quickly increasing, due to high mechanical properties of fibres, which allow FRP strips to be effectively applied for increasing tensioned resisting area or as hooping confinement elements of compressed members.

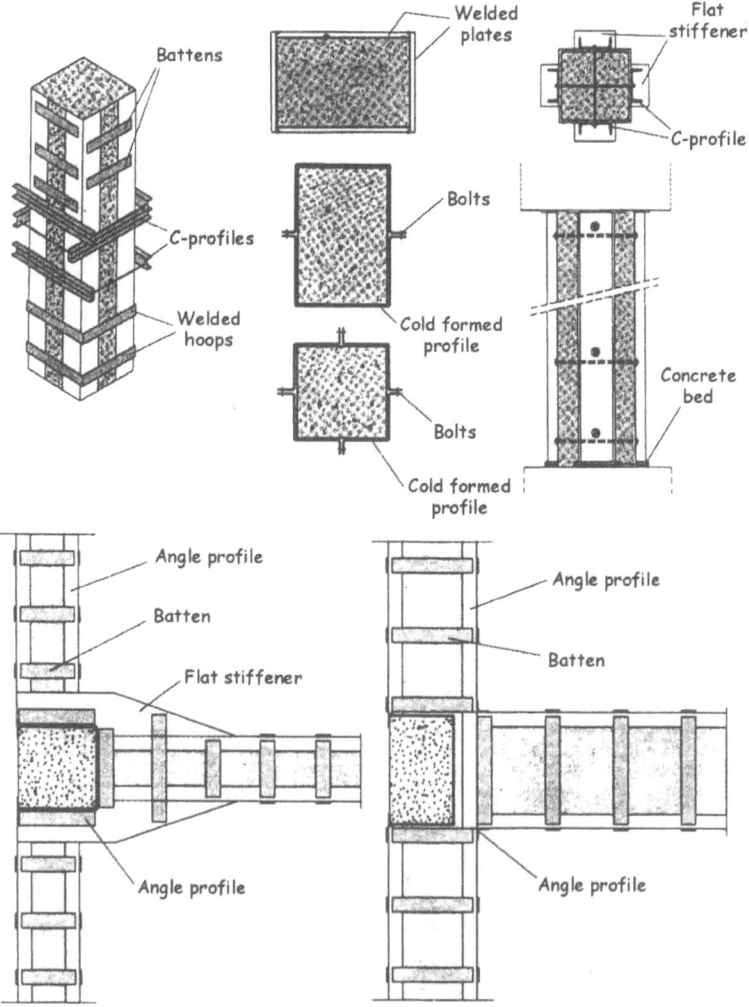

Figure 61. Bonded steel elements for the strengthening of r.c. members.

5.2 Interventions on r.c. beams and columns

Concerning practical interventions on r.c. elements, many options are possible. In general, a good performance is obtained combining the above techniques with each other. For example epoxy resins can be conveniently used for sticking steel elements, in the form of flat and angles profiles, together with r.c. columns or beams (Figure 61). In such a case resins advantageously substitute for mechanical fasteners (e.g. expansion bolts), that could involve high stress concentration into *in-situ* concrete.

Alternatively, an additional concrete cast can be fastened to the existing structure by means of steel connecting devices, as shown in Figure 62, so as to obtain a composite structure. Similarly, resins can be used in combination with inert material to prepare concrete or mortar with improved mechanical features, to be used in the restoration of damaged parts or for increasing the resisting section of existing elements.

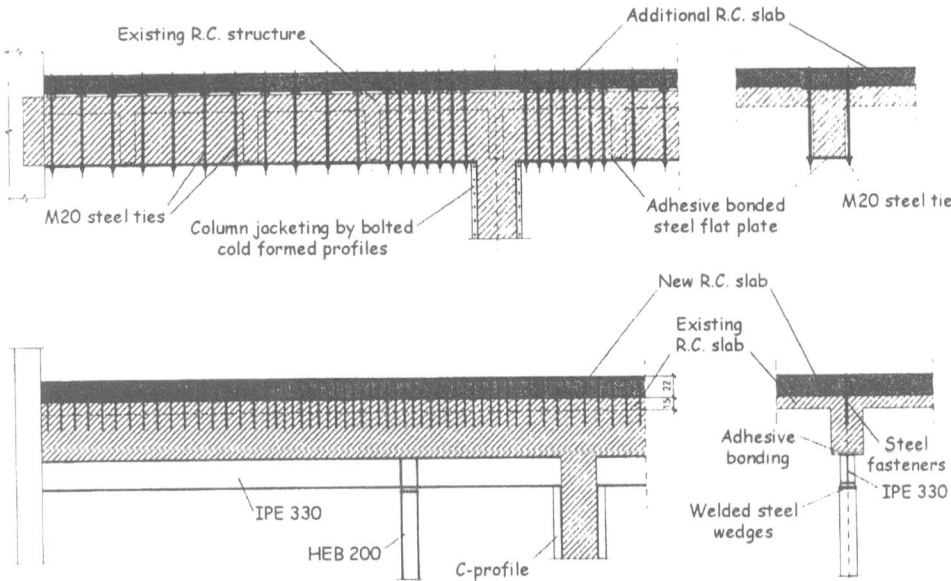

Figure 62. Strengthening options for increasing the load bearing capacity of r.c. beams.

In case of r.c. columns, goals of the intervention can be either restoring the original load-bearing capacity of the column (in case of damage) or increasing the load bearing capacity of the column, when a higher structural performance is required. Damage in r.c. elements can be purely external, when it is limited to the outer perimeter without involving the inner core of the member, but also internal, when deep degradation of the element body is present. When damage is limited to the external concrete cover and no increase of acting load is planned, the cross section is usually kept as it is and intervention is concerned with the local repairing of damaged parts with expansive concrete and/or resins (Figure 63a). As an alternative, confinement with expansive concrete reinforced with steel network may be applied, after removal of damaged parts. Confinement with steel plates, according to the technique of so-called "*beton plaqué*" can be used as well.

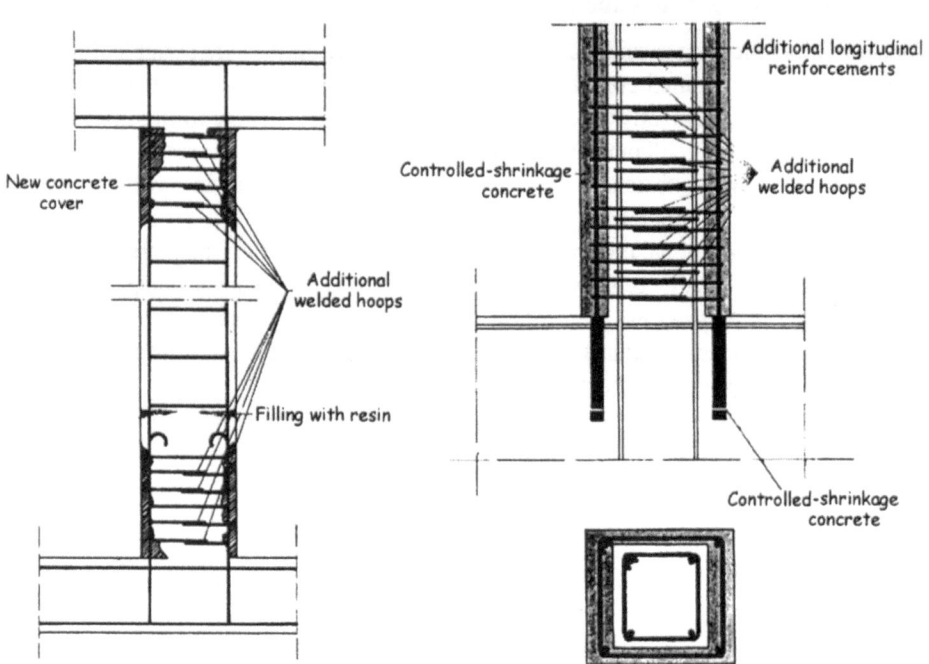

Figure 63. Repairing (a) and strengthening (b) of r.c. column by means of concrete re-covering (a) or r.c. jacketing (b).

When the original cross section has to be increased due to severe damage of the inner core and/or strong increase of acting load, then the existing column can be coupled with additional r.c. or steel columns. Such new elements can be placed either aside the existing member or can be encircling around it. Column jacketing by means of new r.c. skin may represent a good provision, which involves just a slight modification of global stiffness distribution in the building (Figure 63b). An ultimate solution in case of extremely severe damage can be the complete elimination of the column and the construction of a new one.

In all above cases great attention should be paid to the connection between the existing column and the additional elements. This is highly important in order to ensure the best structural effectiveness of the intervention, as well as to reduce the risk of instability of outer concrete or steel skin. Also, the effect of shrinkage should be kept into account, as it can strongly affect the final distribution of load between existing and additional elements. Similarly, when the column is severely damaged, or when a great increase of acting load is planned, a careful evaluation of its residual stiffness and load bearing capacity should be done, so that an accurate estimation of actual load distribution between members can be achieved.

In the end, as for interventions on masonry buildings, caution is recommended when using bracing systems for increasing resistance against horizontal actions (Figure 64) or integrative structures as a substitution of eliminated members (Figure 65). This additional elements, in fact, involve a radical modification in the global behaviour of the building, in particular under horizontal actions. Their effect, therefore, should be carefully evaluated and, most of all, adequate strengthening of foundations must be provided at bracing location.

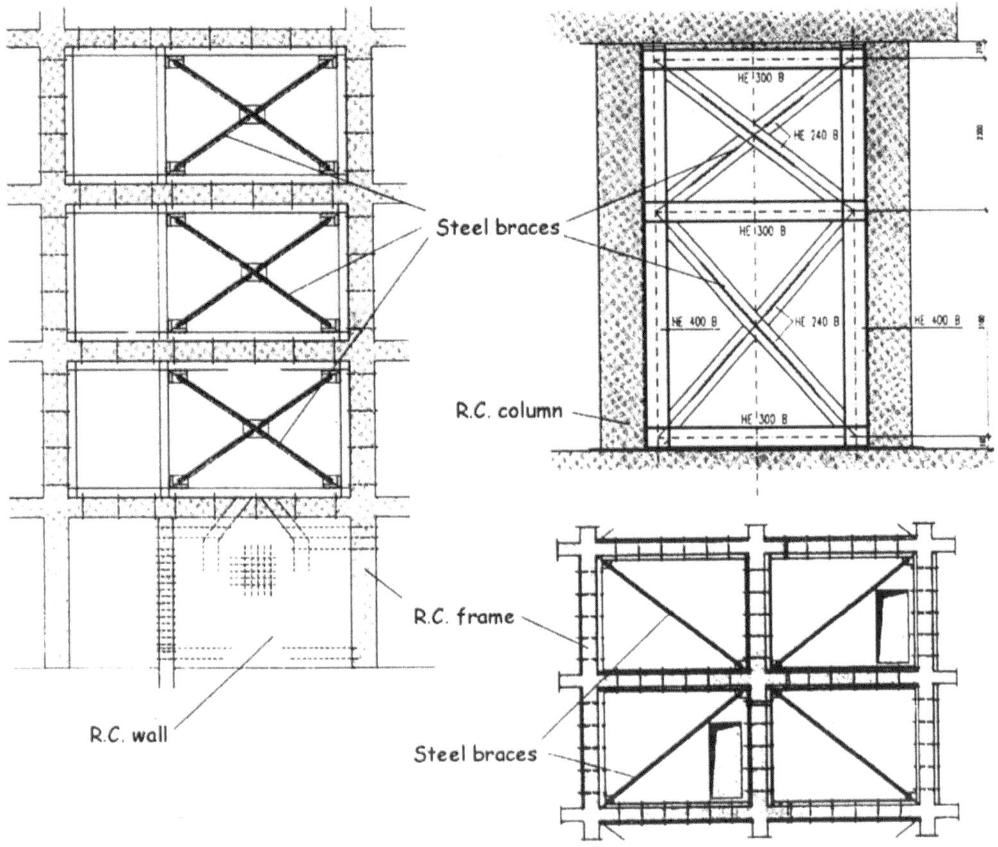

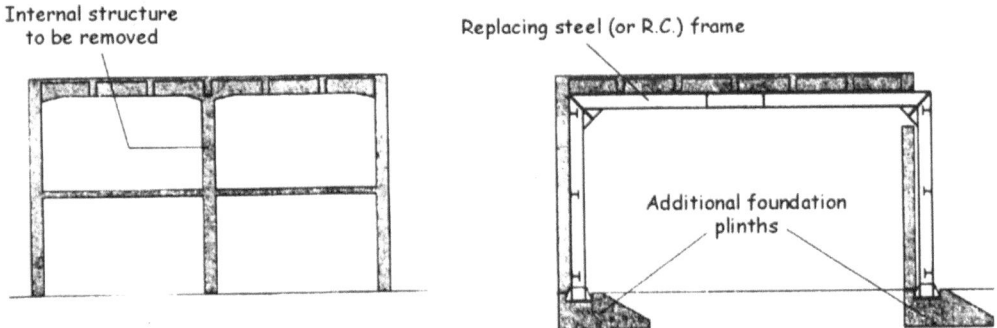

Figure 64. Additional bracing structures for improving resistance and stiffness against horizontal loads.

Internal structure
to be removed

Replacing steel (or R.C.) frame

Additional foundation
plinths

Figure 65. Transformation of a r.c. frame involving the substitution of the central column with a replacing steel or r.c. one-bay frame.

References

Note for the reader

As mentioned in the introduction, Italy is very advanced in the field of structural rehabilitation. This caused a great amount of existing papers, books and manuals on this topic to be issued in this country in italian language. Literature on this topic is enormously wide and, as a consequence, can not be listed completely herein. For this reason just a few, but highly representative titles are mentioned in the first two of the following reference lists. Because of the general character of this chapter, these books are not explicitly referred to in the text, as they would be worth being cited almost everywhere. Their reading is highly recommended to whomever may be interested in the structural restoration practice, from both design and execution points of view. The first list contains references to textbooks and manuals issued in Italy. Even tough the language may result not understandable to foreign readers, such books are able to provide a valuable and helpful information about all strengthening techniques, including interventions on historical and monumental buildings, by means of the rich contents of images and illustrations, only. They can be considered, as a matter of fact, as the today state-of-the-art of this field. Reference to some recent international conference proceedings (in English) is given in the second list, as they can be regarded as a comprehensive guide to the latest trends in the structural rehabilitation. A third list (Cited references) contains some more specific titles, all of them referred to in the chapter text as appropriate.

Textbooks and manuals

AA.VV. (1991). L'acciaio nel recupero edilizio e nel restauro – Repertorio delle soluzioni tecniche. Consorzio CREA.

Albi Marini, A.: Corso di consolidamento degli edifici, notes from the lectures held at the Engineering Faculty of the University of Naples Federico II.

Avramidou, N. (1990). Criteri di progettazione per il restauro delle strutture in cemento armato. Liguori, Napoli.

Barbarito, B. (1993). Collaudo e risanamento delle strutture. UTET, Torino.

Defez, A. (1990). Il consolidamento degli edifici. Liguori, Napoli.

Del Piero, G. (1983). Il consolidamento delle costruzioni. CISM, Udine.

Del Piero, G. (1984). Le costruzioni in muratura. CISM, Udine.

Fera, C. (1986). Principi di progettazione e rassegna di interventi. Collana Acciaio e Riuso Edilizio, Nuova Italsider, Genova.

Guerrieri, F. (Editor) (1999). Manuale per la riabilitazione e la ricostruzione postsismica degli edifici – Regione dell'Umbria. DEI, Roma.

La manna, L.F., and Bellicini A.L. (2000). Tecniche di risanamento degli edifici. Carocci, Roma.

Mandara, A., and Mazzolani, F.M. (1992). Nuove strategie di protezione sismica per edifici monumentali: il caso della Collegiata di San Giovanni Battista in Carife", Soprintendenza ai B.A.A.A.S., Salerno - Avellino.

Mastrodicasa, S. (1980). Dissesti statici delle strutture edilizie. Hoepli, Milano.

Mazzolani, F.M., and Mandara, A. (1991). L'acciaio nel consolidamento. ASSA, Milano.

Mazzolani, F.M., and Mandara, A. (1992). L'acciaio nel restauro. ASSA, Milano.
Pasta, A. (1999). Restauro conservativo e antisismico. Flaccovio.
Sarà, G. (Editor) (1989). Restauro strutturale. Liguori, Napoli.

Conference Proceedings

International Seminar on Structural Analysis of Historical Constructions, Barcelona, 1995.

2nd International Seminar on Structural Analysis of Historical Constructions, Barcelona, 1998.

International IABSE Conference on Savings Buildings in Central and Eastern Europe, Berlin, 1998 (on CD-ROM).

2nd International Congr. On Science and Technology for the Safeguard of Cultural Heritage in the Mediterranean Basin, Paris, 1999.

5th International Congress on Restoration of Architectural Heritage, Florence (Italy), 2000 (on CD-ROM).

3rd International Congress On Science and Technology for the Safeguard of Cultural Heritage in the Mediterranean Basin, Alcalà de Henares (Spain), 2001(on CD-ROM).

3rd International Seminar on Historical Constructions, Guimaraes, Portugal, 2001.

Cited references

Ballio, G., and Calvi, G.M. (1993). Strengthening of masonry structures by lateral confinement. *Proc. of IABSE Symp. on Structural Preservation of the Architectural Heritage*, Rome.

Croci, G., Bonci, A., and Viskovic, A., (2000). The use of shape memory alloys devices in the basilica of St Francis in Assisi. *Proceedings of the Final Workshop of ISTECH Project*, Ispra (Italy), 117-140.

Dolce, M., Cardone, D., and Nigro, D. (2000). Experimental tests on seismic devices based on shape memory alloys. *Proceedings of the 12th World Conference on Earthquake Engineering*, Auckland, (on CD-ROM).

Ghersi, A., Landolfo, R., and Mazzolani, F.M. (2002). Design of metallic cold-formed thin-walled members. Spon Press, London – New York.

Giuffrè, A., and Martines, G. (1989). Impiego del titanio nel consolidamento del capitello della Colonna Antonina. *Proceedings of the Convegno e Mostra A.N.I.A.SPE.R.* Rome, 45-50.

Indirli, M. (2000). The demo-intervention of the ISTECH Project: the Bell Tower of S. Giorgio in Trignano (Italy). *Proceedings of the Final Workshop of ISTECH Project*, Ispra (Italy), 141-153.

Landolfo, R., and Mazzolani, F.M. (1989). Indagine sperimentale su lamiere grecate tipo TRP 200. (in Italian) Acciaio n. 1.

Landolfo, R., and Mazzolani, F.M. (1990). Sistemi misti acciaio-calcestruzzo con l'impiego di lamiere grecate di terza generazione: indagine sperimentale. (in Italian) Acciaio n. 1.

Mandara, A., and Mazzolani, F.M. (1998) Confining of masonry walls with steel elements. *Proceedings of the International IABSE Conference on Savings Buildings in Central and Eastern Europe*, Berlin (on CD-ROM).

Mandara, A. (1992) L'uso dell'acciaio nel consolidamento di elementi murari verticali. L'Edilizia", 11.

Mazzolani, F.M., and Mandara, A. (1999). Methodology for the structural rehabilitation of the Main Hall of "Mercati Traianei" in Rome", *Proceedings of the 2ⁿᵈ Int. Congr. On Science and Technology for the Safeguard of Cultural Heritage in the Mediterranean Basin*, Paris.

Mazzolani, F.M., and Mandara, A. (2000). Advanced metal systems in structural rehabilitation of monumental constructions. *Proceedings of The International Conference on Structural Engineering, Mechanics and Computation*, Cape Town, Vol. 1, 75-86.

Mazzolani, F.M., and Mandara, A. (1989a). Esame critico degli interventi di consolidamento su edifici in muratura danneggiati dal bradisismo flegreo. Acciaio n.9.

Mazzolani, F.M., and Mandara, A. (1989b). L'uso dell'acciaio negli interventi di restauro nell'edilizia monumentale dell'Italia Meridionale. Acciaio n.10.

Pegon, P., Armelle, A., Pinto, and A., Renda, V. (2000), The ELSA Laboratory and the protection of cultural heritage. *Proceedings of the 5ᵗʰ International Congress on Restoration of Architectural Heritage*, Florence (on CD-ROM).

Pinto, A., Ferreira, S., and Barros, V. (2001). Underpinning solutions of historical constructions. *Proc. of the 3ʳᵈ Int. Seminar on Historical Constructions,* Guimaraes, Portugal, 1003-1012.

Spoelstra, M., and Monti, G. (1999). FRP-confined concrete model. *Journal of Composite for Constructions,* ASCE 3(3), August 1999, 143-150.

Triantafillou, T.C. (1998). Strengthening of masonry structures using epoxy-bonded FRP laminates. *Journal of Composite for Constructions,* ASCE 2(2), 96-104.

Valluzzi, M.R., Valdemarca, M., and Modena, C. (2001). Experimental analysis and modeling of brick masonry vaults strengthened by frp laminates. *Journal of Composite for Constructions,* ASCE, August 2001.

Acknowledgements

The author wishes to acknowledge the contribution taken from italian books and manuals listed above. Some concepts and illustrations shown in the text have been, in fact, inspired by these books.

Similarly, an important source for many pictures concerning the use of steel in structural rehabilitation was the outcome of the research issued by the Italian Ministry of University and Scientific-Technological Research (MURST) and carried out in the years 1988-90 by CO.RI.R.E. (COnsorzio di RIcerca per il Recupero Edilizio) under the coordination of Prof. F.M. Mazzolani.

Last, but not least, the author is very grateful to Prof. F.M. Mazzolani himself for his precious and helpful advices and for his relentless encouragement in the study of this and other research topics.

CHAPTER 5

USE OF SPECIAL TECHNIQUES IN REFURBISHMENT

J-P. Muzeau

Blaise Pascal University, Clermont-Ferrand, France

Abstract. This chapter deals with the use of some special techniques, which are possible to be used in refurbishment. It contains the description of the main methods available for partial demolition, the presentation of non-classical fasteners as well as of some techniques that allow a connection to be improved or upgraded. Then, two examples of building refurbishment are detailed and the upgrading of a suspension bridge is described.

Very often, light, moderate or severe demolition may be the initial task to be decided before really starting the refurbishment of a construction. In the first part of this chapter, some of the available techniques, those that can be used in the limited case of partial demolition, are presented. Choice criteria with regards to environmental or site conditions are provided to help in choosing the most efficient method.

In refurbishment activities, the accessibility conditions are sometimes very restrictive and may influence the technical solutions to be chosen for connecting new elements to existing ones. The second part of this chapter describes several non-classical fasteners, which can be efficient in specific site conditions.

Refurbishment frequently requires to reinforce existing connections or to create new supports. Some proposals are presented in the third part. They deal mainly with beam-to-column connections in the case of columns made of concrete or steel.

Two examples of building refurbishment are described in the fourth part. They concern the transformation of two rather old buildings into more convenient structures at Blaise Pascal University, Clermont-Ferrand, France.

Finally, the fifth part details the upgrading of the "25th of April Bridge" in Lisbon, Portugal, which has been refurbished recently to welcome a new railway traffic inside its truss beam and to increase its road traffic capacity on the upper deck.

1   Techniques for Partial Demolition[1]

Refurbishment is generally associated with more or less important changes in the use of constructions. So, it may be required to modify some of their parts in order to create new stairs, new walls or new floors, to extend their capacity by addition of extra storeys or to reinforce partially their strength.

To adapt or to update an existing construction, it is often necessary to carry out some transformations and to demolish some of its parts. Hence, the choice of the most efficient demolition technique, as function of the structural material, becomes one of the first steps in a refurbishment. It allows the whole structure, or only some of its parts, to be prepared for their new functions.

In this chapter, the main methods existing in the domain of partial demolition are detailed (Mur J. & Muzeau J.P., 1979), (Muzeau, 1996), (Farinha, 1998). Then tables of comparison are established. They allow the most efficient techniques to be chosen amongst mechanical, blasting and thermal processes.

1.1   Mechanical techniques

Hand demolition. The use of hand demolition may be necessary to avoid any significant or severe mechanical actions to be developed in the existing structure, mainly when its stability could be affected prior to its reinforcement.

With masonry, taking off blocks, layer by layer, with hand tools (pick and hammer), is sometimes required when salvage material is valuable. It is a very slow method that is only realistic for masonry but not for concrete what ever could be its reinforcement.

Hand dismantling may also be efficient for steel or timber structures when the elements may be re-used or when large deformations at the connections must be absolutely avoided.

Processes using percussion. Pneumatic picks are compressed-air tools designed for high frequency impacts (1000 to 2000 hits/min) produced by a pick or a blade. Relatively light (3 to 16 daN), they can be used in any position. The pick diameters range from 20 to 35 mm with lengths from 350 to 450 mm.

Concrete breakers (Photo 1) are heavier (18 to 40 daN). They can be used only in vertical position. Most are pneumatic but some are hydraulic or electric.

These tools are similar but much smaller than rock breakers always installed on an excavator. Hydraulic or pneumatic, rock breakers are about twenty times more powerful than pneumatic picks, and this is generally too much for partial demolition. The tool diameters range from 80 to 115 mm.

All these tools use points and spades or even burst wedges. They produce small fragments, which can be easily extracted.

Internal openings in floors or walls may be carried out by this way because these tools do not create falls of large elements, which could affect the stability of the structure. If necessary, larger structural members can be supported by a crane or by temporary works to avoid critical damages.

---

[1] This paragraph is co-authored with Dr. António Baptista, Senior Researcher, LNEC, Lisbon, Portugal.

Noise, dust and vibrations are some of the problems associated with these tools and it is rather difficult to control the crack spreading.

Photo 1. Example of concrete breakers

Alligator shear. An alligator shear (Figure 1) may be installed on a hydraulic excavator to cut steel members or reinforcing bars. Nevertheless, its use creates generally some damages in the nearest parts of the structure due to the relatively large displacements involved during the shear process.

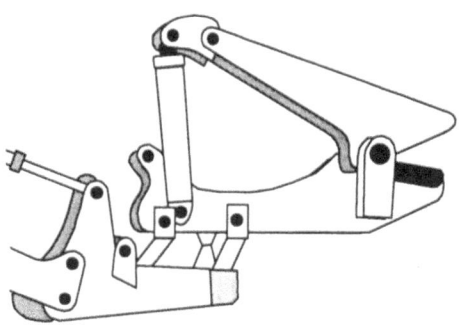

Figure 1. Alligator shear

Hydraulic bursters. Two main types of hydraulic bursters exist. The first one, called Roc-Jac burster, is a steel cylinder with a number of pistons that are forced radially outwards against the surrounding structure (Figure 2). The second type only contains one radial wedge piston. Darda, Roc-Pak and Gullick-Dobson bursters are of that kind (Figure 3). The difference between these tools is that the Darda burster has its piston forced outwards while, for the others, the piston moves inwards.

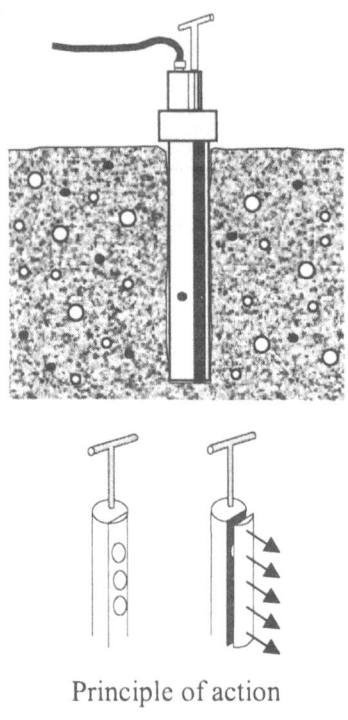

Principle of action

Figure 2. Roc-Jac hydraulic burster

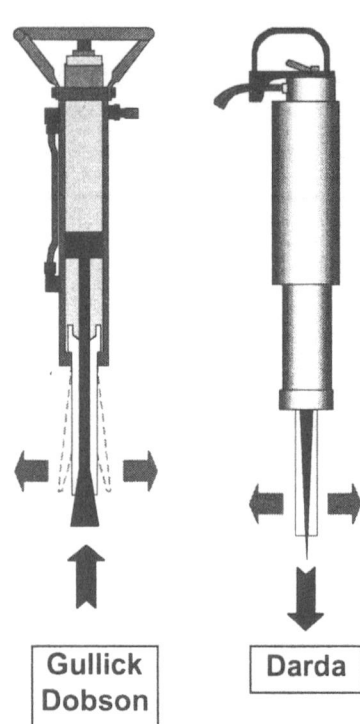

Figure 3. Darda and Gullick Dobson hydraulic bursters

Being founded on the principle of a wedge action, these tools are only efficient with hard and brittle materials (photo 2). All of them require pre-drilled holes.

Roc-Jac develops a radial load of about 1750 kN while Darda and Roc-Pac are able to produce a bursting load of 3500 kN. Less powerful, the Gullick-Dobson burster is designed to develop 1000 kN.

These tools are economical. They allow the cracks to be controlled and they are fast in their use. They are not noisy and they do not create dust, projections or vibrations. They are light and easy to handle.

Their main drawback is that they require a pre-drilled hole of a rather good quality and, in the case of reinforced concrete, the reinforcing bars limit their action. It is to be noted that Roc-Pac is able to work in a hole obtain with a thermal lance (§ 1.3).

Photo 2. Hydraulic burster in action (Courtesy Zschokke)

Diamond tools. A large range of diamond tools is available: from drills to cores, including circular, linear or reciprocating saws as well as diamond coated wires.

Circular saws, with diamonds set in the blade, can cut concrete to a depth of about 600 mm (photo 3). The diameter of the tools varies from 300 mm to 1.80 m with thickness from 1.2 to 20 mm. A diamond saw is a thin steel disc set with diamonds on the tungsten-carbide matrix of the rim so as to cut stone, concrete, steel or other hard material. Automatic advance is generally provided to obtain an efficient and safe displacement in the cutting.

Photo 3. Diamond circular saw in action (Courtesy Clipper Norton)

It is to be noted that a reciprocating saw exists which can cut up to 1.2 m concrete depth (photo 4). It provides a very high quality of linear cutting. To operate, it requires only a preliminary hole to be drilled before its installation, whose diameter must be larger than the width of the linear saw. A detail of the blade is shown on photo 5.

Photo 4. Reciprocating saw in action (Courtesy Zschokke)

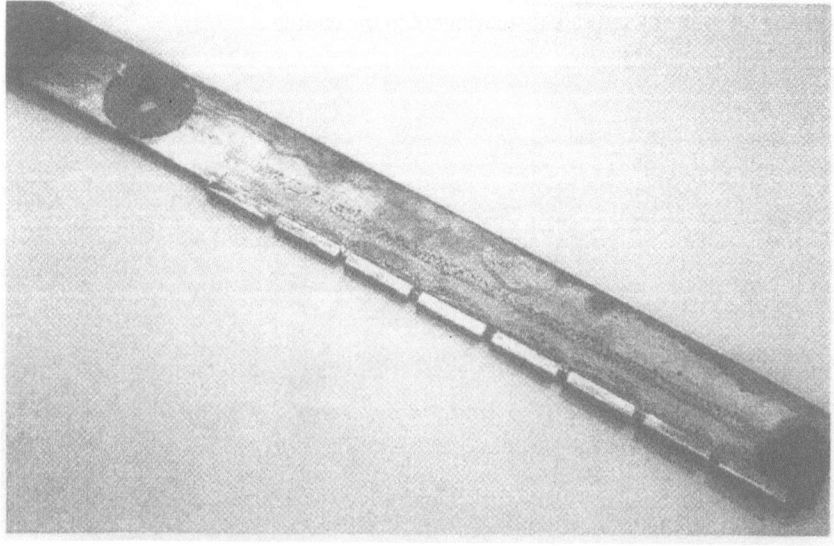

Photo 5. Detail of the linear reciprocating saw blade (Courtesy Zschokke)

Diamond drilling, with cores of which diameters are up to 600 mm and lengths from 150 to 850 mm, allow any attack angle (photo 6). Using this method, it is possible to cut rectangular openings obtained by tangent holes (photo 7). The equipment must be driven automatically. The depth of drilling may be more than 3 m.

Photo 6. Drilling with a severe attack angle (Courtesy Zschokke)

Photo 7. Drilling of an opening in a concrete floor, with a diamond core (Courtesy Zschokke)

Nowadays, diamond coated wires are more and more used and preferred to the reciprocating saw. This equipment consists in a steel wire supporting regularly spaced-coated diamonds (Figure 4). It is moved by a powerful main driving wheel and the cutting is obtained by a system maintaining the wire in sufficient tension (Figure 5).

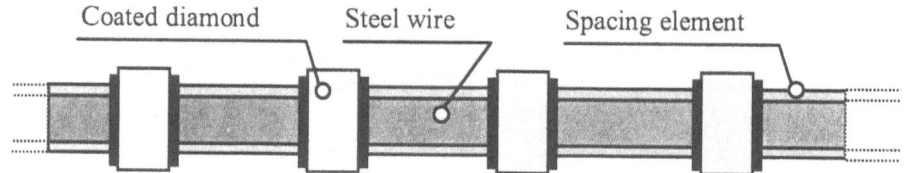

Figure 4. Diamond coated wire (AFFSB, 1995)

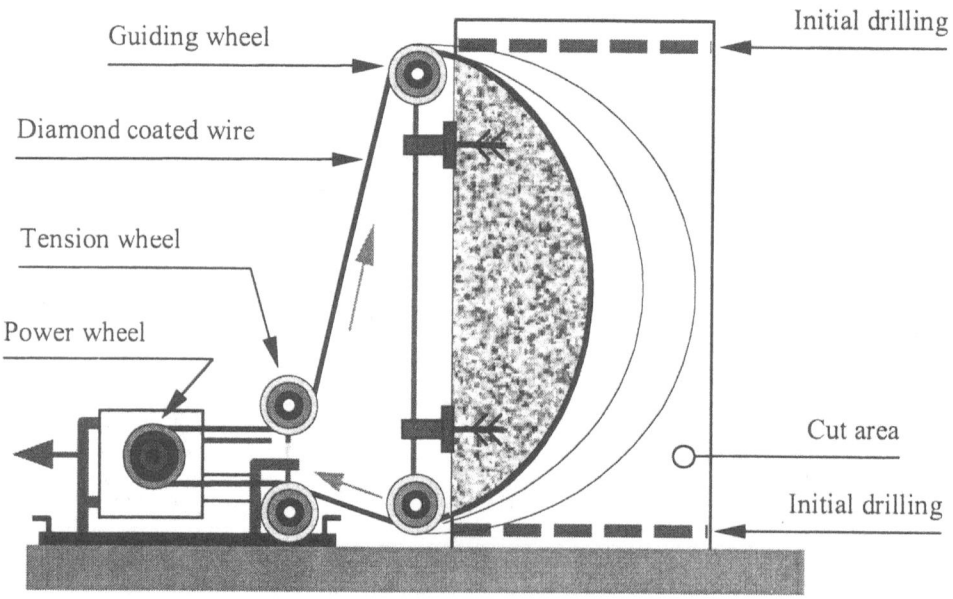

Figure 5. Diamond coated wire system (GE Superabrasives, 1991)

This process allows structures to be cut at high speed. Founded on the use of a wire which can be as long as necessary, there are practically no limitations in the depth of the cutting or in the dimensions of the structures which can be treated in this way. The installation of several intermediate wheels allows the direction of the wire to be changed. So, the cutting can be carried out in any positions (Figure 6).

Diamond tools can cut steel bars as well as concrete itself but all require cooling by a regular water spray (about 5 to 20 l/min). This process is safe, fast and efficient but also rather expensive. It is not noisy and it produces a work of high quality without vibrations or impacts.

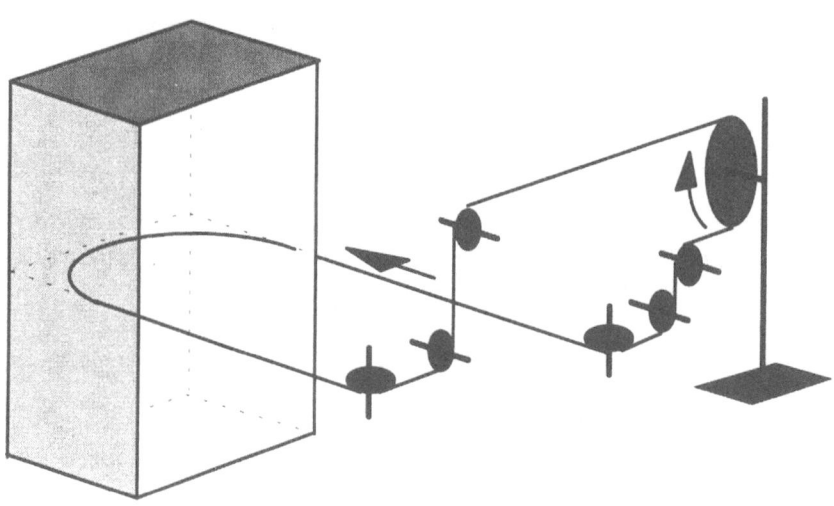

Figure 6. Example of allowed possibility with diamond coated wire (AFFSB, 1995)

It is possible to choose the size of the debris depending on the transport possibilities. It does not change the mechanical characteristics of the remaining part of the structure. A careful choice of the tools is required to avoid a rapid loss of efficiency and a rigid connection to a stable support is needed. Most of diamond tools are considered to be very efficient in partial demolition.

Jet blasting. Jet blasting with very high water pressure is a more and more popular process in demolition. Without any additives, a pressure of about 200 to 250 MPa allows water to cut concrete. An addition of quartz or abrasive sand allows reinforced concrete to be demolished.

High precision, no vibrations, no dust and no noise are the main advantages of this process. Obviously, by nature, it requires availability of water on site.

1.2 Blasting technique: Gas-expansion burster

If blasting is frequent in total demolition, one explosive process only can be used in the field of partial demolition: the gas expansion burster.

This method consists in a cylinder inserted into a prepared cavity. Its name is "Cardox process" (Figure 7). The cylinder contains chemicals or liquefied gas, which expand rapidly with a considerable force when ignited by an electrical charge. It causes fracture and fragmentation of the surrounding structural mass.

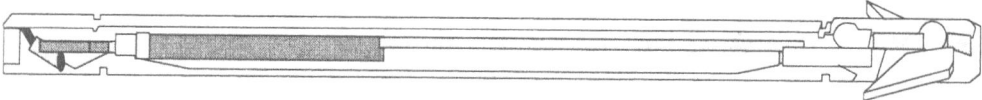

Figure 7. Cardox cylinder

The action being very fast (20 to 40/1000 s), the Cardox cylinder does not create vibrations. It can be used in urban areas but is limited to concrete with light reinforcement.

1.3 Thermal processes

Oxyacetylene torch. This method, well kNown in the steel industry, is used to cut steel sheets or profiles. An oxyacetylene torch is a metal-cutting flame obtained from compressed oxygen and acetylene or other fuel gas in separate steel cylinders.

Thermic lance. Concrete melting between 1500 and 1700°C, it is possible to cut structural concrete and steel members by very intensive heat.

The thermic lance consists in boring holes into concrete using the high temperature of a burning steel tube (lance) packed with iron rods through which oxygen is passed (photo 8). With the oxygen/iron reaction, the temperature is able to reach 2000 to 2500 °C.

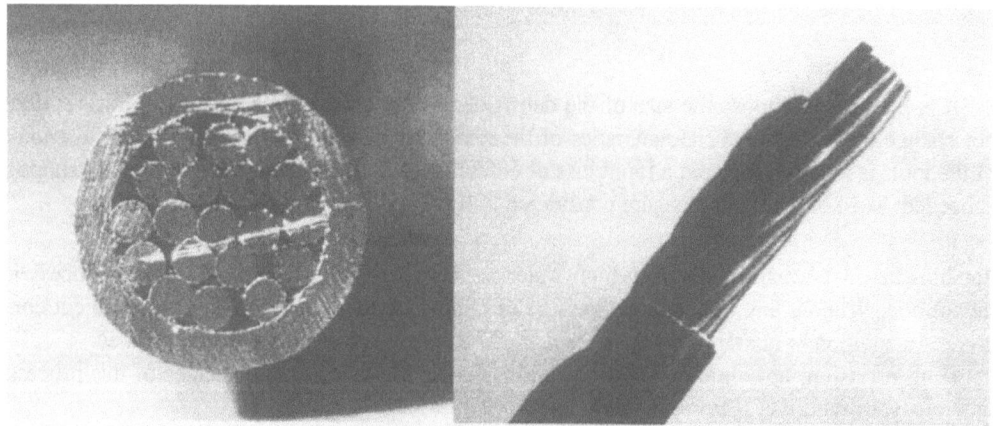

Photo 8. Thermic lance

To start the process, oxyacetylene or similar fuel gas mixture is passed to ignite the end of the lance. After ignition, oxygen is fed, to keep the lance burning.

If concrete is heavily reinforced, it can be cut more easily than non-reinforced, since the reinforcement burns in the oxygen jet (photo 9).

The most common lances are 3 to 4 m long with diameters equal to 13, 17 or 21 mm. The bored holes have diameters between 30 to 60 mm.

This technique is very attractive in the field of partial demolition (photo 10). No noise, no impacts, no vibrations are produced. It does not create cracks in the surrounding concrete. This is a fast process, usable in urban sites without problem. Its allows large thickness of material to be bored. It is to be noted that this technique can also be used under water.

Photo 9. Thermic lance cutting (Courtesy Zschokke)

Photo 10. Result of an opening cut with a thermic lance (Courtesy Zschokke)

1.4 Choice criteria

Following the presentation of the possible techniques available for partial demolition, tables are drawn to help in choosing the most efficient process with regard to different criteria:

- available processes according to the material to be demolished (Tables 1),
- conditions for using particular partial demolition process (Tables 2),
- possibility for using regarding environmental criteria (Tables 3).

Table 1.1. Available processes for masonry, concrete and reinforced concrete

MASONRY	CONCRETE AND REINFORCED CONCRETE	
Low-rise buildings	*Low-rise buildings*	*High-rise buildings*
Hand demolition	Pneumatic hammers	Pneumatic hammers
Pneumatic hammers	Hydraulic burster	Hydraulic burster
Thermic lance	Gas-expansion burster	Thermic lance
Gas-expansion burster	Thermic lance	Diamond tools
Hydraulic burster	Diamond tools	Jet blasting
Diamond tools	Jet blasting	
Jet blasting		

Table 1.2. Available processes for prestressed concrete or steel structures

PRESTRESSED CONCRETE	STEEL STRUCTURES	
	Low industrial buildings	*High rise buildings*
Thermic lance	Hand dismantling	Hand dismantling
	Oxyacetylene torch	Oxyacetylene torch
	Alligator shear	Thermic lance
	Thermic lance	Diamond tools
	Diamond tools	

Table 2.1. Conditions for drilling

DRILLING	Pneumatic drills	Diamond core	Thermic lance
Use	masonry, concrete	masonry, concrete	masonry, concrete, steel
Efficient thickness	up to 2 m	very large	very large
Speed	slow	fast	fast
Cost	economical	expensive	expensive
Labour force	no requirements	very qualified	not very qualified
Manageability	good	medium	good
Possibility of control	possible	excellent	difficult with thickness
Quality of the work	medium	excellent	medium
Requirements	No	water/electricity	no

Table 2.2. Conditions for cutting

CUTTING	Diamond core	Diamond saw	Hydraulic burster
Use	masonry, concrete	masonry, concrete	masonry, concrete
Efficient thickness	very large	up to 1.2 m	variable
Speed	fast	fast	fast after drilling a hole
Cost	expensive	expensive	economical
Labour force	very qualified	very qualified	no requirements
Manageability	medium	medium	excellent
Possibility of control	possible	possible	possible
Quality of the work	excellent	excellent	medium
Requirements	water/electricity	water/electricity	no

Table 2.3. Conditions for cutting (continued)

CUTTING	Oxyacetylene torch	Thermic lance
Use	steel	masonry, concrete, steel
Efficient thickness	rather small	very large
Speed	fast	fast
Cost	economical	expensive
Labour force	no requirements	low qualification
Manageability	good	good
Possibility of control	possible	difficult with thickness
Quality of the work	medium	medium
Requirements	no	no

Table 3. Criteria for using regarding environmental criteria

CRITERIA	Pneumatic hammers	Diamond tools	Hydraulic bursters	Oxyacetylene torch	Thermic lance
Use in urban areas	Possible	Possible	Possible	Possible	Possible
Risk for workforce	no	no	no	no	no
Risk of fire	no	no	no	yes	yes
Smoke production	no	no	no	no	high
Dust production	yes	no	no	no	low
Projections	low	no	no	fluid slag	fluid slag
Vibrations	high	no	no	no	no
Noise	significant	negligible	no	negligible	negligible

1.5 Conclusion

Refurbishment generally requires some modifications to the existing structure to allow for the change of use (openings, modification of the position of walls or of floors, or extensions for example).

Techniques of partial demolition depend on the kind of material to be demolished or simply cut. Three main types can be distinguished (mechanical, blasting techniques and thermal processes). Each of them is closely linked to different specifications and it is necessary to kNow their advantages and disadvantages to choose the most efficient available process.

2 The Use of Special Fasteners in Refurbishment[2]

In refurbishment activities, the accessibility conditions are sometimes very restrictive and may influence the choice of technical solutions. It is particularly the case for connections.

On the one hand, the use of high-strength bolts under torque control may be difficult. As an alternative, a new kind of fastener is available, easier to be installed. Its name is Huck-Fit bolt. It possesses main geometrical and mechanical properties equivalent to those of high-strength bolts grade 10.9 (noted HS) but with a completely different installation procedure: it is obtained by swaging a collar on a pin. It has to be noted that Huck-Fit bolts, as HS bolts, can also be used as classical shear or tension bolts of high steel grade (ignoring the preload).

On the other hand, blind bolts can be advantageous to connect structural elements accessible from one side only and, obviously, for hollow sections. Some of them are also based on an installation sequence involving a swaging process and they offer relatively high strengths. These are named BOM and HSBB fasteners. To complete this description, the Ultra-Twist system, being an efficient improvement of the HSBB bolt, is also detailed.

All these fasteners are described in this chapter paying attention to their technological characteristics. All offer real potential for use in refurbishment as well as in new structures where their use is already well established in a number of countries.

---

[2] This paragraph is co-authored with Dr. Michel Dréan, Lecturer, IFMA, and Dr. Abdelhamid Bouchaïr, Senior Lecturer, CUST, Blaise Pascal University, both located at Clermont-Ferrand, France.

2.1 The Huck-Fit bolts

Huck-Fit bolts are swaged bolts constituted by two different elements: a pin, with a head (Figure 8), and a collar (Figure 9). Only the main types are shown here. More detailed information are given in (Czarnomska *et al.*, 1993).

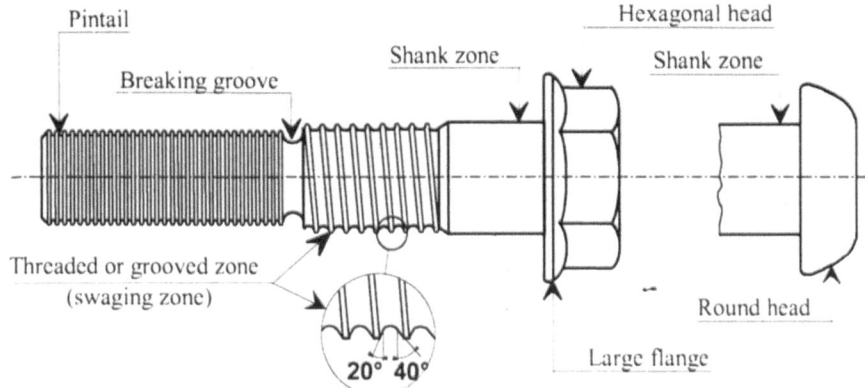

Figure 8. Huck-Fit bolt components: pin

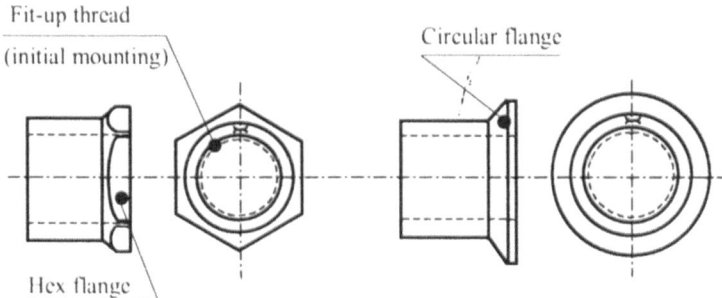

Figure 9. Huck-Fit bolt components: collar

The available diameters are Ø 12, 16, 20, 24 and 27 mm. Helical grooves, hexagonal head of the pin and of the flange of some collars provide the possibility of removing the bolt later, if required, using standard bolting tools.

Installation process. The installation process consists in swaging the collar on the pin with a special hydraulic tool system. The internal diameter of the nose assembly being smaller than the external diameter of the collar, it forces the collar material to flow into the pin grooves. The sequence of the installation process can be divided into four steps represented in Figure 10.

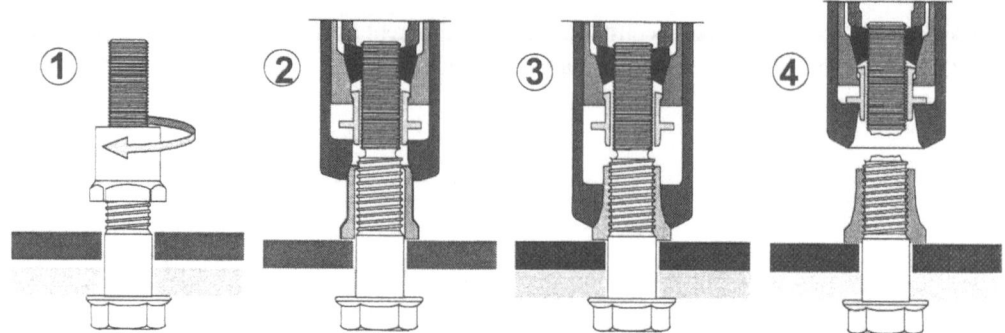

Figure 10. Installation process

1. The pin is inserted through a prepared hole. The collar is threaded onto the pin.
2. The nose assembly is placed over the pintail. It pulls on it, drawing the work pieces together. The anvil pushes on the collar.
3. Continued pulling on the pintail moves the anvil forward, swaging the collar into the lock grooves. Controlled pulling and swaging lengthens the pin and the collar to develop a permanent clamp force in the pin.
4. When swaging is completed, the pintail separates from the pin and the tool disengages the swaged collar from the anvil.

The tool is light and easy to handle. It is more compact than many powered tightening tools of similar form. Installation takes only a few seconds, so that much time is saved as compared to traditional torque tightening.

Properties of installed Huck-Fit bolts. The general properties of swaged bolts are the followings:

- swaging of the collar on the pin creates the second head of the system. Lock grooves in the swaging area lead to a resistant and durable link.
- the ductile steel of the collar is swaged into the locking grooves. The installed fastener is comparable to a screw-nut system without gap. So, in the case of vibrations, the behaviour of swaged bolts is much more efficient than the one of HS bolts: the rotation of the collar being quite impossible.

The helical grooves look like a thread but with a shape constituted by a succession of linear and elliptical parts (fairly different from a classical bolt). The groove shape with angles equal to 20° and 40° is not as severe as in a classical thread (Figure 11). As the resistant section is greater and the stress concentration factor is lower, it provides an excellent behaviour in fatigue.

The ultimate resistance of the constitutive steels (collar ≈ 450 MPa and pin ≈ 1000 MPa) associated to cross-sections of the shear areas such as: $Y_1 \approx 2\,Y_2$ give a shear resistance of the collar slightly smaller than the one of the pin. So, in a bolt loaded in tension, the ultimate resistance is the shear stripping resistance of the collar grooves and not the tension resistance of the threaded cross-section of the pin. This behaviour leads to a reasonably ductile behaviour and not to the rather brittle one of classical HS bolts.

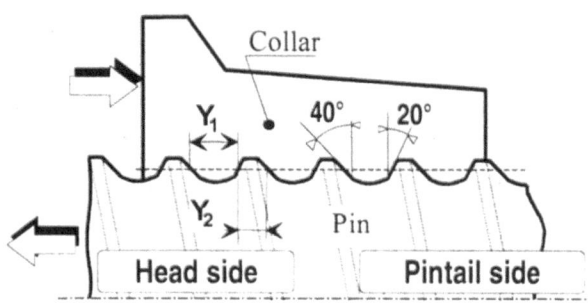

Figure 11. Groove shape

Because of the groove shape, the resistant cross-section of the pin is larger than the one associated with a classical thread. The stripping resistance of the collar is also greater than that of the threaded section of a classical bolt. So, with the same steel grade, the tension resistance of the Huck-Fit bolts is larger than the one of HS bolts. Moreover, the Huck-Fit bolt is such that shear of the bolt occurs almost always in the shank (unthreaded) zone. All these properties give a mechanical behaviour to the Huck-Fit bolts which is always larger than the one of high-strength bolts grade 10.9 of the same diameter.

Finally, because of the installation procedure is controlled by the failure of the breaking groove acting as a fuse, and because of the shape of the bolt itself, it is possible to check if the swaging action is correctly performed only by a visual verification.

Numerous tests have been carried out to clarify, on the one hand, the mechanical properties of Huck-Fit bolts and to analyse, on the other hand, their behaviour in shear, tension and moment (end-plate) connections (Czarnomska et al., 1993), (Baptista et al., 1997), (Muzeau et al., 1995), (Aribert et al., 1994).

Design and characteristic resistance. Both Huck-Fit bolts (Ryan, 1994) and HS bolts can be used as:

1. in shear:

   - either as pre-loaded bolts in slip-resistant connections;
   - or as classical bolts grade 10.9 designed to resist to a shear action on the pin (considered as non-preloaded);

2. in tension:

   - either as a preloaded bolt;
     or considered as a non-preloaded bolt in tension;
   - in combined shear and tension as in moment connections for instance.

A comparison of the cross-sectional area of Huck-Fit (HF) and high-strength (HS) bolts is given in Table 4 where A is the cross-section area of the bolt shank and A_s the one of the threaded or grooved part.

Regarding the resistance according to Eurocode 3, the design pre-loading force is :

- HF bolts: value guaranteed by the manufacturer for special fasteners (Huck, 1990);
- HS bolts: $F_{p.Cd} = 0.7\ f_{ub}\ A_s$;

and the characteristic shear resistance:

- threaded part: $F_{v.Rk} = 0.5\ f_{ub}\ A_s$;
- non threaded part: $F_{v.Rk} = 0.6\ f_{ub}\ A.$

Table 4. Cross-sectional areas of HF or HS bolts (mm²)

Bolt	A	A_s^{HF}	A_s^{HS}
M12	113	98	84.3
M16	201	176	157
M20	314	275	245
M24	452	396	353
M27	572	503	459

Table 5. Compared pre-loading and characteristic shear resistance of HF or HS bolts (kN)

Bolt	$F_{p.Cd}^{HF}$	$F_{v.Rk}^{HF}$ shank zone	HF bolts $F_{v.Rk}^{HF}$ grooved zone	$F_{p.Cd}^{HS}$	$F_{v.Rk}^{HS}$ shank zone	HS bolts $F_{v.Rk}^{HS}$ threaded zone
M12	62.5	70.1	50.7	59.0	67.8	42.1
M16	112.5	124.8	91.1	109.9	120.6	78.5
M20	184.5	195.0	142.3	171.5	188.4	122.5
M24	254.0	285.2	204.9	247.1	271.2	176.5
M27	330.5	355.2	260.3	321.3	343.2	229.5

The value of f_{ub} for the HF bolt is taken as 1035 MPa, the value guaranteed by the manufacturer. For the high-strength 10.9 bolt, the value of 1000 MPa is used as required by Eurocode 3 (Table 5).

In the same way, the characteristic tension resistance is (Table 3):

- HF bolts: value guaranteed by manufacturer (Huck, 1990);
- HS bolts: $F_{t.Rk} = 0.9\ f_{ub}\ A_s$.

Table 3. Characteristics tension resistance of HF or HS bolts (kN)

Bolt	HF $F_{t,Rk}$	HS $F_{t,Rk}$
M12	88.2	75.9
M16	163.6	141.3
M20	255.4	220.5
M24	368.2	317.7
M27	478.2	413.1

The design resistance is given by $F_{Rd} = F_{Rk}/\gamma_{Mb}$. While Eurocode 3 proposes a value of the partial safety factor $\gamma_{Mb} = 1.25$, each country may define its own one. For resistance, in France for instance, a value of $\gamma_{Mb} = 1,5$ is required for bolts in tension, while the value $\gamma_{Mb} = 1.25$ is maintained for bolts in shear.

2.2 Blind swaged fasteners

Two kinds of blind bolts which can be used in structural steel design, have been developed as swaged bolts. They are named BOM and HSBB (Korol *et al.*, 1993).

The blind oversized mechanically fastener. Up to now, this type of fastener has been used mostly for the mechanical, road, rail and other transport industries. Nevertheless, it offers real potential for use in steel construction.

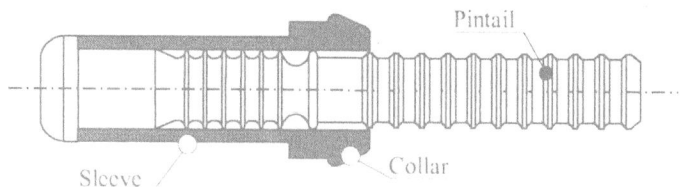

Figure 12. Shape of a BOM fastener

The blind oversized mechanically fastener (BOM) has been developed many years ago. It is formed by a pin and a sleeve (Figure 12). Its installation procedure is shown on Figure 13. After the insertion of the bolt into the prepared hole from the accessible side (a), through the pulling action of the installation tool, the sleeve collapses and a bulb is formed in it on the blind side (b). The pulling action of the tool continuing, the collar is swaged (c) and the pintail breaks (d).

This bolt is available in different diameters, in inch sizes, from about 5 mm to 19 mm. Its tension capacity is close to 80% of the one of a high-strength bolt grade 8.8, the failure occurring by shear off of the bulb created in the sleeve. Due to the heat treatment of the pin, its shear resistance is much higher (about twice that of a HS bolt grade 8.8). As a consequence, this bolt is potentially well suited for shear connections.

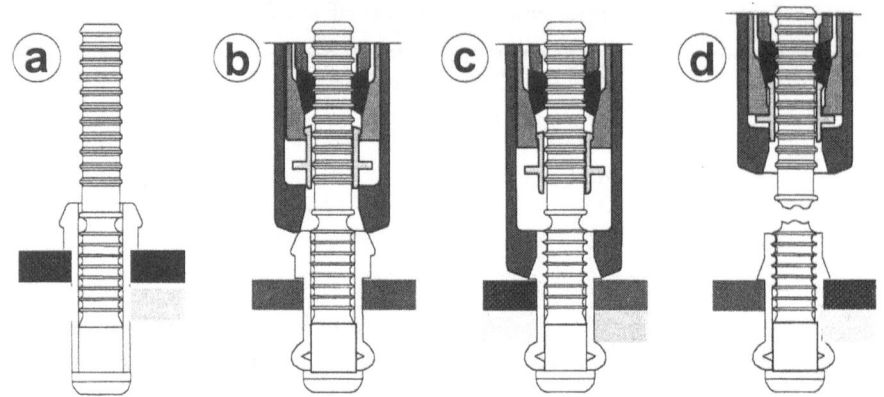

Figure 13. Installation of a BOM bolt

The high strength blind bolts. The high strength blind bolt (HSBB) has been developed more recently and is designed to provide a clamp load action close to the one of a HS bolt, grade 8.8. Its different components are shown on Figure 14.

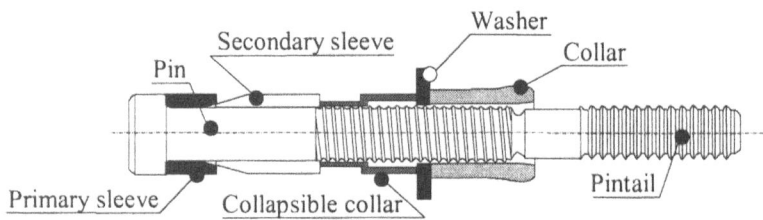

Figure 14. Components of a HSBB

The installation process of HSBB leads to the initial and final configurations shown on Figure 15. It consists in forcing the primary sleeve to cover the secondary one and to create the "head" of the blind side of the fastener. Then, the collar is swaged on the grooves before failure of the pintail.

Table 4. Comparison of characteristics tension resistance (kN)

Bolt	Ø	$F_{p.Cd}$	$F_{t.Rk}$	$F_{v.Rk}$
HS grade 8.8	M20	130	168	110 - 126
BOM	M19	40	116	171
HSBB	M20	130	173	78

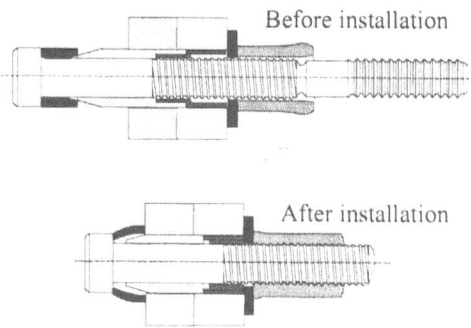

Figure 15. HSBB before and after installation.

The main results of tests carried out on that bolt (Korol et al., 1993) are given in Table 4 in comparison of those provided by similar HS bolts grade 8.8 (according to Eurocode 3) and BOM. As it can be observed, this bolt provides an efficient resistance in tension but that its shear capacity is reduced due to a smaller pin diameter.

2.3 The Ultra-Twist blind bolt

As an efficient improvement to the previous fastener, a new device rather close to the previous one was recently developed (Sadri, 1995), (Huck, 1994). Its general shape is shown on Figure 16. The swaged collar of the BOM fastener is replaced by a nut on the pin.

The installation sequence is presented on Figure 17. The bolt is installed from the accessible side in the prepared hole (a). It is engaged in a standard electric torsion wrench tool used for installation of Twist-Off control (T-C) type fasteners. The backside bulb is formed (b). As the installation load increases, the shear washer shears allowing the backside bulb to come into contact with the work surface of the back side (c). Torquing of the installation tool completes the installation (d). It creates the clamp force and the torque pintail shears off.

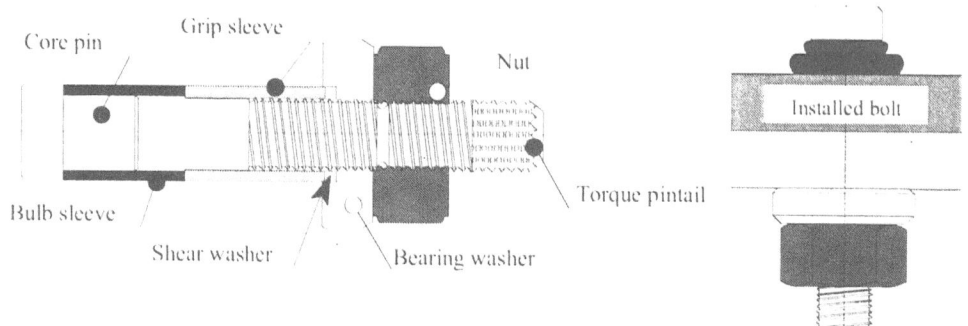

Figure 16. Ultra-Twist blind bolt.

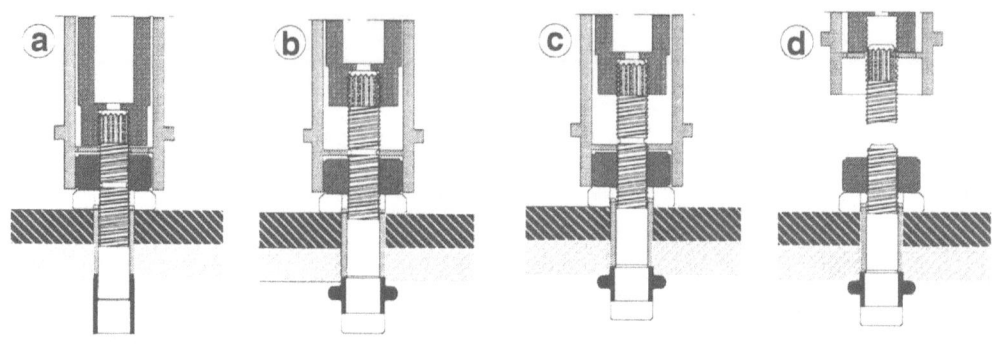

Figure 17. Ultra-Twist blind bolt installation.

Table 5. Characteristics tension resistance of Ultra-Twist blind bolts

Ø (mm)	A_s (mm²)	$F_{p.Cd}$ (kN)	$F_{t.Rk}$ (kN)	$F_{v.Rk}$ (kN)
20	169	138.04	182.7	121.8
24	235	191.76	253.8	169.2
27	321	261.80	346.5	231.0

This bolt is not a swaged bolt but as it retains the essential characteristics of the BOM fastener, it appeared interesting to include it here. It is a blind bolt whose mechanical characteristics calculated with the relationships in § 1.3, are given in Table 5. A comparison with Table 4 shows that this new kind of fastener represents an efficient improvement both in tension and in shear resistance, over the BOM version. Being based on the use of a standard tool used for the installation of T-C bolts, one of the other advantages provided by this new blind bolt is that it does not require the special tool for swaging.

2.4 Conclusion

Swaged bolts can be an efficient alternative solution for bolted connections in refurbishment. Their installation being based on the failure of the breaking groove, it is usually relatively easy to install in difficult access conditions.

Blind bolts present mechanical characteristics close to the ones of high-strength bolts grade 8.8. Some of them are swaged, while another kind is installed under torque-control effect. All of them are an interesting alternative to classical fasteners for connection of hollow sections and for any situation where access is possible from one side of the work surface only.

They can be used as classical sheared and/or tensioned bolts and they may also be used as pre-loaded bolts. In case of fatigue or seismic actions for example, this possibility is to be considered as a very attractive way to avoid significant variations in the axial load in the bolts.

When refurbishment of steel structures is to be designed, it is important to be aware of the capacity, limits and potential difficulties for all the available fasteners so as to permit the choice of the most efficient device. Numerous technical problems due to site conditions are always better solved when the possibilities of the choice are as large as possible.

3   Improvement of Existing Connections[3]

Refurbishment may require to reinforce existing connections, or to create new supports. That can be achieved by transforming the joint category (hinged to fixed...), by adding new elements (higher resistance...) or by changing the type of fasteners (upgrading the bolts...).

The creation of new supports depends strongly on the type of structure to be refurbished.

If it is generally possible to transform a hinged connection into a fixed one, it must be strongly emphasised that when the bending resistance of the connection becomes higher, that creates a deep modification in the load transfer requiring a careful check of the whole structure.

It is often possible to reinforce a connection by addition or by modification of existing elements. For instance, new stiffeners may be installed or additional plates may be used to increase the thickness of a flange or a web.

Two main solutions exist to improve the fasteners: to use devices with improved characteristics (larger diameter, higher grade...) or to use other kinds of fasteners (HS bolts instead of black bolts or rivets for instance, injection bolts...). The gluing of steel elements to reinforce a joint is not considered here but it is a very valuable solution!

This paragraph contains a discussion about the type of fasteners, which can be used to reinforce a connection. Then, some proposals of connections to be used in the case of existing walls or columns made of concrete or steel, are given and only connections to existing members are considered (this presentation does not refer to a new self-bearing structure included behind or inside an old one). Refurbishment of masonry structures is analysed in Chapter 4 and is not considered here.

3.1   Which kind of fasteners can be used?

Most of classical fasteners are efficient in refurbishment:

- rivets (if they are needed but preferably installed at the workshop),
- black bolts (or ordinary bolts),
- HS bolts (but the friction grip cannot be considered if the surface of the elements cannot be treated) or injection bolts,
- welding on site as well as at the workshop,
- TC bolts,
  self-drilling screws for thin-walled members.

The type of fasteners or techniques described in the previous paragraph may also be an alternative:

[3] This paragraph is co-authored with Prof. Miklós Iványi, Dr. László Hegedûs & Balázs Vásárhelyi, Department of Structural Engineering, Budapest University of Technology & Economics, Hungary.

1. Huck-Fit bolts,
2. blind bolts:

   – BOM,
   – HSBB,
   – Ultra-twist bolts,

On the one hand, the choice of the fastener type depends on the existing material. A connection to a concrete element is only possible with threaded elements connecting new supports for a new structure. A connection to steel members may be welded or bolted but it is generally easier on site to weld new elements on a steel member than to drill holes.

On the other hand, the choice of the fastener type depends also on accessibility. If a full accessibility exists, any kind of fasteners can be used depending on the material. In case of restricted accessibility, welded elements may be a valuable solution but cares have to be taken against fire. Blind bolts are efficient if it is possible to drill a hole and if the material behind the accessible element allows the deformation of the second head to be developed. Blind bolts are also interesting in case of hollow sections for instance.

3.2 Proposal for connections

Some proposals for typical beam-to-column or beam-to-wall connections are given here. They concern only the cases where a steel beam is connected to a concrete or steel column (or wall).

Reinforcements of existing connections always need to be designed. Care has to be taken to avoid to mix different kinds of fasteners (welds and bolts for instance) of which mechanical behaviours are not compatible.

Connections to a concrete element (wall or column). Because the reinforcing bars of an existing reinforced concrete element are difficult to be measured and because it is rather difficult to increase them, hinged supports are generally acceptable to connect a new beam. So, five hinged supports are presented against only one transmitting bending forces.

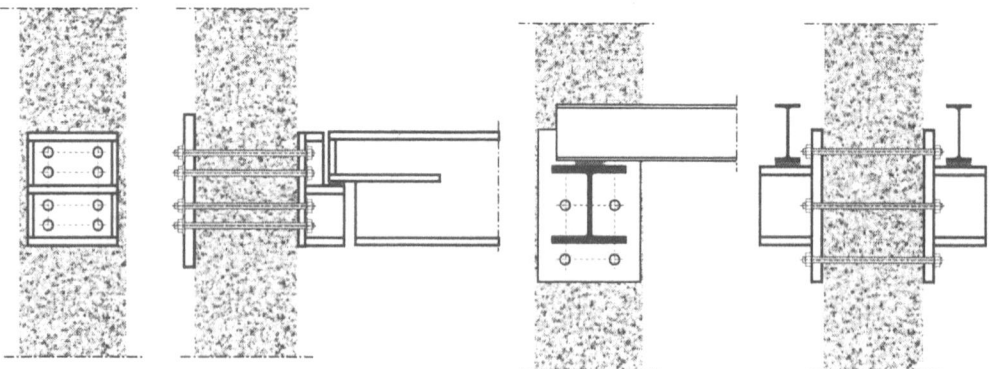

Figure 18. Hinged connection to a concrete wall.

Figure 19. Hinged connection to a concrete wall.

Figure 18 represents a hinged connection to a concrete wall or a concrete column. In this case, the steel beam is then simply supported. The special bolted corbel acts as a simple support. This type of connection needs sufficient resistance in the concrete. This needs also access to the opposite face of the concrete element.

A slightly small bending moment exists so the concrete element has to be checked against this load case. The height of the corbel has to be chosen close to the beam depth.

Figure 19 shows another kind of hinged connection attached to a concrete column. Here, two steel beams are used and they are simply supported.

This connection needs sufficient resistance of the concrete column under the added axial load.

This design may be required when it is not possible to access to the external face of the concrete column or if the access to the internal face is also not possible.

Because of the double beam set-up, the beams are smaller than when using only one profile; the column is symmetrically loaded and no bending moment has to be considered.

The lower I section support allows easy installation, and easy access. The height of the connection is much higher than the beam depth itself and it is visible. That must be acceptable.

If required, the height of the support can be increased to create a stronger bearing.

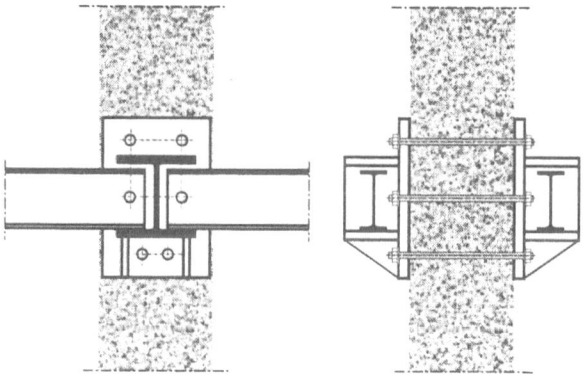

Figure 20. Hinged connection to a concrete wall.

Figure 20 presents a third type of hinged connection attached to a concrete column.

Compared to the previous one, the total height of the connection is smaller because the new steel beams are installed on the bottom flange of the corbel. The lower stiffeners are generally necessary to increase the bearing capacity of the supporting element.

As previously, because of the double beam set-up, the beams are smaller than when using only one profile and the column is symmetrically loaded. So, no bending moment has to be considered.

Figure 21 represents a fifth hinged connection to a concrete column or wall.

It requires a rather good quality in the concrete because of the large number of holes to be drilled. This connection transmits mainly shear forces.

When rather small loads have to be connected, a single angle may be sufficient but the symmetry is lost).

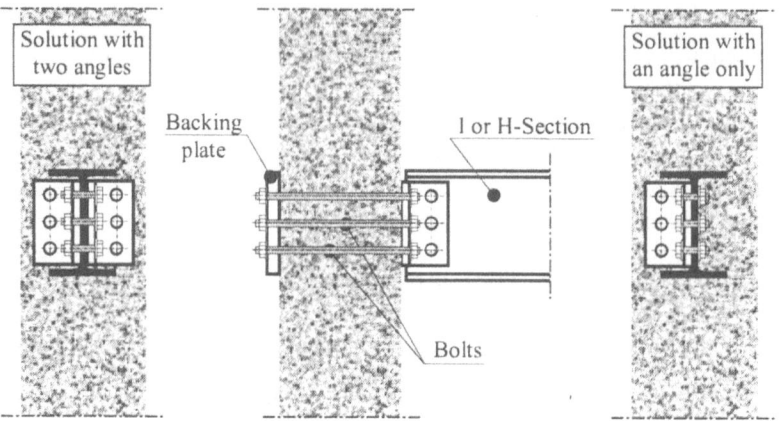

Figure 21. Hinged connection to a concrete wall.

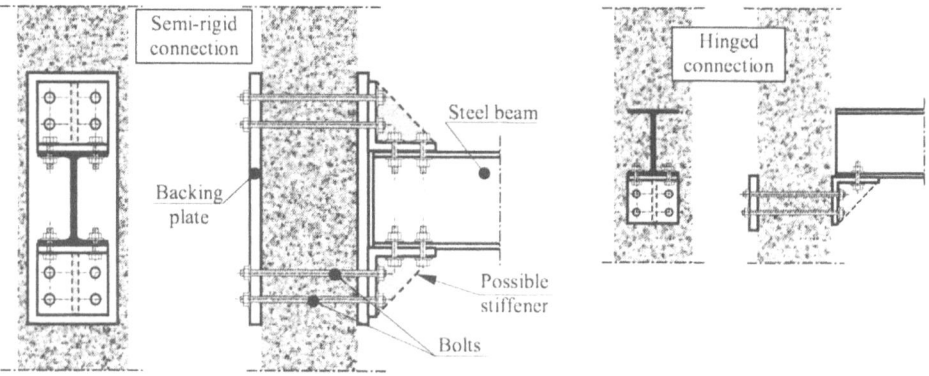

Figure 22. Semi-rigid connection to a concrete wall.

Figure 22 shows a semi-rigid connection of a beam to a concrete wall or a concrete column. It represents an easy solution and its design and installation are rather simple.

The height of the connection depends mainly on the beam depth. Its capacity may be limited to the angle rigidity, which can be improved with its thickness or with stiffeners.

This solution requires the access to the opposite face of the column. Connections where only the lower angle exists are simple connections.

Connections to a steel column. The design of new connections to existing steel members is not very different from classical connections used in new structures. In the case of refurbishment, restrictions due to accessibility may influence the choice of the connection. Possibilities to weld or to drill holes are also criteria to be taken into account to design the connection.

Figure 23 presents a support welded to the existing column. This is a hinged connection where stiffeners may be required in the column

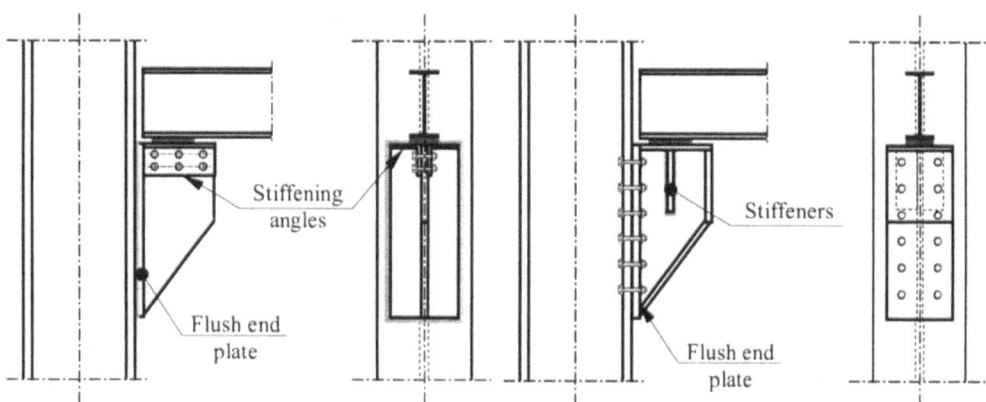

Figure 23. Support welded to an existing column.   **Figure 24.** Support bolted to an existing column.

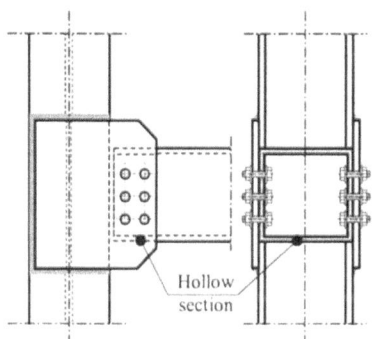

Figure 25. Semi-rigid beam-to-column connection.

Figure 24 shows a type of connection similar to the previous case but when the support is bolted to the existing column. This is also a hinged connection.

Figure 25 shows a semi-rigid connection. Blind bolts are required if the beam flanges are not opened to access for the nuts and the tightening of the bolts. The connecting plates are welded to the existing column. It is indeed possible to design plates bolted to the column.

Figure 26 represents the improvement of the stiffness of a connection by welding of stiffeners in the column web. Figure 27 proposes another type of solution when the column web is not sufficient.

Figure 28 presents a bolted semi-rigid connection using angles.

If it is more common to use an end-plate welded to the beam and bolted to the column, accessibility conditions may lead the present type of connection. It is the case, for instance, when the distance between two columns has to be adjusted. When a gap is required, the present solution is acceptable.

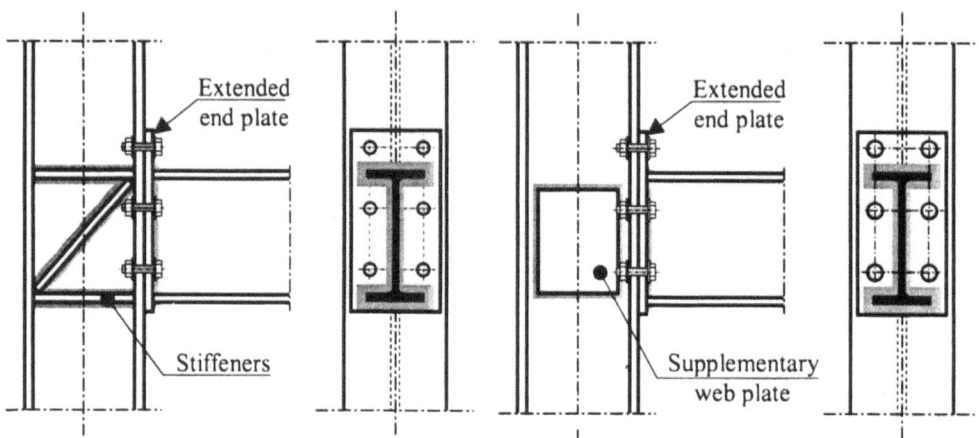

Figure 26. Improvement of a connection stiffness.   **Figure 27**. Improvement of a web resistance.

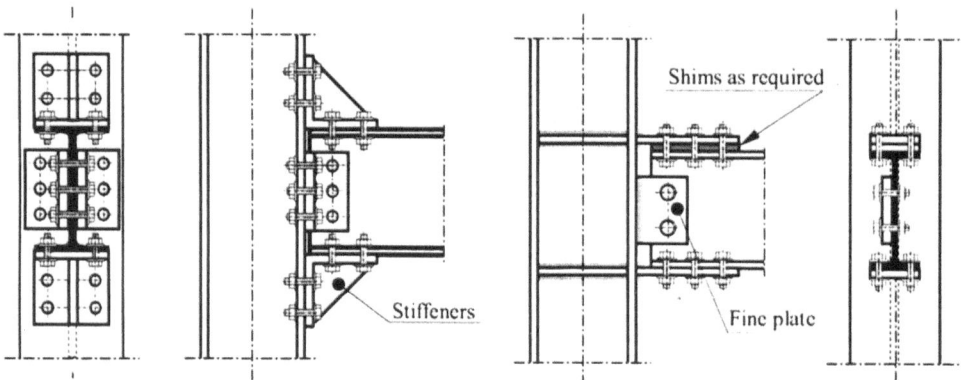

Figure 28. Bolted semi-rigid beam-to-column connec-   **Figure 29**. Welded semi-rigid beam-to-column
tion.   connection.

This solution represents also the case where the angle web existed (the connection was previously a simple hinged connection). In that case, the angles would create a more rigid connection.

Figure 29 is another solution to transform a hinged connection (web angle only) to a rigid connection. Again, depending on the thickness of the flange plates, this connection may act as a semirigid connection.

3.3 Conclusion

In all cases, when connections are modified, the structure must be carefully checked, locally and globally, to insure that it is able to resist to these modifications: to increase the resistance of a connection acts on its stiffness and the load distribution may be strongly affected. In other words, the improvement of a connection may strongly affect the mechanical behaviour of the whole structure and it may not necessarily be an improvement for the structure itself!

Many solutions exist to create or to reinforce connections in existing structures. The main difficulty is to be sure that the structure is able to accept any additional internal forces. If not, a solution with a self-supported structure has to be designed.

4   Two Examples of Refurbishment at Blaise Pascal University[4]

Blaise Pascal University owns different buildings built in several places in Clermont-Ferrand. Its main location contains administration but also several facilities for students, some laboratories, libraries, lecture theatres and teaching rooms. To increase its capacity, it was necessary to transform two rather old buildings into more convenient structures. The "Manège" has been refurbished to create teaching and meeting rooms and to extend a restaurant created some years ago. The second one, the "Paul Collomp" building has been refurbished and completely reorganised with a new lecture theatre, a library, an extra floor, new spaces to circulate and to stand and lecture rooms. Both were designed with the smallest possible weight and with a high level of transparency.

4.1 The "Manège" building

Before. The "Manège" building is a dressage arena built in the 1860s. It was one of the main constructions of the old military quarter in Clermont-Ferrand. This large building (1320 m^2) had different uses. At first, its name proves that it was designed for riding. Later, it was transformed into a ballroom. After the Second World War, it was given to the university. Then, the large covered surface was divided into different teaching rooms and, recently, its East part has been transformed to be used as a restaurant.

The roof, covered with tiles, is supported by several pre-cast "Polonceau" trusses provided originally. The main rafters are fabricated from timber. The hangers and all the tension members, including the main ties, comprise steel bars. The compressed members are made with cast-iron possessing sufficient cross-section to resist to buckling (Figure 30). Most of the ends of the steel bars are threaded and connected to the frame with a nut allowing a precise adjustment to be obtained. Details of connections ① and ② are shown on photos 12 and 13.

---

[4] This paragraph is co-authored with Dr. Abdelhamid Bouchaïr, CUST, Blaise Pascal University, Clermont-Ferrand, France.

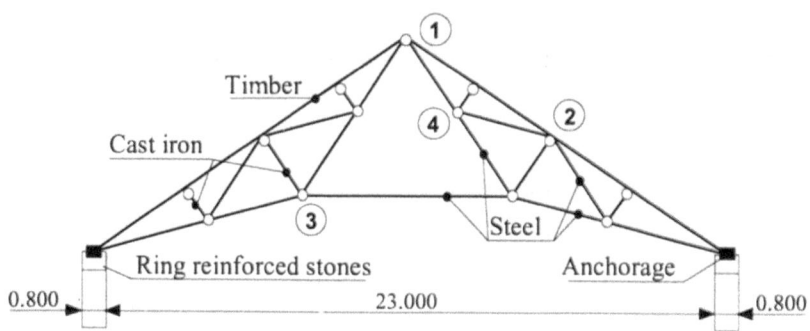

Figure 30. Polonceau truss

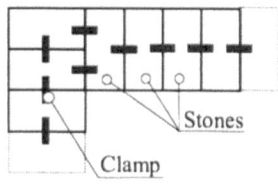

Figure 31. Reinforcement at ordinary courses

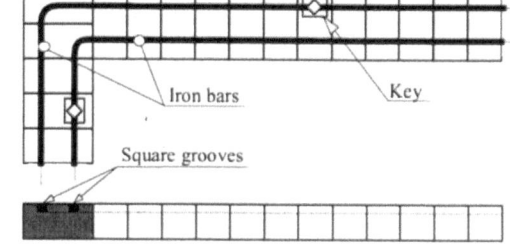

Figure 32. Ring reinforced stones

Photo 11. External view of the "Manège"

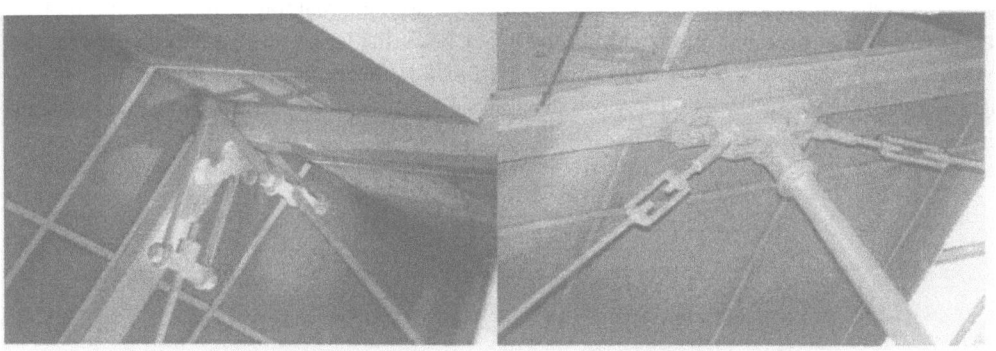

Photo 12. Detail of connection ①   **Photo 13.** Detail of connection ②

The method of construction follows a very old masonry technique using stone reinforcement. If clamps are provided at some ordinary courses (Figure 31), much stronger reinforcement are installed to transform the main stone layers into *ring reinforced stones* acting like ring beams (Figure 32). Square grooves were cut in the stones (hard andesite volcanic lava) at its upper face and iron bars were forged into the grooves. This whole behaves like a corset giving a high resistance to the building even in case of settlement for example.

The façades including very large windows (photo 11), this method of construction provides a strong and efficient resistance to the structure which appeared difficult to change without any risk of cracks or failure.

The refurbishment. The "Manège" has been refurbished to create new and modern teaching rooms (1900 m^2) and to renovate and extend the restaurant (700 m^2). Figures 33 and 34 give the general drawings of the new scheme of the refurbishment.

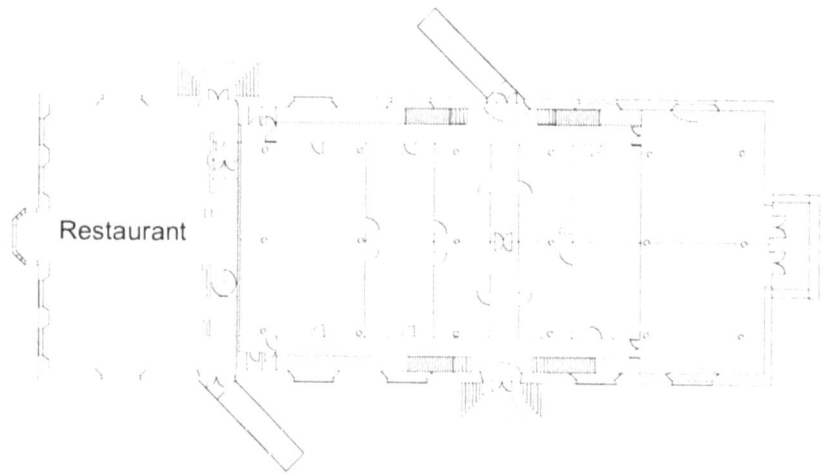

Restaurant

Figure 33. Level 0

The architects have proposed to build a set of superposed internal transparent modules on two levels. A floor has been created to support the upper level, the lower one being organised on the existing one. Because of the system of construction with reinforced stones, it was decided to design the internal structure without any action on the external walls to avoid any changes in their global mechanical equilibrium.

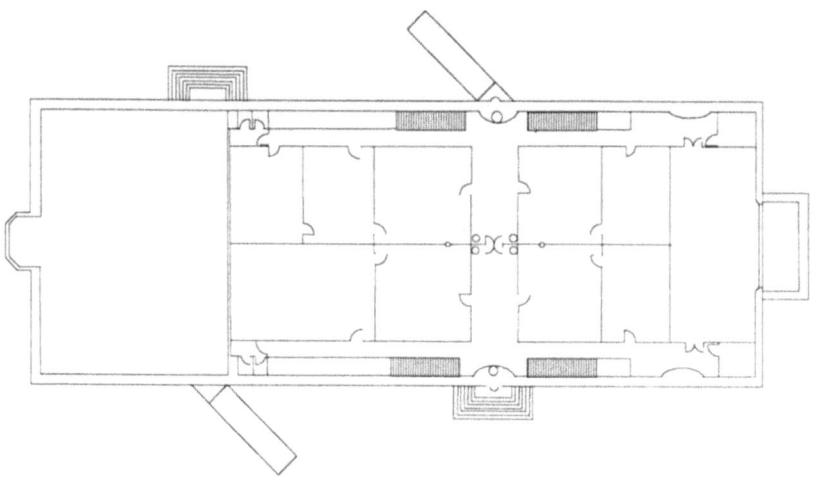

Figure 34. Level 1

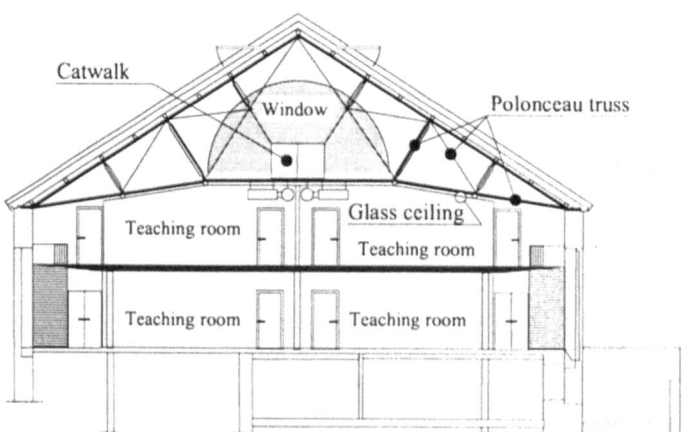

Figure 35. Elevation of the structure

For a steel designer, one of the most interesting ideas is that the "Polonceau" trusses are underlined and kept visible. Acting on the nut of each threaded tension bar, the trusses were readjusted to provide two perfect planes to the roof. The walls of all the teaching rooms and the ceiling of the ones at the upper level are built with glass to create a real transparency of the internal structure (photo 14). So, it is possible to have a general view of the bearing structure from inside the building (Figure 35). This conception provides a real feeling of freedom and it emphasises the lightness of the structure.

Long skylight windows, located in the pitch roof, increase the natural lighting inside the construction.

Internal light steel stairs disposed close to the external walls permit the passage from one level to the other one. Balustrades fabricated in glass panels framed with steel tubular sections are provided, keeping clear the required transparency (photos 15 and 16).

Photo 14. Internal view of the "Polonceau" trusses and upper teaching rooms.

A catwalk was created at the level of the main tie of the "Polonceau" truss and was hung from it. It supports the services including ventilation and smoke extractors. It allows the roof windows to be reached for cleaning purpose. The 780 m² glass ceiling of the internal structure is also supported by the "Polonceau" trusses. Details of modified connections ③ and ④ (Figure 30) are shown on Figure 36 and photos 17 and 18, emphasizing the extra elements designed to support the new structure.

Photo 15. Stairs and lower level.   **Photo 16.** The upper level

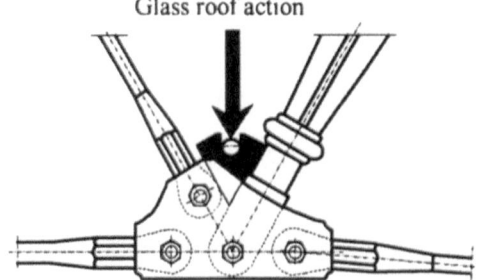

Glass roof action

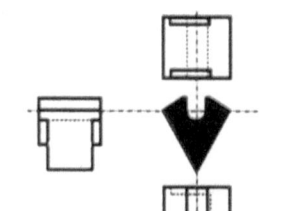

Glass roof supporting element

Figure 36. Detail of connection ③

In this refurbishment, the use of steel associated with glass is an efficient solution to obtain lightness and transparency. The efficiency of the "Polonceau" trusses allows all the support system to be organised without any interference with the existing walls.

Photo 17. Connection ③

Photo 18. Connection ④

4.2 The "Paul Collomp" building

How it was built. The "Paul Collomp" building was built in the 1930s and it was heightened by one storey in 1960. At the beginning, it was designed following *Le Corbusier* and *Mies van Der Rohe's* concept of "the free plane" theory (a skeleton structure constituted by frames supporting floors allowing all the necessary further changes of the internal use to be possible without any major modifications).

This is a reinforced concrete building with four levels including the basement and a flat roof (Figure 37). The whole building is supported by rather light and shallow foundations.

The refurbishment. The project consists of constructing an intermediate floor between the first and the second original ones with a large horizontal opening to provide a new lecture theatre (Figure 38). An additional structure covering the "Cour anglaise" is erected to extend the existing building with its circulation areas (stairs and horizontal distribution with large standing zones). All the existing façades are reorganised with the maximum possible amount of glazing (photo 19).

Because the construction has already been heightened in 1960, it was necessary to minimise the weight of the new construction to limit the extra loads supported by the foundations. That explains why a steel structure with composite floors was chosen.

The architects argued: "Steel design leads to easier connections. It allows site adaptability and the possibility of creating longer spans by welding together shorter elements" (Panthéon J.F. and Saveau J., 1991).

The bearing structure of the lecture theatre is composed of trussed beams supported by reinforced concrete beams of the original structure (Figure 39). The new intermediate floor, carried out by a grid of steel beams, is constructed in the main building (Figure 40). It is connected to the columns by the help of special bolted corbels (Figure 41). The general arrangement of this new intermediate floor is shown on Figure 42.

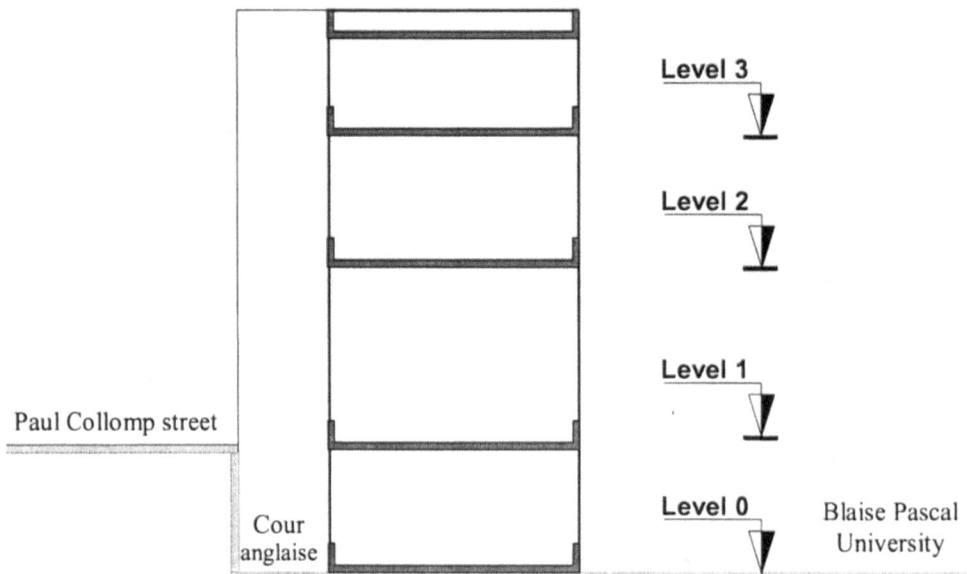

Figure 37. The existing structure

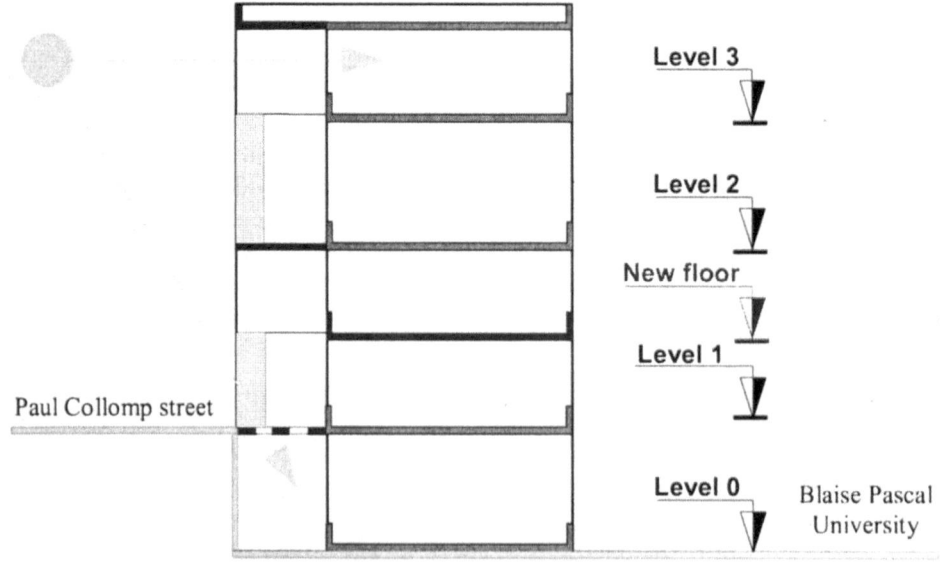

Figure 38. The refurbished structure

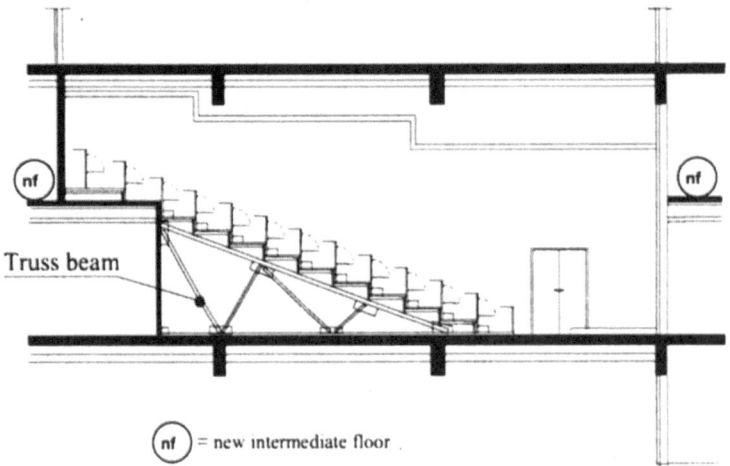

Figure 39. Supporting structure of the lecture theatre

Photo 19. The East façade with its large windows

In the West, above the "Cour anglaise" which, at the beginning, has been designed to provide light to the basement, a new structure with composite floors and tubular columns filled up with concrete is erected, supporting steel stairs and large communication floors (photo 20). It was decided to create a new light and transparent façade with glass panels, built independently from the structure to keep free the different floors. A detail is shown on photo 21.

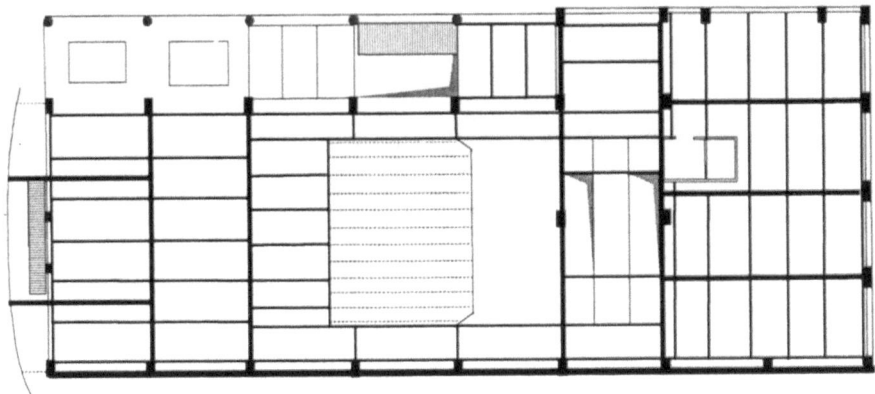

Figure 40. Supporting structure of the intermediate floor

This structure is a self-bearing one and it is connected to the old one. Automatic or printed shades are disposed into the glass walls to protect the building from the sunshine. To keep transparency and to lighten the basement of the original building, some parts of the first floor contain glass panels. It is to be noted that it was the first time in France that glass floors were designed as one hour fire-stop floor.

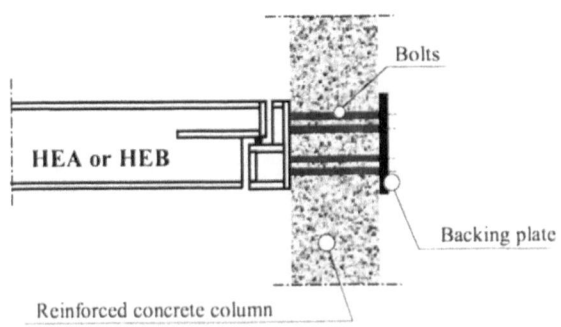

Figure 41. Connections to the existing columns with special corbels

In the Southern part, a vertical lath acting like a light filter (or a shield) was installed close to the wall of the old building (photo 22). It protects the construction against the sunshine and shelters the exit steel stairs (photo 23).

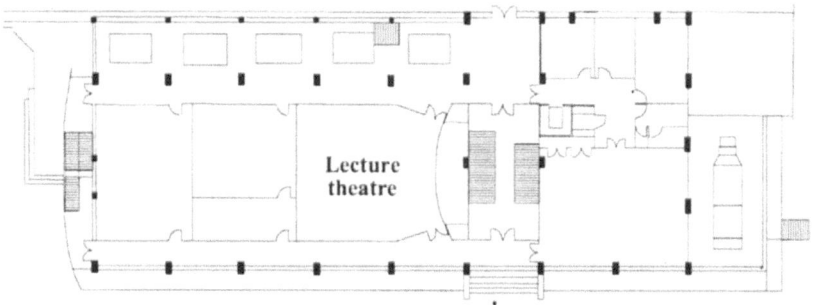

Figure 42. Intermediate floor at level + 8.00

Photo 20. Internal view with stairs, glass wall and floor above the "cour anglaise"

4.3 Conclusion

These examples show successful refurbishment with steel at Blaise Pascal University. The structures have been chosen with the idea of keeping maximum transparency in the new rooms associated with very light constructions. This choice allows the previous structure to be improved in the "Manège". It also allows a new lecture theatre and large circulation areas to be created in the "Paul Collomp" building.

Photo 21. Detail of the glass wall supports

Photo 22. The South façade   **Photo 23.** Exit stairs of the South façade

5   Upgrading of the "25th of April Bridge" in Lisbon[5]

The "25th of April Bridge" in Lisbon is a crucial civil engineering structure in Portugal. It allows a fast access to the South part of this country across the large estuary of the Tagus River. It has been used as a road bridge for about thirty years, but at the beginning, it was designed to be able to accommodate a railway crossing inside its truss steel girder as well as a larger number of road lanes on the upper deck. The train loads becoming larger than it was expected thirty years ago, it was necessary to reinforce the whole structure to fit the actual traffic and train loads (Muzeau & Baptista, 2001).

In this paragraph, the original design is briefly described, including the main figures of this elegant suspension bridge. Then, the improvements in the main steel bridge and in the approach concrete viaduct are presented.

It should be reminded that Lisbon is very close to the Atlantic Ocean coast and that it is located in a seismic zone.

5.1   The initial bridge

Started in November 1962, the "25th of April Bridge" in Lisbon, originally named "Salazar Bridge", was opened to road traffic in August 6, 1966. It spans over the Tagus River and it was the main bridge linking the capital of Portugal to the South Tagus bank.

Photo 24. Initial state of the 25th of April Bridge

[5] This paragraph is co-authored with Dr. António Baptista, Senior Researcher, LNEC, Lisbon, Portugal.

It is a suspension bridge (Photo 24) whose shape reminds strongly the suspension bridges in San Francisco. It was build by American companies (United States Steel Export Company & Morrison kNudsen Company), chosen after an international competition. It has been designed by D.B. Steinman, Boynton, Gronquist & London.

The main span of the bridge is 1013 m (Figure 43). It is completed by two side spans of 483 m and bank spans of 99 m, one in the South and two in the North. The total length of the suspension bridge is 2277 m. The towers are 190 m high.

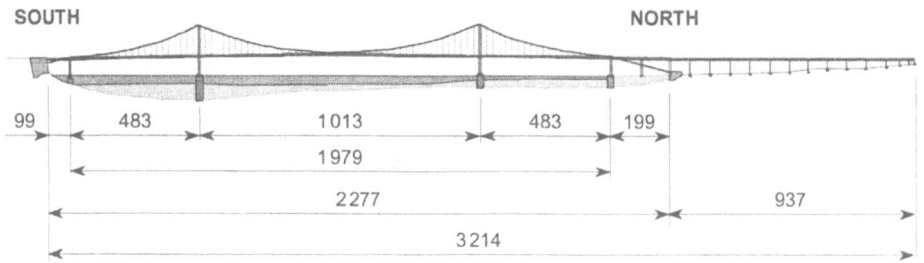

Figure 43. Elevation view of the Bridge

On the South side, it is directly linked to the rock bank of the river but on the North side, it is extended by a concrete viaduct of a total length equal to 937 m, whose spans vary around 75 m. Therefore, the total length of the bridge in 3214 m.

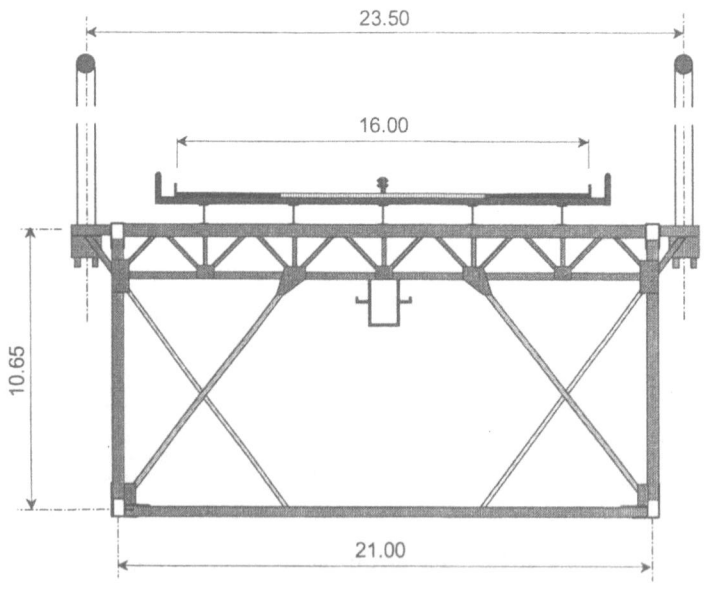

Figure 44. Initial cross-sections of the truss girder

The steel truss beam main dimensions are 10.65 m deep and 21.00 m wide (Figure 44). It is a continuous beam made of bolted elements. Due to the marine atmosphere, special attention has been brought to the protection against corrosion, which is mainly based on galvanisation and painting.

The spacing between the original suspension cables is 23.50 m. They are made of 37 strands constituted by 11248 parallel steel wires of 5 mm in diameter. Their external total size is 586 mm.

At the first stage, the cross-section was designed to accommodate four traffic lanes (Figure 45a). In order to decrease the dynamic wind effects, the two central lanes were made of steel grids.

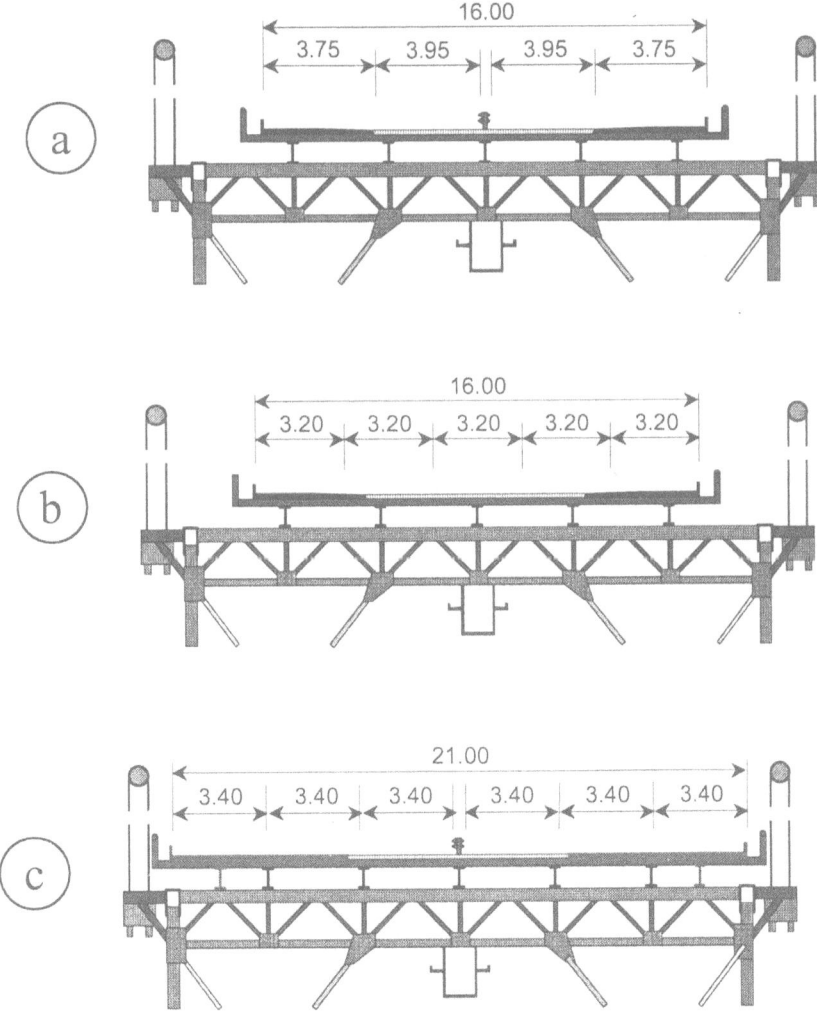

Figure 45. Initial, intermediate and final cross-section of the upper deck with four, five and six lanes

The evolution of the traffic becoming larger, on July 1990, by removing the central barrier, the upper part of the steel deck has been improved to accommodate a fifth lane (Figure 45b). It has to be focused that at the initial stage, the project was designed to be able to support a total of six traffic lanes (Figure 45c). So, no special reinforcements were required at that stage to extend the bridge capacity from four to five lanes.

One of the main characteristics of the project is that, since the beginning, the possibility to include a railway platform inside the truss girder was considered. That explains why, on the South bank of the River, a tunnel was carried out at the initial stage of the project (Photo 25).

Photo 25. South bank tunnel (Canto Moniz et al., 1966)

5.2 The improvement

Due to a very large increase of the road traffic and public requirements concerning the railway link, it was decided, in 1989, to achieve the completion of the project. So, the upper deck was extended to six traffic lanes (opened in November 1998) and a railway platform was built at the lower level of the truss girder (inaugurated in July 1999).

If an important road traffic evolution was assessed at the initial stage of the project, this increase was 2.5 times greater than expected (for 1980, the assessment was 20 000 cars/day as the actual figure was 53 500 cars/day). The measured traffic was more than 120 000 cars/day in the year 1994. Nevertheless, it must be focused that the corresponding loads did not increase consequently: they remain practically stable from four lanes to six lanes.

On the other hand, the evolution in frequency and in train capacities (double-deck trains for example), creates higher loads than those considered in the initial design. Therefore, the main structure required a strong improvement regarding both the new traffic loads and the fatigue resistance. However, no actions have been required regarding the towers and the foundations.

The improvement of the suspension bridge: the main suspension structure. Different options have been considered to improve the suspension bridge. At the initial phase, it was proposed to reinforce the supporting structure with three cable stays fixed at the main towers (Figure 46). Another solution raised in self-anchored secondary cables whose compressive axial loads would have been supported by the new internal structure built to support the railway tracks (Figure 47). The final solution consists in an externally secondary suspension cable anchored close to the previous one (Figure 48). Steinman was again the designer.

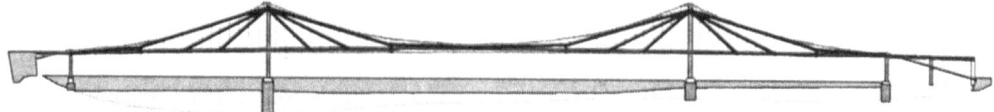

Figure 46. Initial solution with cable-stays

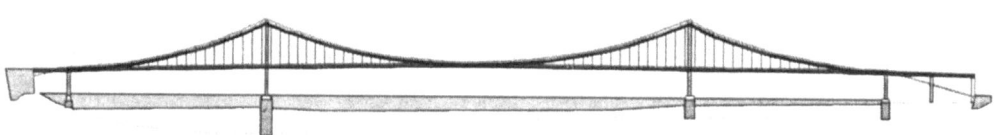

Figure 47. Other proposal with secondary self-anchored cables

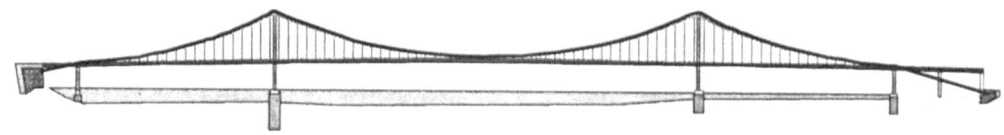

Figure 48. The final solution

The total costs of the three proposals were very close. The third solution was chosen because it leads to less maintenance costs and to a higher fatigue resistance. Figure 49 represents this option. In that case, the truss beam is supported each 11.5 m creating a more uniform load distribution.

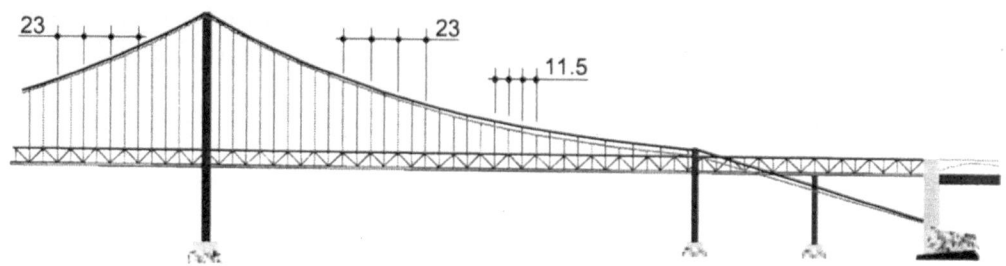

Figure 49. Detail of the North part of the suspension bridge

The new anchorage, made of separate concrete blocks, have been located close to the existing ones (Photo 26).

Photo 26. North bank anchorage blocks

In order to accommodate the new secondary cables at the top of the towers, special saddles have been installed as an extension to the original supports (Photos 27 and 28).

Photo 27. Main tower saddle (Fernandes et al., 1966)   **Photo 28.** Intermediate support saddle

The installation of a secondary cable over the initial one is shown in Photo 29. The new cables are made of 19 strands constituted by 4104 parallel steel wires of 3.5 mm in diameter. Their external total size is 354 mm. They are located at 3.70 m above the existing cables.

Photo 29. Secondary cables located above the existing ones

The improvement of the suspension bridge: the upper deck. The width increase of the upper deck has been carried out by addition of steel cantilevers bolted to the existing structure (Photo 30). So it was possible to welcome six traffic lanes according to the transverse view represented on Figure 50.

Photo 30. Extension of the transverse beams (Fernandes et al., 1966)

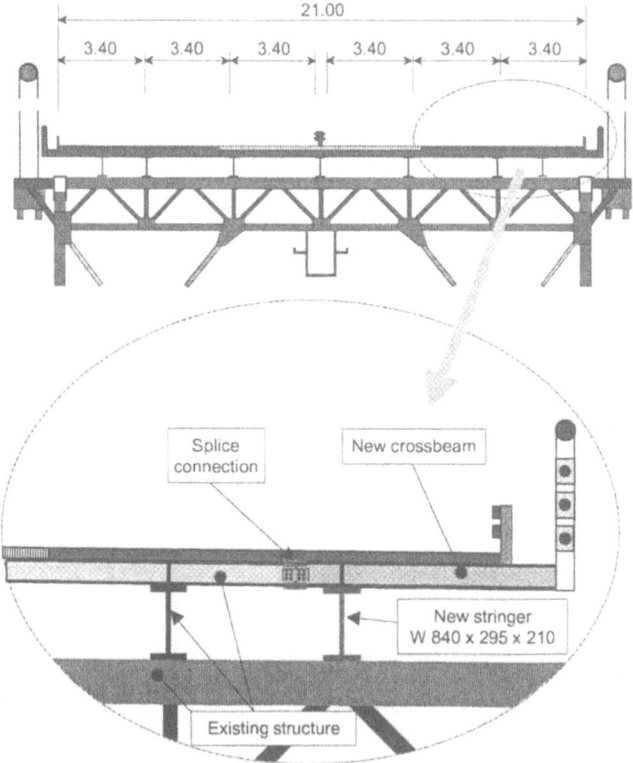

Figure 50. Transverse section of the upper deck

The improvement of the suspension bridge: the railway track included into the truss girder.
To accommodate the railway track in the caisson truss girder, the structure had to be straighten.

Photo 31. The stiffening truss

Photo 32. The bottom chord

The main railway supporting structure is composed of four stringers (1 m deep) associated to a bracing system carried out from 80×80×8mm angles. It is associated to the lower elements of the caisson as an orthotropic system. The horizontal bracing is a K bracing system as it is shown on Figure 51. Photos 31 and 32 (Fernandes et al., 1966) represent the general design of the main railway supporting structure. Photos 33 and 34 (Fernandes et al., 1966) represent the railway tracks after completion.

Photo 33. Railway tracks & evacuation railway

Photo 34. Railway tracks after completion

In order to fulfil the design requirements, some elements of the initial truss beam had to be reinforced due to increased shear forces and bending moments. Therefore, the advantage to work on a bolted structure was an important factor in the improvement.

The top chord has been reinforced by addition of steel plates, which have been bolted by replacing the initial bolts on the top flange. The truss diagonals have been reinforced by the same way to be able to resist higher axial loads without bending. Figures 52 and 53 show the reinforced elements.

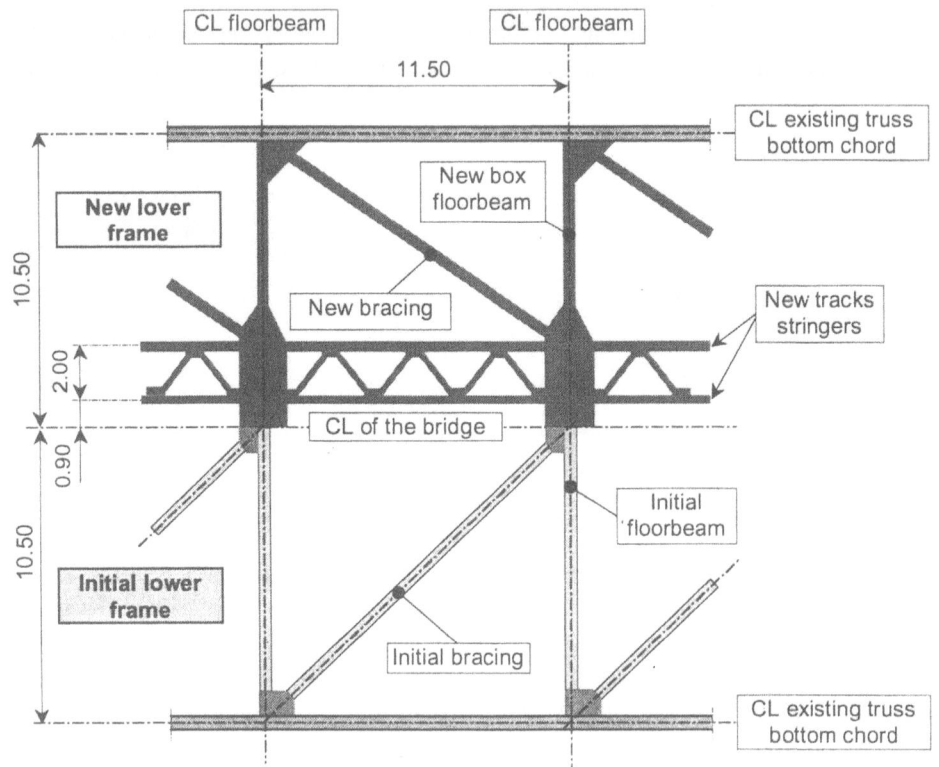

Figure 51. Horizontal bracing system

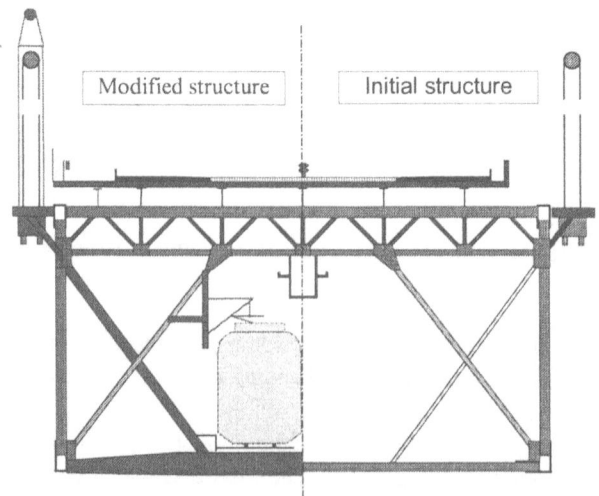

Figure 53. Reinforced elements at the bottom flange

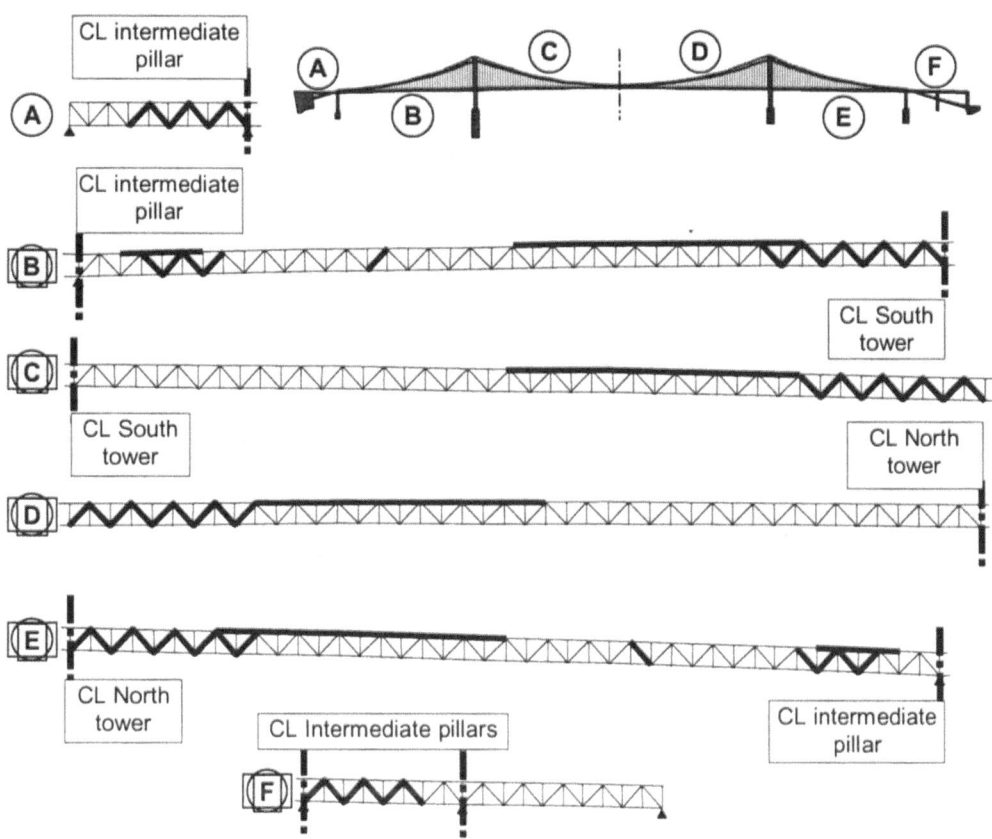

Figure 52. Reinforced elements along the main truss girder

Photo 35 shows one stage of the installation. It is dated on July 1997. In this picture, the general shape of the truss beam has to be detailed. Due to the additional weight of the reinforcing elements and to their localisation, the beam deflection at mid span was relatively important (here about 3 m) compared to the standard situation. Figure 54 shows the evolution of the deflection which, obviously, comes back to zero at the end of the improvement. Photo 36 is a picture of the truss beam after the improvement.

Photo 35. The bridge during the improvement

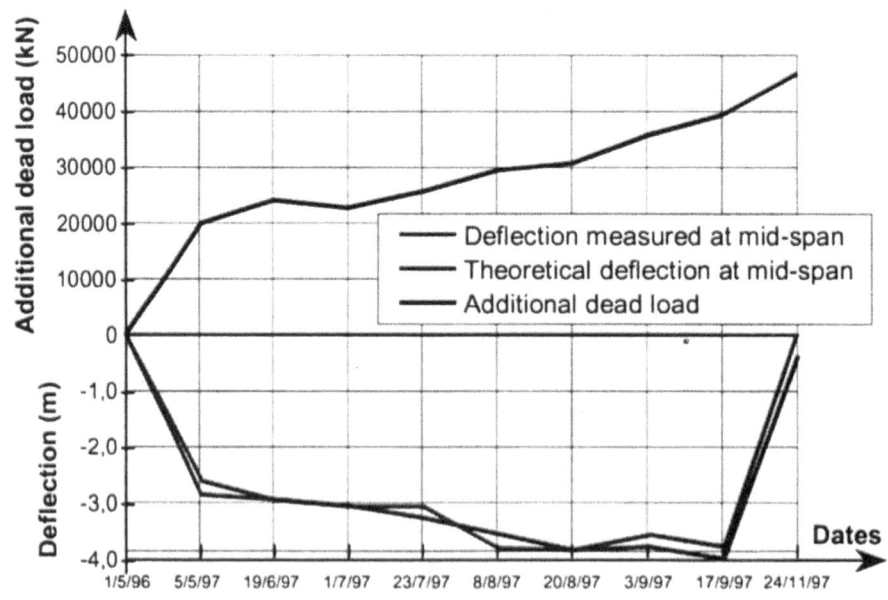

Figure 54. Evolution of the mid span deflection during the improvement

Photo 36. Downside view of the truss girder after completion of the improvement

5.3 The improvement of the approach viaduct

The approach viaduct is located on the North bank of the Tagus river, over the Alcântara urban zone.

The structural solution. At the initial stage, intermediate supports were located in the prestressed concrete structure of the viaduct in order to bear the future railway tracks (Photo 37).

Photo 37. Structure of the approach viaduct (Canto Moniz et al., 1966)

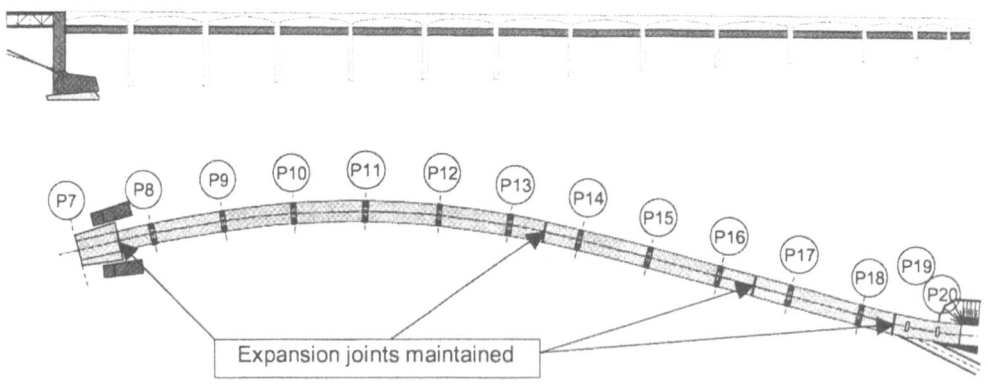

Figure 55. Final expansion joints

The initial concrete structure was built as a succession of cantilever elements. Expansion joints were displayed between them at each mid-span but they required frequent reparation. Therefore, it was decided to create a structural continuity in the structure and only four expansion joints are kept (Figure 55). In that goal, additional prestressing has been carried out by means of external prestressed bars and cables, and steel plates have been located to fit the gap between the elements (Figure 56).

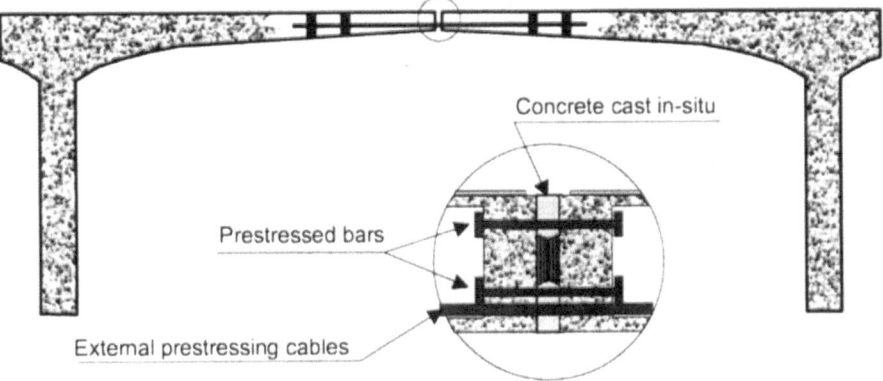

Figure 56. Creation of continuity in the cantilever elements

Photo 38. Steel girder installed in the approach viaduct

At the origin of the project, a pure steel solution was expected. Thirty years later, a composite solution appeared to be the most efficient. It is made of a pair of uniform main plate girders (4.0 m deep) horizontally braced at the bottom flange level with transverse diaphragms at about 6.40 m apart (Photo 38 & Figure 57). The concrete slab is 350 mm thick and it is connected by studs welded to the main girders.

The steel structure is represented on Figure 58

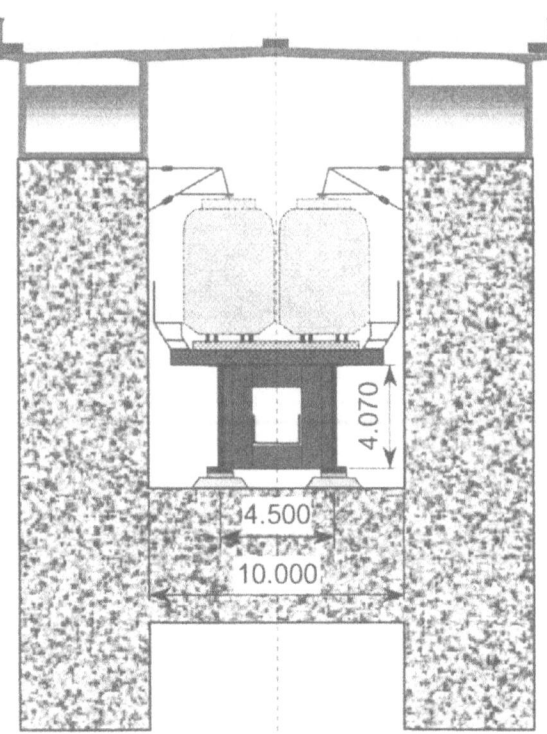

Figure 57. Transverse view of the approach viaduct

Photo 39. Inside view of the steel structure

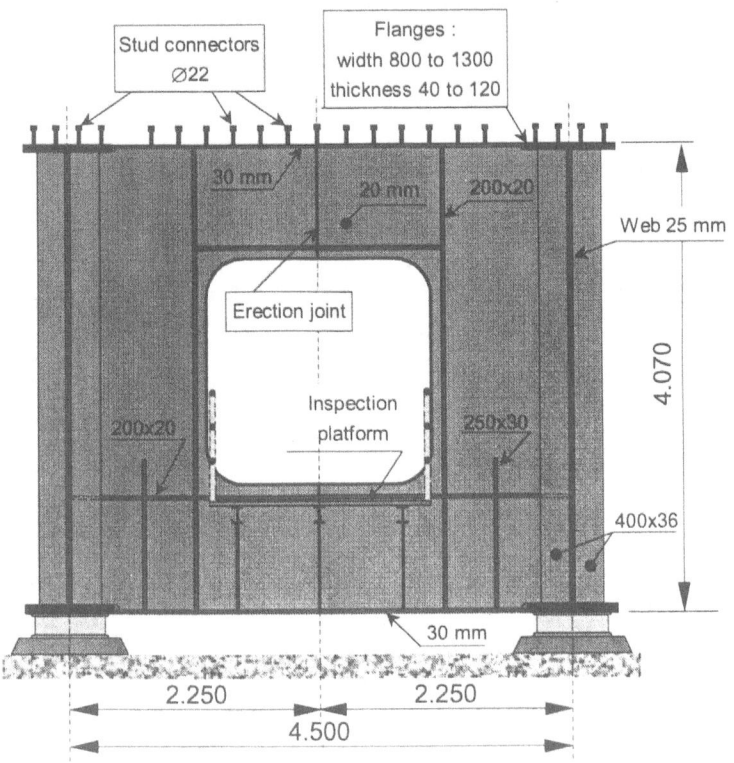

Figure 58. Steel structure of the approach viaduct

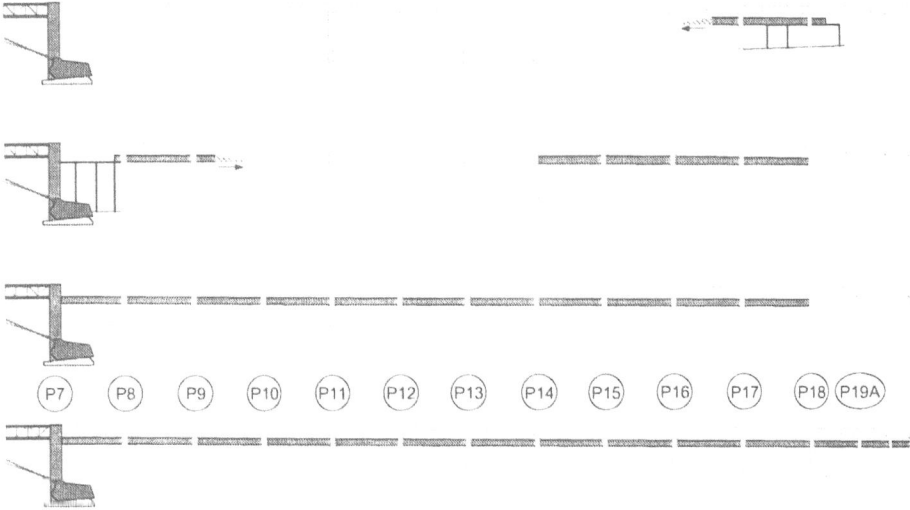

Figure 59. Erection phases

Photo 40. Temporary structure and launching of the railway track steel girder

The different steps of the installation process are shown on Figure 59.

A temporary steel truss (Photo 40) beam was used to support the first section of the steel plate girder before launching the structure, element by element. Each of them had a length of 20 m. They were assembled by welding. The launching started from both ends of the viaduct. Photo 41 represents a view of the approach viaduct after completion of the refurbishment.

5.4 Conclusion

The improvement of the "25th of April Bridge" was designed since the very beginning of the project but some improvements and some reinforcements were necessary to take into account the real state of both the road and train traffics.

A secondary cable was added in the suspension bridge and the truss steel beam required to be stiffened to welcome the rail track girders.

A composite solution was used in the approach viaduct supported by the columns of the existing prestressed concrete structure which has been previously turned into continuous structure.

Another new bridge, the "Vasco de Gama" bridge, has been built recently in another area in Lisbon (close to the site of the international Expo'98) but the "25th of April Bridge" remains the most important bridge in Lisbon for the suburban traffic. The upgrading described in this chapter represents thirty years of dreams for more than 200 000 Portuguese crossing it every day. It improved significantly their quality of life.

Photo 41. The new composite structure installed in the approach viaduct

6 References

AFFSB (1995). Association française des foreurs et scieurs de béton (1995) *La Gazette. n°1*, AFFSB.

Aribert J.M., Lachal A., Muzeau J.P. and Racher P. (1994). Recent Tests on Steel and Composite Connections in France. Proceedings of the *Second State of the Art Workshop, COST C1 "Semi-rigid Behaviour of Civil Engineering Structural Connections"*, pp. 65-74, Prague, Czeck Républic.

Baptista A., Fournely E., Muzeau J.P. and Ryan I. (1997). Parafusos Huck-Fit: Descrição e comportamento mecânic (Huck-Fit Bolts: Technical Description and Mechanical Behaviour). *Revista Portuguesa de Engenharia de Estruturas*, RPEE, n°41, pp. 21-28.

Canto Moniz J., Silva Lima C. and Teixeira C. (1966). A Ponte Salazar. Ministério das Obras Públicas, *Gabinete da Ponte sobre o Tejo*, Lisbon.

Czarnomska m., Muzeau J.P. and Racher P. (1993). Etude expérimentale comparative du comportement d'assemblages par boulons sertis huck-fit ou par boulons hr. *Revue Construction Métallique*, n°2, pp. 9-29.

Farinha J. (1998). Técnicas de demolição. *Encontro Nacional sobre Conservação et reabilitação de estruturas*, LNEC, Lisbon, Portugal, p. 785-794.

Fernandes m., Canto Moniz L. and Viana J. N. (1999) A Ponte 25 de Abril. Construção da 2a fase – Instalação do Caminho de Ferro. *Ministério do Equipamento, do Planeamento e da Administração do Território, Rede Ferroviária Nacional* – REFER, E.P..

GE Superabrasives (1991). Technologie du sciage au fil diamanté. *GES 91.960 F1995*.

Huck (1990). Metric Huck-Fit fastening system, Doc. N°UKB-865 11-90 5M, *Huck Manufacturing Company.*

Huck (1994). The Ultra-Twist fastening system, Doc. N°HWB-892 9-94 2M, *Huck Manufactering Company.*

Korol R.M., Ghobarah A. and Mourad S. (1993). Blind bolting W-shape beams to HSS columns. *Journal of Structural Engineering,* vol. 119, n°12.

Mur J. and Muzeau J.P. (1979). Etude comparative des divers procédés de démolition. Critères de choix. *Annales de l'ITBTP,* n°377.

Muzeau J.P. and Baptista A. (2001). Récentes évolutions du pont du 25 avril à Lisbonne, *Bulletin Pont Métalliques,* n°22, OTUA, Paris, France (to be published).

Muzeau J.P., Missoum A., Faugeras J.C. and Ryan I. (1995). Huck-Fit fasteners: Available preload in plate connections with lack of flatness. *IXth International Conference Metal Structures,* Vol. 1, pp. 261-270, Kraków, Pologne.

Muzeau J.P. (1996). Techniques for partial demolition. International Seminar "Refurbishment of structures", Tihany, Hungary, June 1996, *TEMPUS SJEP 09524-95,* published by the Technical University of Budapest, ISBN 963-421-543-2, pp. 171-183

Panthéon J.F. and Saveau J. Architects (1991). Université Blaise Pascal. Réhabilitation du bâtiment Paul Collomp à Clermont-Ferrand. Avant-projet détaillé, Mémoire, *Ministère de l'Education Nationale,* France.

Reis A. and Melo L. (1997). The incremental launched railway deck for the Tagus Bridge viaduct, *Int. Conf. "New Technologies in Structural Engineering",* NEW TECH'97, IABSE-FIP, Lisbon, Portugal, pp. 869-876.

Reis A., Moura R., Pedro J.O. and Farinha J. (1999). Reabilitação e Reforço do Viaduto de Acesso Norte à Ponte Sobre o Tejo em Lisboa, *Jornadas Portuguesas de Engenharia de Estruturas, JPEE'98,* L.N.E.C., Lisbon, pp. 587-596.

Ryan I. (1994). Boulons sertis Huck-Fit - Caractéristiques, spécifications techniques et guide d'utilisation. *Rapport CTICM n°9.005-2,* version 1.0, CTICM, France.

Sadri m.S. (1995) Advanced technique for joining of hollow steel structures. *Nordic Steel Construction Conference* '95.

CHAPTER 6

STRENGTHENING OF STEEL FRAMES IN SEISMIC RESISTANCE

H. Akiyama
Nihon University, Tokyo, Japan

Abstract

Based on the balance of energy between the seismic input and the structural resistance, the basic formulation for strengthening buildings is derived. The loading effect of an earthquake can be grasped as the energy input, while the seismic resistance of an building can be considered to be the energy absorption capacity of the building.

In order to improve the seismic resistance of existing buildings, the following methods are effective;

1) to dispose energy absorbing structural elements in addition to existing structural skeletons
2) to equip an energy absorbing story with the energy absorption capacity large enough to be able to absorb the total energy input, leaving the other stories undamaged
3) to dispose energy absorbing elements over the building evenly, keeping the structural elements which sustain gravitational loadings elastic

1. Introduction

In earthquake-prone countries, the strengthening of existing buildings raises problems more serious and troublesome than the designing of new buildings do. The seismic engineering is developing day by day and is still far from completion. Therefore, any existing buildings must be checked the necessity of strengthening in the light of advanced technology. The necessity of strengthening is determined by its emergency and cost-benefit. The most emergent situation is created under the real occurrence of an earthquake. As has been exhibited in the Ilyogoken-nanbu earthquake (1995), causes for structural damage are ascribed to:

- poor materials
- poor constructional works
- poor design
- excessive seismic input

Not only the deterioration of materials by time, but also poor properties of original materials were disclosed in the Hyogoken-nanbu earthquake. Some steel structures indicated brittle fracture without no traces of plastic deformation.

Again, it was strongly recognized that the ductility of steels depends totally on the toughness of materials against brittle fracture. Poor constructional works are always found in every occasion of earthquake. In steel structures, defects in welding and anchor bolts in column bases are prevailing. Poor designs in steel structures are found in connections. The stress concentration in connections in members' end causes a premature failure of connections, which prevents fully development of adjacent members' plastification.

The seismic input depending strongly to the epicentral distance and local ground conditions reach sometimes far greater than the design level. In the Hyogoken-nanbu earthquake, it reached locally a level about two times as great as the level of the national building code.

The actual damage is very persuasive for the decision of repairing and strengthening. When a building suffers damage, the recovery of its function and the removal of clients' anxiety are of primary importance.

Reparations are usually made crudely and rashly. Broken connections are rewelded incompletely. Broken column bases are covered by new concretes. Buckled diagonal bracings are simply replaced by new ones. These treatments become inevitably emergent to save time, space and cost. Therefore, while the experiences of various damages are invaluable in developing seismic engineering, practices of reparation taken immediately after an earthquake are generally not so suggestive as proper measures of strengthening.

Decision for strengthening is made based on the following items.

- diagnosis on seismic resistance
- selection of practical measures
- analysis on cost

In this lecture, a main focus is brought into practical measures based on the estimate of seismic resistance.

The seismic resistance of a structure can be generally described in terms of the balance of energy between the input energy exerted by an earthquake and the energy absorbed by a structure.

In ordinary structures, structural members are expected to absorb the input energy by means of plastification. Therefore, not only the strength but also the inelastic deformation capacity must be simultaneously considered in strengthening.

The advanced technique such as base-isolation provides a more promising approach. The advanced structural type including base-isolated structures is identified to be the flexible-stiff mixed structure. The structural performance of this type of structure is superior to the ordinary structures which are categorized to be stiff structures.

By adding flexible elements, which remain elastic, the structural responses are stabilized.

The laminated rubber bearings used as flexible elements enable the set-up of the energy-absorbing story, which can absorb the total energy input to a structure. By applying the energy absorbing story, not only strengthening of the existing building but also its expansion in scale becomes possible.

Any structures on the earth must sustain the gravitational loading. Basically, the structural skeletons are so designed to remain elastic with a considerable marginal strength under the gravitational loading. Therefore, the structural skeletons can be the flexible elements in the flexible-stiff mixed structure, as far as the rigid elements capable of absorbing the total energy input are evenly distributed over a structure. This type of structure in which the damage concentration can be almost dismissed is categorized to be the damage-dispersing type of the flexible-stiff mixed structure in contrast with the damage-concentrating type of the flexible-stiff mixed structure (the base-isolated structure).

2. Basic Structural Behavior of Structures during Earthquakes

2.1 Total energy input exerted by an earthquake

Fig.2.1 shows a single-mass oscillatory system subjected to an uni-directional horizontal ground motion. The equation of motion for the system is expressed as follows.

$$M\ddot{y}+C\dot{y}+F(y) = F_e \tag{2.1}$$

where M : mass

$C\dot{y}$: damping force

F_e : seismic force $(-M\ddot{z}_0)$

z_0 : horizontal ground motion

y : displacement of the mass relative to the ground

Multiplied by $dt = \dot{y}dt$ on both sides, and integrated over the entire duration of an earthquake, t_0, Eq.(2.1) is reduced to

$$M \int_0^{t_0} \ddot{y}\dot{y}dt + C \int_0^{t_0} \dot{y}^2 dt + \int_0^{t_0} F(y)\dot{y}dt = \int_0^{t_0} F_e\dot{y}dt \tag{2.2}$$

The right-hand side of the above equation expresses the total amount of energy exerted by an earthquake, E.

The second term of the left-hand side expresses the energy consumed by the damping mechanism, W_h. The first term of the left-hand side, which can be reduced to $M\dot{y}_{t=t_0}^2/2$, expresses the kinetic energy at the instant when the earthquake motion vanishes. The third term expresses the strain energy deposited in the spring system, which consists of cumulative plastic strain energy, W_p, and elastic strain energy at the instant when the earthquake motion fades away. The kinetic energy and the elastic strain energy constitute the elastic vibrational energy, W_e. Therefore, Eq.(2.2) becomes as follows.

$$W_p + W_e + W_h = E \tag{2.3}$$

Eqs. (2.2) and (2.3) can be extensively applied to multi-degree of freedom systems by expressing relevant quantities with maxtrices and vectors.

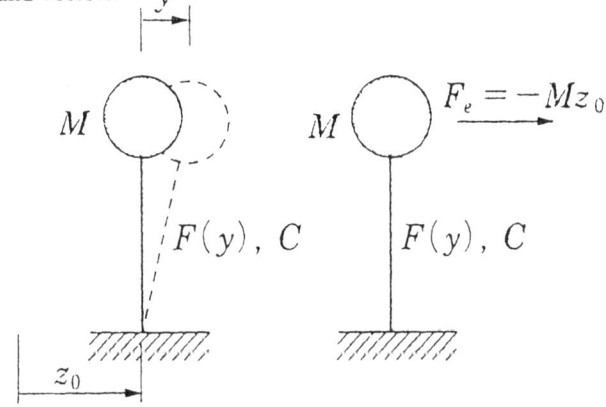

Fig.2.1 One-Mass Vibrational system

Eq.(2.1) expressing the equilibrium of forces at the instant can afford a real response of a structure over elastic and inelastic ranges through its integration. Obtained information, however, is of scattering nature and does not tell us a general trend of structural responses.

On the other hand, Eqs.(2.1) and (2.2) imply an integral of structural responses and make it possible to quantify energy constituents such as W_e, W_p and W_h for wide varieties of design parameters. Moreover, the constancy of the total energy input, E enhances their applicability to the seismic design problem:(Akiyama, 1985):

> The total energy input into a structure by an earthquake depends mainly on the total mass and the fundamental natural period of the structure, scarcely influenced by the strength distribution, the mass distribution and the stiffness distribution.

The energy spectrum for an earthquake is defined to be a relationship between the equivalent velocity, V_E and the fundamental natural period of the structure, T. V_E is converted from the total energy input, E through the following equation:

$$V_E = \sqrt{\frac{2E}{M}}$$

Under strong earthquakes, which should be considered in the design process, structures exhibit inelastic behavior characterized by a strong nonlinearity. Therefore, the energy spectra for the design use must involve the effect of structural nonlinearity. The structural nonlinearity was found to be equivalent to a nonlinearity due to damping,   and it was found that the energy spectra for plastified nonlinear structures can be enveloped by those of the damped elastic systems with 10% of critical damping, i.e. $h = 0.1$. Based on the energy spectrum of the elastic system, the spectrum for design use can be constructed as shown in Fig.2.2. As is shown in the figure, the design spectrum is an envelope of the energy spectrum of the elastic system and is formed by two line segments.

One is the line, which passes through the point of origin, and another is the line parallel to the abscissa. Thus, the design spectrum is characterized by the level of the spectrum, V_{Em} and the corner period, T_G which corresponds to the intersection point of two line segments.

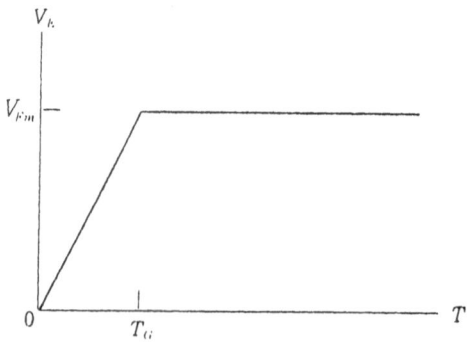

Fig.2.2 Energy Spectrum for Design Use

2.2 Energy input attributable to damage

The structural damage is directly related to the strain energy (Housner, 1956, 1959). Therefore, it is essential for the structural design purpose to know the energy input attributable to damage, E_D as shown by the following equation.

$$E_D = W_e + W_p = E - W_h \qquad (2.4)$$

Through numerical analyses, it was found that a rough estimate is possible by the following empirical formula.

$$E_D = \frac{1}{(1+3h+1.2\sqrt{h})^2} \qquad (2.5)$$

The equivalent expression of E_D in the velocity, V_D is obtained by applying the similar conversion as follows.

$$V_D = \sqrt{\frac{2E_D}{M}} = \frac{V_E}{1+3h+1.2\sqrt{h}} \qquad (2.6)$$

The $V_D - T$ relationship was found to be very close to the velocity spectrum for the elastic system.

2.3 Expression of structural damage

Taking the elastic-perfectly plastic type of restoring force characteristics, the structural damage is defined as follows.

a) cumulative plastic deformation

The cumulative plastic deformation δ_p is defined as

$$\delta_p = \sum_i \Delta\delta_{pi} = \delta_p^+ + \delta_p^- \qquad (2.7)$$

where $\Delta\delta_{pi}$: plastic deformation increment

δ_p^+ : sum of the plastic deformation increments in the positive direction

δ_p^- : sum of the plastic deformation increments in the negative direction

By using the cumulative plastic deformation, W_p is expressed as

$$W_p = Q_Y \delta_p = \eta Q_Y \delta_Y \qquad (2.8)$$

where Q_Y : yield strength

δ_Y : yield deformation

η : cumulative inelastic deformation ratio

η is defined as

$$\eta = \frac{W_Y}{Q_Y \delta_Y} = \eta^+ + \eta^- \qquad (2.9)$$

where η^+ : cumulative inelastic deformation ratio in the positive direction

η^- : cumulative inelastic deformation ratio in the negative direction

Eq.(2.9) can be a general definition for the restoring force characteristics different from the elastic-perfectly plastic type.

(b) maximum plastic deformation

The maximum plastic deformation is defined as follows.

$$\delta_{pm} = \text{Max}\left\{\delta_{pm}^+, \delta_{pm}^-\right\}$$

(2.10)

$$\text{where} \quad \delta_{pm}^+ = \delta_m^+ - \delta_Y$$

$$\delta_{pm}^- = \delta_m^- - \delta_Y$$

δ_m^+ : maximum deformation in the positive direction

δ_m^- : maximum deformation in the negative direction

Non-dimentionalized deformations are defined as follows.

$$\mu_m = \text{Max}\left\{\mu^+, \mu^-\right\}$$

(2.11)

$$\text{where} \quad \mu^+ = \frac{\delta_{pm}^+}{\delta_Y}, \mu^- = \frac{\delta_{pm}^-}{\delta_Y}$$

μ^+ : maximum plastic deformation ratio in the positive direction

μ^- : maximum plastic deformation ratio in the negative direction

μ_m : maximum plastic deformation ratio, the bigger of μ^+ and μ^-

(residual deformation

1... residual permanent deformation left as an ostensible damage is the residual deformation and is expressed as follows.

$$\delta_r = \left|\delta_{pm}^+ - \delta_{pm}^-\right|$$

(2.12)

The cumulative plastic deformation can be quantified by knowing the energy absorbed through inelastic deformations. The development of the maximum plastic deformation and the residual deformation depend on the structural type.

The most influential factor is the presence of the elastic element. Generally, the structural element can be recognized to be a flexible-stiff mixed structure. The flexible element is the element that remains elastic. The stiff element is the element that behaves inelastically. Its restoring force characteristics are shown in Fig.2.3.

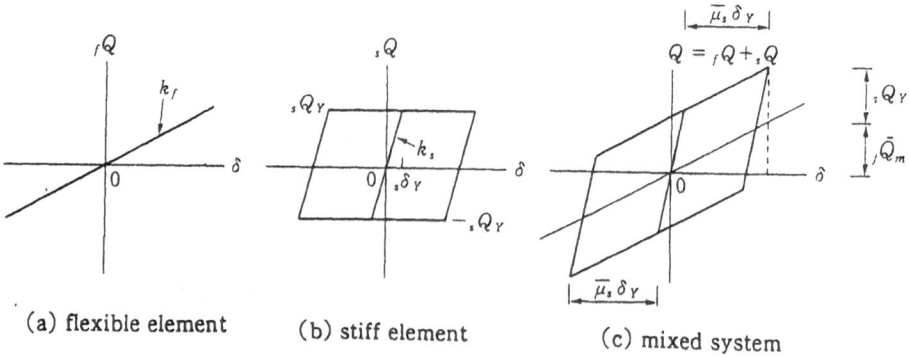

(a) flexible element   (b) stiff element   (c) mixed system

Fig.2.3 Flexible-Stiff Mixed Structure

By the presence of flexible element, hysteretic behaviors tend to orient the point of origin. As a result, the residual deformation is reduced and the maximum deformation is restrained under a fixed amount of energy input. The flexible-stiff mixed structure is characterized by the shear force ratio, r_q defined by

$$r_q = \frac{{}_f\bar{Q}_m}{{}_sQ_Y} \tag{2.13}$$

where $\quad {}_f\bar{Q}_m$: average of maximum shear forces developed in the positive and negative loading domains in the flexible element.

$\quad {}_sQ_Y$: yield shear force of the rigid element.

Ordinary structures, which are not intentionally equipped with flexible elements, are identified to be structures singly equipped with stiff elements (ordinary structure).

Through numerical analyses, a conservative estimate for η/μ_m was obtained as follows.

$$\left. \begin{array}{l} \text{for } r_q \leq 1.0, \ \dfrac{\eta}{\mu_m} = 4+4r_q \\[2em] \text{for } r_q > 1.0, \ \dfrac{\eta}{\mu_m} = 8.0 \end{array} \right\} \tag{2.14}$$

For the residual deformation, the medium responses can be expressed as

$$\left. \begin{array}{l} \text{for } r_q \leq 1.0, \ \dfrac{\delta_r}{\delta_m} = \dfrac{1}{2+3r_q} \\[2em] \text{for } r_q > 1.0, \ \dfrac{\delta_r}{\delta_m} = 0.2 \end{array} \right\} \tag{2.15}$$

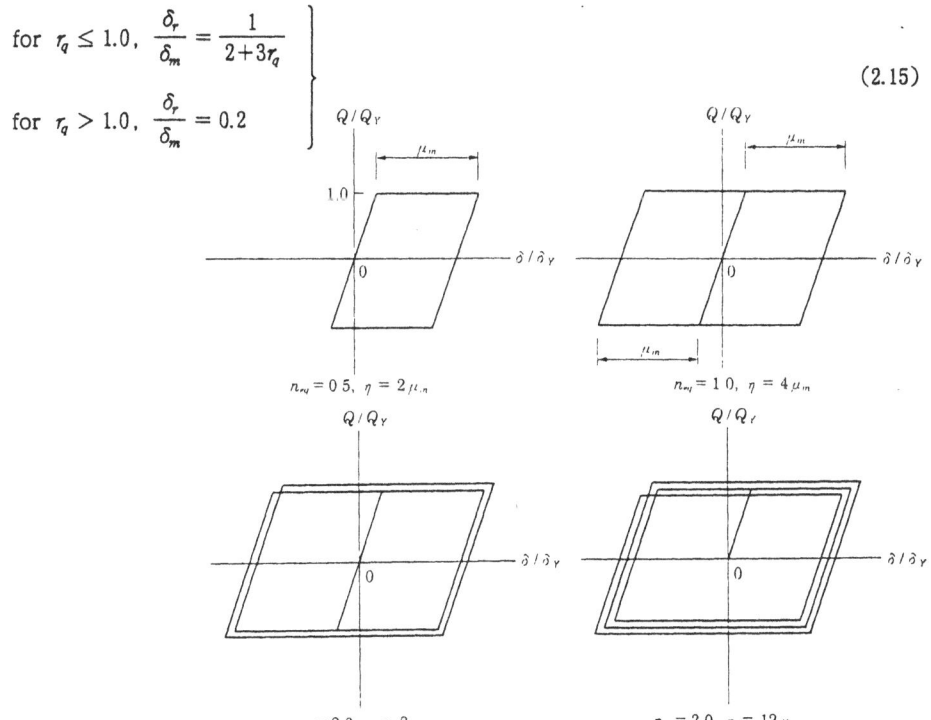

Fig.2.4 Equivalent Hysteretic Loop

Based on the value of η/μ_m, a fictitious hysteretic loop can be envisaged. Referring to Fig.2.4, the number of cycles in a fictitious hysteretic loop, n_{eq} and η/μ_m can be related as follows.

$$n_{eq} = \frac{\eta}{4\mu_m} \tag{2.16}$$

In ordinary structures which are characterized by $r_q = 0$, the fictitious hysteretic loop is the one-cycle loop with the amplitude of μ_m. Similarly, for the flexible-stiff mixed structure characterized by $r_q > 1.0$, the fictitious hysteretic loop is the two-cycle loop with the amplitude of μ_m. Such a relationship may hold for the case of restoring force characteristics other than the elastic-perfectly plastic type, as far as the difference is not so remarkable.

2.4 Basic damage distribution low applied to shear-type systems with elastic-perfectly plastic restoring-force characteristics

(a) standard damage distribution

The yield strength of ith story is denoted by Q_{Yi} and the elastic limit deformation under Q_{Yi} is denoted δ_{Yi}. Using the cumulative inelastic deformation ratio of ith story, η_i, W_{pi} is expressed as

$$W_{pi} = \eta_i Q_{Yi} \delta_{Yi} \tag{2.17}$$

The spring constant of ith story is denoted by Q_{Yi}. Using the total mass, M and the fundamental natural period, T, the spring constant of an equivalent one-mass system, k_{eq} is defined as

$$k_{eq} = \frac{4\pi^2 M}{T^2} \tag{2.18}$$

k_i is related to k_{eq} by the following expression

$$k_i = \kappa_i k_{eq} \tag{2.19}$$

Then, Eq.(2.17) is rewritten as

$$W_{pi} = \frac{Mg^2 T^2}{4\pi^2} c_i \alpha_i^2 \eta_i \tag{2.20}$$

$$\text{where} \quad c_i = \left(\frac{\sum\limits_{j=i}^{n} m_j}{M} \right)^2 \frac{1}{\kappa_i} \tag{2.21}$$

$$\alpha_i = \frac{Q_{Yi}}{\sum\limits_{j=i}^{N} m_j g} \quad : \text{yield shear force coefficient}$$

As a standard distribution of damage, the damage distribution under an uniform distribution of η_i is taken as follows.

$$\frac{W_{pk}}{W_p} = \frac{c_k \alpha_k^2}{\sum\limits_{i=1}^{N} c_i \alpha_i^2} \tag{2.22}$$

The yield shear force coefficient distribution, which realizes the standard damage distribution, is defined to be the optimum yield shear force coefficient distribution and is expressed as

$$\bar{\alpha}_i = \frac{\alpha_i}{\alpha_1}$$ (2.23)

Eq.(2.22) is also written as

$$\frac{W_{pk}}{W_p} = \frac{s_k}{\Sigma s_j}$$ (2.24)

$$\text{where} \quad s_i = c_i \kappa_i \bar{\alpha}_i^2 = \sum_{j=i}^{N} \left(\frac{m_j}{M}\right) \bar{\alpha}_i^2 \left(\frac{k_1}{k_i}\right)$$ (2.25)

(b) optimum yield shear force coefficient distribution

If the optimum yield shear force coefficient distribution is unchanged, irrespective of the quantity of η_i, $\bar{\alpha}_i$ can be given by the shear force coefficient distribution of the elastic system which corresponds to the case of infinitesimally small constant damage.

The above-mentioned inference has been checked to be applied to practical cases; and an unified expressi ⁔. for the El Centro (1940) record, is expressed as follows.

$$\left. \begin{array}{l} \text{for} \quad x \geq 0.2 \\ \qquad \bar{\alpha}_i(x) = 1+1.5927x-11.8519x^2+42.5833x^3-59.4827x^4+30.1586x^5 \\ \text{for} \quad x < 0.2 \\ \qquad \bar{\alpha}_i(x) = 1+0.5x \end{array} \right\}$$ (2.26)

$$\text{where} \quad x = 1- \frac{\displaystyle\sum_{j=i}^{N} m_j}{M} = \frac{i-1}{N} \quad (\text{for } m_i / m_1 = 1.0)$$

(c) damage distribution low

Based on numerical analyses for systems equipped with strength distribution different from $\bar{\alpha}_i$, it was found that the damage distribution in general case can be expressed by a following expression continuous to Eq.(2.24).

$$\frac{W_{pn}}{W_p} = \frac{s_i p_i^{-n}}{\displaystyle\sum_{j=1}^{N} s_j p_j^{-n}} = \frac{1}{\gamma_i}$$ (2.27)

$$\text{where} \quad p_j = \frac{\alpha_j}{\alpha_1 \bar{\alpha}_j}$$

n : damage concentration index
γ_i : damage dispersion factor

p_j expresses the extent of discrepancy of α_j / α_1 from the optimum distribution. When $p_j = 1$, it is obvious that Eq.(2.27) is reduced to Eq.(2.24). γ_i is termed the damage dispersion factor, since γ_i indicates the extent of damage dispersion into stories other than ith story. n is a positive exponent. As n increases, the dependency of damage distribution on p_j increases. In case of $n = 0$, the damage concentration does not take place. In stories with $p_j > 1$, damage is reduced as n increases. Reversely, in stories with $p_j < 1$, damage concentration is emphasized, as n increases.

(d) damage concentration index

When the damage distribution is given by Eq.(2.27), the value of n can be obtained by observing the change of the damage distribution due to a decrease of strength in the observed story as follows.

First, the damage distribution in kth story under an arbitrary strength distribution is obtained as follows.

$$a = \frac{W_{pk}}{W_p} = \frac{s_k p_k^{-n}}{\sum_{j=1}^{N} s_j p_j^{-n}}$$

(2.28)

Next, the damage distribution under another strength distribution in which the strength in kth story is modified by multiplying p_d is obtained us follows.

$$b = \frac{W_{pk}}{W_p} = \frac{s_k p_k^{-n} p_d^{-n}}{\sum_{j \neq k} s_j p_j^{-n} s_k p_k^{-n} p_d^{-n}}$$

(2.29)

From Eqs.(2.28) and (2.29), the value of n is obtained as follows.

$$n = -\ell_n \left\{ \frac{b(1-a)}{a(1-b)} \right\} / \ell_n p_d$$

(2.30)

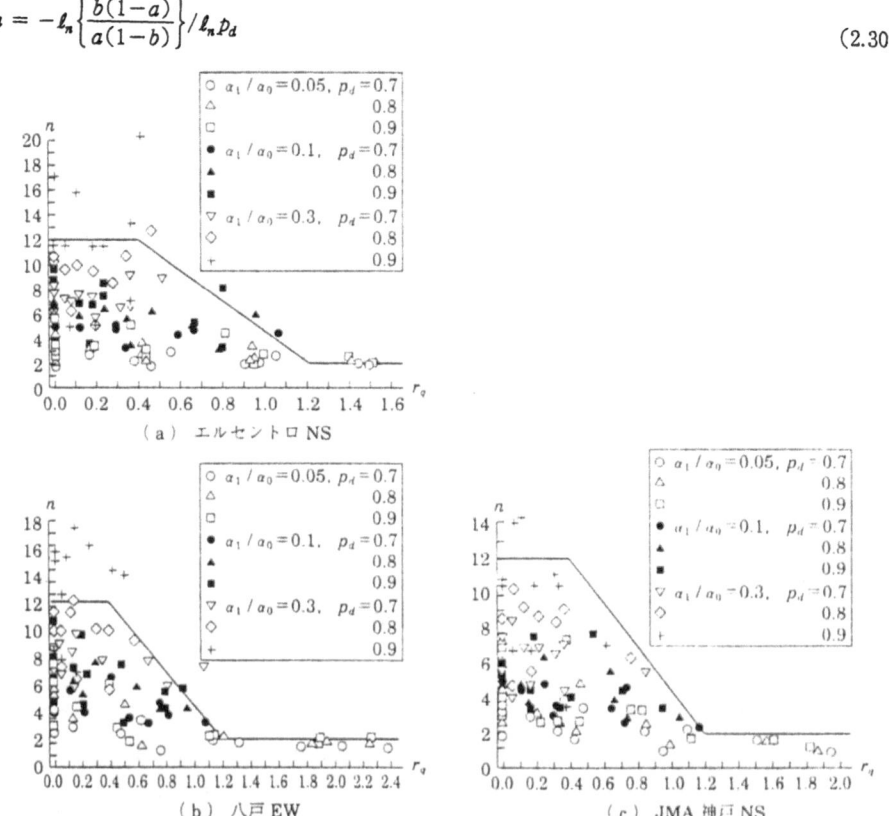

Fig.2.5 n-Values for Flexible-Stiff Mixed Structure

According to the above-mentioned procedure, damage concentration indices were obtained as follows.

weak-column frames   : $n = 12.0$
diagonally braced frames : $n = 12.0$
weak-beam frames   : $n = 6.0$
flexible-stiff mixed frame : $n \leq 6.0$ (for $r_q > 0.8$)

In flexible-stiff mixed frames, the presence of the flexible element produces a remarkable effect in reducing the damage concentration index as shown in Fig.2.5.

2.5 Structural parameters

Structural parameters which govern structural responses are shown in Fig.2.6 under the following conditions.

$$\frac{\alpha_i}{\alpha_1} = \tilde{\alpha}_i$$

$$\frac{m_i}{?} = 1.0$$

$$\frac{\ell}{\delta_{Y1}} = 1.0$$

$\bar{Q}_{Yi}$ is the yield shear force distribution given by the following equation.

$$\bar{Q}_{Yi} = \frac{Q_{Yi}}{Q_{Y1}} = \left(\frac{\sum\limits_{j=1}^{N} m_j}{M} \right) \tilde{\alpha}_i \tag{2.31}$$

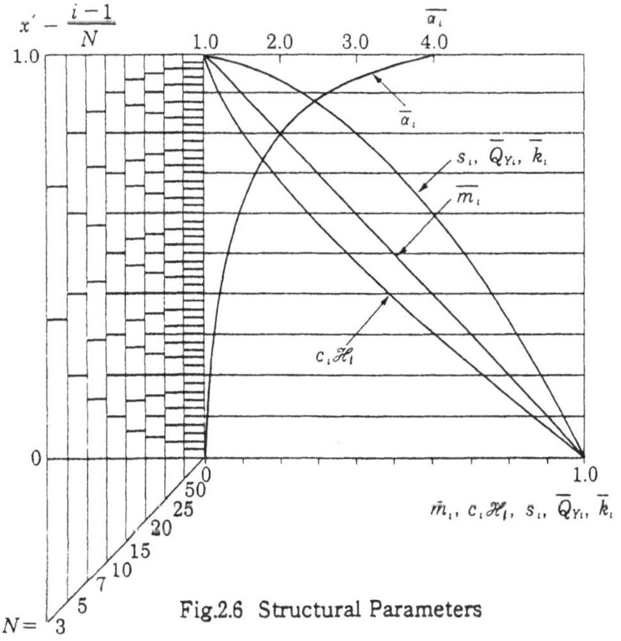

Fig.2.6 Structural Parameters

Considering the constancy of the yield deformation, the spring constant distribution, $\bar{k}_i$ is given by

$$\bar{k}_i = \frac{k_i}{k_1} = \bar{Q}_{Yi} \qquad (2.32)$$

Considering that $s_1 = 1.0$ and $s_i / s_1 = s_i$, obviously, the following relationship holds.

$$s_i = \left(\frac{\sum_{j=i}^{N} m_j}{M}\right)^2 \bar{\alpha}_i^2 \frac{k_1}{k_i} = \frac{\bar{Q}_{Yi}^2}{\bar{k}_i} = \bar{Q}_{Yi} \qquad (2.33)$$

$c_i \kappa_1$ is expressed as

$$c_1 \kappa_1 = \left(\frac{\sum_{j=i} m_j}{M}\right)^2 / \bar{k}_i \qquad (2.34)$$

The next approximate expressions are very useful in order to grasp the general tendency of the structu responses.

$$\kappa \quad 0.48 + 0.52N \qquad (2.35)$$

$$\sum_i s_i = 0.36 + 0.64N \qquad (2.36)$$

2.6 Effective period

In the inelastic systems, the overall stiffness is softened as the plastic deformation develops. Thus, the instantaneous period changes between the initial period, T_0 and the longest period, T_m. The energy spectrum for the purely elastic system, $_0V_E^{(T)}$ is strongly dependent on the natural period, T_0. On the other hand, the inelastic system is characterized by a wide range of variety of the natural period. Therefore, the energy input in the inelastic system can be expressed as follows.

$$E = \frac{\int_{T_0}^{T_m} {}_0V_E^2(T)dT}{2(T_m - T_0)} \qquad (2.37)$$

The design energy spectrum is such as shown in Fig.2.2 and in the shorter period range, the spectrum is written as follows.

$$V_E = aT \qquad (2.38)$$

Substituting Eq.(2.38) into $_0V_E$ in Eq.(2.37) E is obtained as

$$E = \frac{M}{2}\left(a\sqrt{\frac{T_0^2 + T_0 T_m + T_m^2}{3}}\right)^2 = \frac{M(aT_e)^2}{2} \qquad (2.39)$$

$$\text{where} \quad T_e = \sqrt{\frac{T_0^2 + T_0 T_m + T_m^2}{3}} \qquad (2.40)$$

Eq.(2.39) implies that the energy input in the inelastic system can be expressed by applying the effective period, T_e in place of T_0.

T_e can be also approximated by

$$T_e = \frac{T_0 + T_m}{2} \tag{2.41}$$

When the restoring force characteristics are to be described precisely, a standard load-deformation relationship is required. To be standard is same as to be well-definable. The best well-definable load-deformation relationship is the load-deformation relationship under the monotonic loading. Thus, the monotonic load-deformation curve is indispensable to describe the restoring force characteristics.

In Fig.2.7, a monotonic load deformation curve is schematically shown. Referring to a maximum response, $\bar{\mu}$ and the monotonic load-deformation curve, the instantaneous rigidity of a system, k_0 can be defined by the secant modules as follows.

$$k_0 = \frac{q'}{(1 \cdot} \quad \overline{\delta_Y} \tag{2.42}$$

Where $_q J_Y$: yield level associated with $\bar{\mu}$.

The period of vibration which corresponds to k_s, T_s is defined as follows.

$$T_s = 2\pi \sqrt{\frac{M}{k_s}} = T_0 \sqrt{\frac{1 + \bar{\mu}}{q}} \tag{2.43}$$

The maximum instantaneous period of vibration, T_m can be evaluated on the basis of T_s and is expressed in a following formula.

$$T_m = a_T T_s$$

Where a_T . modifying constant

a_T is obtained as follows, according the type of resting force characteristics.

for the elastic-perfectly plastic type:

$$a_T = \frac{1 + \frac{\bar{\mu}}{8}}{\sqrt{1 + \bar{\mu}}} \fallingdotseq 0.8 \tag{2.44}$$

for the origin-orienting type and degrading type :

$$a_T = 1.0 \tag{2.45}$$

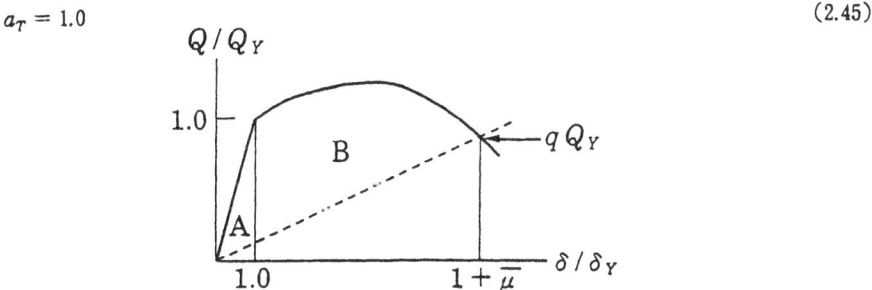

Fig.2.7 Monotonic Load-Deformation Diagram

3. Ordinary Method for Strengthening

3.1 Basic formulation

Ordinary multi-story frames can be identified to be stiff structures consisting of rigid elements. Herewith, a basic formulation is derived by taking the elastic-perfectly plastic type of restoring force characteristics.

The equilibrium of energy is given by Eq.(2.3) or Eq.(2.4)

The elastic vibrational energy is expressed as follows (Akiyama, 1985).

$$W_e = \frac{Mg^2 T^2}{4\pi^2} \cdot \frac{\alpha_1^2}{2} \tag{3.1}$$

W_p consists of cumulative strain energy in every story W_{pi}. Then,

$$W_p = \sum_{i=1}^{N} W_{pi} \tag{3.2}$$

W_{pi} is expressed as follows.

$$W_{pi} = \frac{Mg^2 T^2}{4\pi^2} \cdot c_i \alpha_1 \eta_i \tag{3.3}$$

The total damage can be related to the damage of ith story as follows, referring to Eq.(2.27).

$$W_p = \gamma_i W_{pi} \tag{3.4}$$

where $\quad \gamma_i = \dfrac{\sum\limits_{j=1}^{N} s_j p_j^{-n}}{s_i p_i^{-n}}$

Substituting Eqs.(3.1) and (3.2) and using Eq.(3.3), the following relationships are obtained.

$$_i\alpha_1 = \frac{1}{\sqrt{1 + 2c_i \bar{\alpha}_i^2 \gamma_i \eta_i}} \cdot \frac{2\pi V_D(T_e)}{Tg} \tag{3.5}$$

or

$$_iV_D(T_e) = \frac{\alpha_1 Tg}{2\pi} \cdot \sqrt{1 + 2c_i \bar{\alpha}_i^2 \gamma_i \eta_i} \tag{3.6}$$

where $\quad _i\alpha_1$: required strength level of 1st story determined by the energy absorption capacity of ith story

$\quad\quad _iV_D$: permissible level of seismic input determined by the energy absorption capacity of ith story

In deriving the above equations, the following relation is postulated.

$$\alpha_i = \alpha_1 \bar{\alpha}_i \tag{3.7}$$

In the stiff structures characterized by the elastic-perfectly plastic restoring force characteristics, η_i does not become so large that T_e can be assumed to be equal to T ($= T_0$).

Eq.(3.6) can be rewritten as follows, reminding the definitions of c_i, s_i, and γ_i.

$$_iV_D(T) = \frac{\alpha_1 Tg}{2\pi} \sqrt{1 + \frac{2\eta_k \sum\limits_{j=1}^{N} s_j (p_j / p_i)^{-n}}{\kappa_1}}$$ (3.8)

3.2 Principle of strengthening

The permissible level of the seismic input for a structure is expressed in terms of V_D as follows.

$$V_D = \mathrm{Min}\{_iV_D\}$$ (3.9)

In order to raise the level of V_D, there are three methods:

1) to increase α_1

2) to increase γ_i

3) to increase η_k

Methods in 1) and 2) correspond to the strengthening in the lateral shear force resistance. Even in these methods, the level of η_k must be maintained.

The effective way to increase strength, while not impairing the deformation capacity, is to stiffen the original frame with independent closed frames in which the required level of η_k is equipped as shown in Fig.3.1.

In implementing the third method, the cause which impairs the deformation capacity must be made clear.

These causes are:

• flexural buckling

• local buckling

• lateral torsional buckling

• fracture

Causes for various type of buckling can't be easily removed. Fractural modes of failure in connection often totally impair the energy absorption of members as experienced in the Hyogoken-Nanbu Earthquake (1995) and the Northridge Earthquake (1994). In order to make a member develop its full deformation capacity, the connection must be equipped with the strength greater than the member's maximum strength. Thus, the improvement in η_k is in many cases reduced to the repair ot connections.

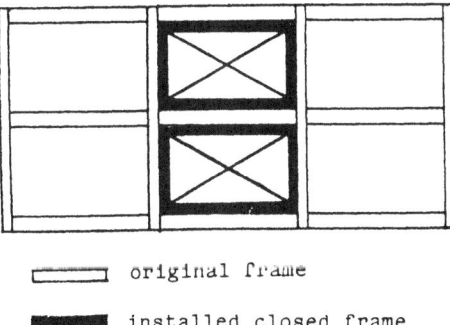

original frame

installed closed frame

Fig.3.1 Installation of Closed Frame

3.3 Illustrative example

Taking a ten-storied shear-type steel frame, the effectiveness of strengthening in strength is demonstrated.

It is possible to estimate p_j in Eq.(3.8) on the basis of the material yield strength prescribed in codes. However, the real strength is influenced by the scatter of material strength. The standard deviations of the yield strength of steels are of the order of 10% of the prescribed yield strength. Considering such a situation, another strength gap p_s is introduced in the observed ith story. Then, p_j applied to Eq.(3.8) is modified to be:

$$\left.\begin{aligned} p_{j \neq i} &= p_{j0} \\ p_i &= p_{i0} p_s \\ p_s &= \frac{1}{1.1} \end{aligned}\right\} \tag{3.10}$$

where $p_{j0} : \dfrac{\alpha_i}{\alpha_1 \bar{\alpha}_i}$ estimated by using prescribed material strength

p_s : strength gap due to scatter of the material

Using p_j in Eq.(3.10), Eq.(3.8) is reduced to

$$_i V_D = \frac{\alpha_1 Tg}{2\pi} \sqrt{1 + \frac{2\eta_i}{\kappa_1}\left\{\frac{\sum_{j \neq 1} s_j p_{j0}^{-n}}{\left(p_{i0} p_s\right)^{-n}} + s_i\right\}} \tag{3.11}$$

Applied parameters are:

$\kappa_1 = 0.48 + 0.52N = 5.68$

$T = 1.2s$

$n = 6$

$\eta_i = 10.0$

$m_i / m_1 = 1.0$

s_i and $\bar{\alpha}_i$ are obtained from Fig.2.6 and are shown in Table 3.1. α_i is assumed as shown in the table.

Table 3.1 Strengthening in Strength

N	s_i	$\bar{\alpha}_i$	initial			primary			secondary		
			α_i	p_{i0}	$_iV_D$ (cm/s)	α_i	p_{i0}	$_iV_D$ (cm/s)	α_i	p_{i0}	$_iV_D$ (cm/s)
10	0.26	2.66	1.02	1.20	371	1.02	1.20	329	1.02	1.20	311
9	0.41	2.01	0.71	1.10	303	0.71	1.10	259	0.71	1.10	(247)
8	0.51	1.70	0.46	0.85	(152)	0.54	(1.00)	(203)	0.60	(1.10)	248
7	0.61	1.52	0.44	0.90	175	0.49	(1.00)	204	0.54	(1.10)	249
6	0.69	1.38	0.45	1.02	240	0.45	1.02	215	0.49	(1.10)	250
5	0.76	1.25	0.44	1.10	225	0.44	1.10	263	0.44	1.10	251
4	0.82	1.15	0.39	1.05	260	0.39	1.05	233	0.41	(1.10)	252
3	0.89	1.10	0.43	1.21	385	0.43	1.21	342	0.43	1.21	325
2	0.91	1.05	0.37	1.11	303	0.37	1.11	260	0.37	1.11	259
1	1.00	1.00	0.32	1.00	232	0.32	1.00	209	0.35	(1.10)	254

Applying Eq.(3.11), V_D is obtained as shown in Table 3.1. p_i is smallest in the eighth floor and the severest damage concentration produces the minimum resistance in terms of V_D in the eighth floor. It can be easily seen that to strengthen relatively weaker seventh and eighth stories is effective to raise the resistance. As a primary strengthening, strength in these stories are increased to the level of $p_{io} = 1.0$.

As a result, V_D is increased by 34% by the 10~15% of strengthening of the strengths in only two stories.

The secondary strengthening is made by strengthening the first story to the level of $p_{i0} = 1.1$ and strengthening other stories, where p_{i0} is smaller than 1.1, to the same level. This time, 10% of strengthening in five stories produces 22% of increase in V_D.

In this manner, not only the increase of strength but also the adjustment of strength distribution to make p_j uniform are very effective in the improvement of the structural resistance against earthquakes.

4. Advanced Method for Strengthening

4.1 General remarks

In ordinary structures, the strengthening is made mainly by increasing strength on the assumption that the deformation capacity of members is unchanged. On the other hand, the strengthening by means of increasing the deformation capacity of structural members is not easy on the reason that to improve drastically members' deformation capacity requires tremendous works and still is not reliable in its effect. However, there can be a promising way of strengthening by changing an ordinary structural type into a more preferable structural type.

More preferable structural types can be generally identified to be the flexible-stiff mixed structure. The practical application of the flexible-stiff mixed structure is divided into two types:

(a) energy-concentrating type

(b) energy-dispersing type

In the energy-concentrating type, the input energy is intended to be concentrated and absorbed in one story where an energy absorbing mechanism is perfectly arranged.

Such a mechanism has been already realized in the base-isolated buildings by introducing laminated rubber bearings and dampers equipped with high energy absorption capacity.

The lack of seismic resistance in existing buildings can be drastically improved by preparing the energy absorbing story based on the energy-concentrating principle. The location of the energy absorbing story should not be restricted in the first story.

What should be checked is not only the energy absorption capacity in the energy absorbing story but also the condition under which the total energy input can assuredly arrive at the energy absorbing story. When the energy absorbing story is located at the elevated position, it was found that the yield shear force coefficient at the base of structure should be grater than a certain limit to pump up the total energy input to the energy absorbing story.

The limit of the base shear coefficient, $_l\alpha_1$ was found to be related to the mass ratio as follows (Akiyama, 1986).

$$_l\alpha_1 = 0.07 \frac{M}{M_2} a_{e1}$$

where M : total mass of system   (4.1)

M_2 : mass which rests on the energy absorbing story

$$a_{e1} = \frac{2\pi V_E}{Tg}$$

As far as the condition of $\alpha_1 > {}_l\alpha_1$ is satisfied, the structure can be made seismic-resistant by making the total energy input absorbed in the energy absorbing story.

In order to make the total energy input concentrate in the energy absorbing story, the energy absorbing story should be weaker relatively than the other stories.

This condition is described as follows, taking the strength reduction factor of 0.5.

$$\alpha_c = \frac{1}{2} \bar{a}_c \left(M_2 / M \right) \alpha_1$$

(4.2)

$$\text{or} \quad \alpha_1 = \frac{2\alpha_c}{\bar{a}_c \left(M_2 / M \right)}$$

where $\bar{a}_i$: yield shear coefficient ratio at the position of the energy absorbing story, determined from $\bar{a}_i(x)$ in Eq.(2.26) by putting $x = 1 - M_2 / M$

Applying this principle, it is possible to strengthen the existing building by extending the scale as far as the existing part has a margin of strength to sustain the gravity load of the extended part.

In the energy-dispersing type, the input energy is intended to be evenly distributed over the whole stories. Compared to the case of (a), the special arrangement for the energy absorbing story is not required, but the energy absorption devices must be distributed all over the structure. This type of structures is found in high-rise or super high-rise buildings in which structural skeletons are intended to remain elastic.

In this type of structure, the energy absorption is made by special devices (dampers) and structural members which support gravity loadings are basically kept elastic or quasi-elastic.

The energy absorbing efficiency becomes greater as the maximum elastic deformation of structural skeletons increases, since the dampers absorbs energy by developing hysteretic behaviors within the limit of the maximum deformation amplitude. Therefore, the energy absorption due to the inelastic deformation is not expected on the part of structural skeletons.

The most important performance of dampers is the high elastic rigidity which enables a quick plastification and the high energy absorption capacity. Various elements for dampers have been already developed as follows.

- extremely mild steels with low yield point
- visco-elastic materials
- friction damper
- lead damper
- non-buckling diagonal bracing
  (core rod with scabbard preventing buckling)
- nicely shaped steel plates and rods arranged collectively

In order to increase elastic rigidity, dampers are installed to form a part of rigid frames as shown in Fig.4.1.

The strengthening by means of dampers is easier than the strengthening of strength in ordinary buildings, since the installation of dampers is made by only adding dampers to them without touching structural skeletons.

4.2 Basic formulation for energy-concentrating type

In highly non-linear systems, it is possible to neglect the elastic vibrational energy, W_e in applying Eq.(2.4). Therefore, Eq.(2.4) is reduced to be

$$W_p = E_D = \frac{M V_D^2}{2} \tag{4.3}$$

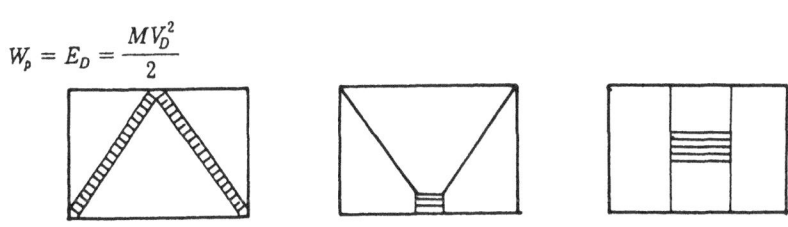

| non-buckling | hysteretic damper | column with |
| diagonal brace | under story shear | shear-yielding |

Fig.4.1 Installation of Rigid Elements fuse metal

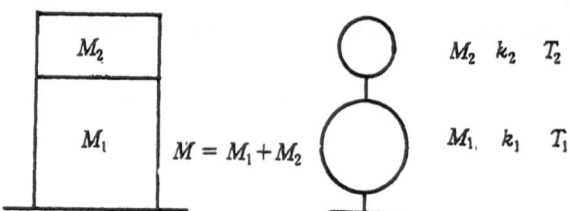

Fig.4.2 Structural Model

The energy absorbing story is assumed to be located in kth story. The structural model is shown in Fig.4.2. The total energy is assumed to concentrat in the kth story. Assumed restoring force characteristics of dampers are of elastic perfectly plastic type.

W_p is expressed as

$$W_p = {_sQ_{Yk}}\,\delta_{pk} \tag{4.4}$$

where δ_{pk} : cumulative plastic deformation in kth story

$_sQ_{Yk}$: strength of dampers in kth story

Neglecting the elastic deformation and applying Eq.(2.14), δ_p is expressed as

$$\delta_{pk} = 8\delta_{mk} \tag{4.5}$$

where δ_{mk} : maximum deformation in kth story

E_D can be converted to the equivalent height by which the total mass is lifted up under the gravitational force as follows.

$$Mgh_{eq} = \frac{MV_D^2}{2}$$

$$\therefore \quad h_{eq} = \frac{V_D^2}{2g} \tag{4.6}$$

where h_{eq} : equivalent height

Substituting Eqs.(4.5) and (4.6) into Eq.(4.3), the following basic formulation is made.

$$_s\alpha_k\delta_{mk} = \frac{h_{eq}}{8}\left(\frac{M}{M_2}\right) \tag{4.7}$$

where M_2 : mass which rests on kth story

It has been already found that the accessibility of the energy into kth story is best when the natural period of the lower structure consisting of the mass M_1 and the spring constant k_1, T_1, and the natural period of the upper structure consisting of the mass M_2 and the spring constant k_2, T_2 are within the following range.

$$\frac{T_2}{T_1} = 0.7 \sim 1.5 \tag{4.8}$$

Herewith, $T_2 = T_1$ is assumed as a practical condition.

The upper structure is a flexible-stiff mixed structure. The spring constant of the elastic element can be expressed as

$$k_f = \frac{f_s Q_{Yk}}{\delta_{mk}}$$ (4.9)

where $f = \frac{{}_f Q_m}{{}_s Q_{Yk}}$

${}_f Q_m$: maximum shear force in the flexible element

The equivalent stiffness of the rigid element which corresponds to the maximum period can be described as follows, referring Eq.(2.44) and neglecting deformations other than in kth story.

$$k_s = \frac{1.56_s Q_{Yk}}{\delta_{mk}}$$ (4.10)

Then, the maximum period of the upper structure, T_m is obtained as

$$T_m = 2\pi \sqrt{\frac{\delta_{mk}}{g(f+1.56)_s \alpha_k}}$$ (4.11)

Approximating the initial period of the upper structure, T_0 to be nullified, the effective period of the upper structure is obtained as follows, referring Eq.(2.40).

$${}_2 T_e = \frac{2\pi}{\sqrt{3}} \sqrt{\frac{\delta_{mk}}{g(f+1.56)_s \alpha_k}} = \frac{1}{\sqrt{3}} \sqrt{\frac{1}{\left(1+\frac{1.56}{f}\right)}} T_f$$ (4.12)

where T_f : natural period of the upper structure in case without dampers

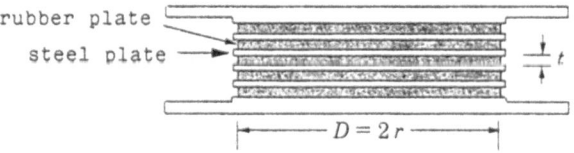

rubber plate

steel plate

$D = 2r$

Fig.4.3 Laminated Rubber Bearing

In applying Eq.(4.8), T_2 should be replaced by ${}_2 T_e$ given by Eq.(4.12). Taking $f = 1$, T_f is obtained to be

$$T_f = 2.77_2 T_e$$ (4.13)

As flexible elements, laminated rubber bearings can be generally applied. When the upper structure is supported by the laminated rubber bearings as shown in Fig.4.3, T_f is expressed as

$$T_f = 2\pi \sqrt{\frac{\bar{\sigma} n_t t}{gG}}$$ (4.14)

where $\bar{\sigma}$: average vertical stress in rubber sheet under the neutral position

$n_t t$: total thickness of rubber

G : shear modules of rubber

The natural period of the total system, T_t is generally expressed by

$$T_t = 2\pi\sqrt{\frac{M}{k}}\, a_r \tag{4.15}$$

where $k = k_1 + k_2$

$M = M_1 + M_2$

$$\frac{1}{a_r^2} = \frac{\left(\dfrac{1}{\mu_1} + \dfrac{r_2}{\mu_2}\right)}{2} - \frac{1}{2}\sqrt{\left(\frac{1}{\mu_1} - \frac{r_2}{\mu_2}\right)^2 + \frac{4r_2^2}{\mu_1\mu_2}} \tag{4.16}$$

$$\mu_1 = \frac{M_1}{M},\ \mu_2 = \frac{M_2}{M},\ \mu_1 + \mu_2 = 1.0$$

$$r_1 = \frac{k_1}{k},\ r_2 = \frac{k_2}{k},\ r_1 + r_2 = 1.0$$

The condition of $T_2 / T_1 = 1.0$ is expressed by

$$\frac{\mu_1}{r_1} = \frac{\mu_2}{r_2} \tag{4.17}$$

Considering $\mu_1 + \mu_2 = 1.0$ and $r_1 + r_2 = 1.0$, Eq.(4.17) is reduced to

$$r_2 = \mu_2 \tag{4.18}$$

Under the condition of Eq.(4.18), a_r is obtained as shown in Fig.4.4. a_r is approximated for $0.1 < M_2 / M < 0.5$

$$a_r = 1.1 + \frac{M_2}{M} \tag{4.19}$$

Then, using T_1 and T_2, T_t is expressed as

$$T_t = a_r\sqrt{r_1 T_1^2 + r_2 T_2^2} = a_r T_1 \tag{4.20}$$

In Fig.4.5, $_s a_k \delta_m - M_2 / M$ relationship expressed by Eq.(4.7) in shown. Also, in the same figure, the optimum yield shear force coefficient distribution is shown on the right-hand ordinate by taking M_i / M on the abscissa. M_i denotes the mass which rests on ith story.

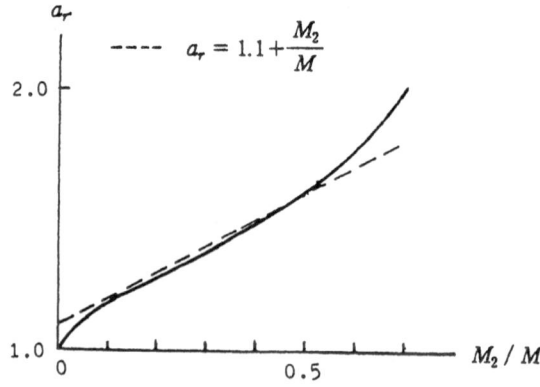

Fig.4.4 $a_r - M_2 / M$ Relationship

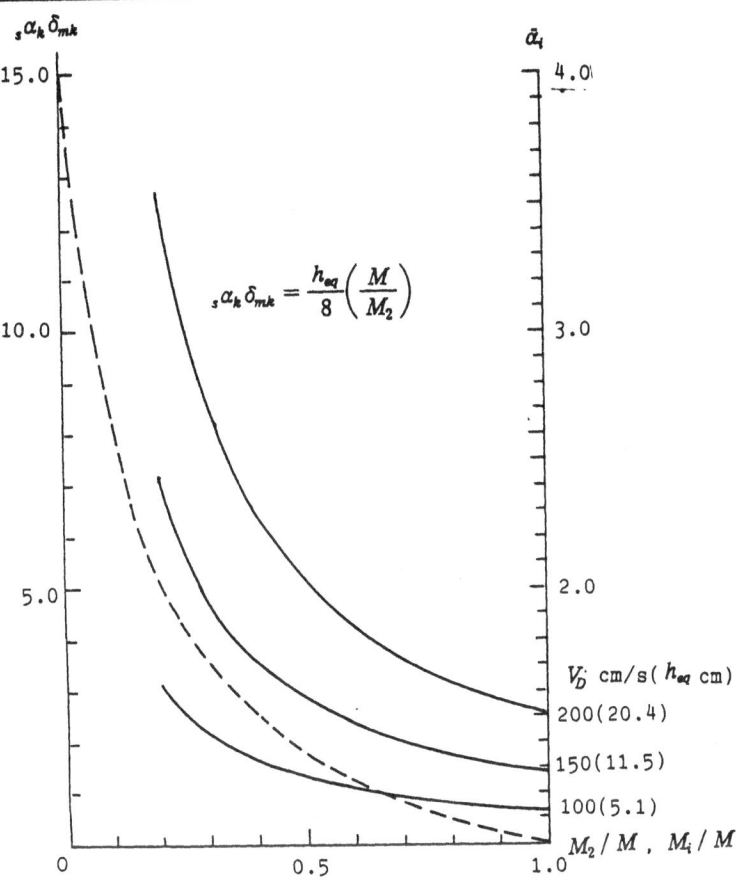

Fig.4.5 $_s\alpha_k\delta_m - M_2/M$ Relationship

The maximum shear force coefficient developed in the kth story, α_{km} is expressed by

$$\alpha_{km} = {}_s\alpha_k(1+f) \tag{4.21}$$

Therefore, in order to prevent plastification in the story other than the kth story, the following condition must be satisfied.

$$\alpha_{i\neq k} \geqq \frac{{}_s\alpha_k(1+f)\bar{\alpha}_i}{\bar{\alpha}_k} \tag{4.22}$$

4.3 Illustrative example for energy-concentrating type

A three story structure is taken as an illustrative example. Structural parameters are assumed to be:

$T_1 = 0.6\text{sec}$

$m_i / m_1 - 1.0, \; \delta_{Yi} / \delta_{Y1} = 1.0$

$\alpha_1 = 0.3, \; \alpha_2 = 0.45, \; \alpha_3 = 0.9$

structural type: weak column type $(n = 1.2) \; \eta_i = 10.0$

Table 4.1 Seismic Resistance of Original Structure

i	S_i	d_i	α_i	$\dot{p}_i$	$\dot{V}_D$ (cm/s)
3	0.56	1.63	0.90	1.84	—
2	0.81	1.18	0.45	1.27	—
1	1.00	1.00	0.30	1.00	93

Similarly to Table 3.1, the structural resistance of the structure is evaluated as shown in Table 4.1. As is shown in the table, a remarkable damage concentration takes place in the first story. This structure is strengthened by applying the energy absorbing story in many ways as shown in Fig.4.6. In the first case, the original structure is totally supported by the base islotion story. In the second case, the third story in supported by the energy absorbing story. In the third and fourth cases, the structure is enlarged by an upwardly extension of one to three stories on the assumption that the original structure has a sufficient strength to support the extended part.

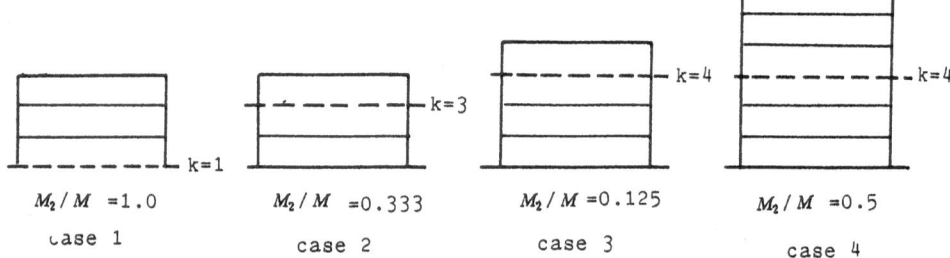

$M_2/M = 1.0$ $M_2/M = 0.333$ $M_2/M = 0.125$ $M_2/M = 0.5$

case 1   case 2   case 3   case 4

Fig.4.6 Position of Energy Absorbing Story

The aimed level of the seismic input is assumed to be

$$V_D = 150 \, \text{cm/s} \; (h_{eq} = 11.5 \, \text{cm}) \tag{4.23}$$

The value of f was assumed to be unify.

In Table 4.2, the required strength for strengthening is shown for each case. Based on the applied deformation, δ_{mk}, the required strength, $_s a_k$ is obtained by Eq.(4.7). Dimensions of the laminated rubber isolator which can develops the maximum deformation, δ_{mk} is also shown.

In the first case, a base-isolated structure is realized and the required strength is considerably smaller than the inherent strength of the original structure.

In the second case, isolators and dampers are installed in the third story. The natural period of shear type structure in generally approximated by

$$T = 2\pi \sqrt{\frac{M\kappa_1}{k_1 g}} \tag{4.24}$$

where M : total mass

$\kappa_1 = 0.48 + 0.52N$

Therefore, the natural period of the original three storied structure is expressed as

$$T = 2\pi \sqrt{\frac{M \cdot 2.04}{k_1 g}} = 0.6 \tag{4.25}$$

Table 4.2 Required Strength in Strengthening

				structure						isolater				
			original			strengthened								
case	i	$\bar{a}_i$	α_{i0}	$\dfrac{\overset{N}{\underset{j=i}{\Sigma}} m_j \alpha_{i0}}{m_1}$	k	α_i	$\dfrac{\overset{N}{\underset{j=i}{\Sigma}} m_j \alpha_i}{m_1}$	$\dfrac{M_2}{M}$	$\dfrac{\delta_{mk}}{,a_k}$	D cm	n_{tt} cm	$\hat{\sigma}$ kg/cm^2	$\dfrac{T_1}{T_f}$	T_t s
1	3	1.68	0.9	—	1	0.235	—	0.1	20cm 0.0715	40	10	100	— 3.17s	1.15
	2	1.38	0.45			0.170								
	1	1.00	0.30			0.143								
2	3	1.68	0.9	—	3	0.432	—	0.333	20cm 0.216	40	10	14.1	0.43s 1.19s	0.61
	2	1.38	0.45			0.354								
	1	1.00	0.30			0.257								
3	4	1.83			4	0.384	0.384	0.25	30cm 0.192	50	15	18.2	0.6s 1.66s	0.81
	3	1.30	0.9	0.9		0.273	0.546							
	2	1.12	0.45	0.9		0.235	0.705							
	1	1.00	0.30	0.9		0.210	0.840							
4	4	1.38			4	0.192	0.576	0.5	30cm 0.096	50	15	18.2	0.6s 1.66s	0.96
	3	1.18	0.9	0.9		0.164	0.668							
	2	1.08	0.45	0.9		0.150	0.750							
	1	1.00	0.30	0.9		0.139	0.834							

Based on Eq.(4.25), the natural period of the lower two storied structure, T_1 is obtained as follows.

$$T_1 = 2\pi \sqrt{\frac{2M \cdot 1.52}{3k_1 g}} = 0.43 \qquad (4.26)$$

In the third case, the strengthened structure becomes four storied structure. Based on the $\bar{a}_i$, the required yield shear coefficient is shown in the Table. The actual yield shear force is proportional to the following quantity.

$$\frac{Q_{Yi}}{m_1 g} = \frac{\overset{N}{\underset{j=i}{\Sigma}} m_j \alpha_i}{m_1} \qquad (4.27)$$

In Table 4.2, the values given by Eq.(4.27) are shown. By comparing these values in the original system and the strengthened system, it is obvious that the lower structure is equipped with sufficient strength to back up the strengthening. In the fourth case, the strengthened structure becomes a six storied structure. Still, the lower structure can back up the strengthening.

4.4 Basic formulation for energy-dispersing type

In the flexible-stiff mixed structure with a higher value of f defined by Eq.(4.9), the damage concentration is mitigated considerably as shown in Fig.2.4. To simplify the problem, the value of n is assumed to be

$$n = 0 \qquad (4.28)$$

The total energy absorption made by the stiff element W_p is expressed as

$$W_p = \tau_1 W_{p1} = \overset{N}{\underset{j=1}{\Sigma}} s_j \cdot {}_s \delta_s Q_{Y1} \delta_{m1} \qquad (4.29)$$

Applying Eq.(4.3), the basic expression similar to Eq.(4.7) is obtained as follows.

$$\alpha_{s1}\delta_{m1} = \frac{h_{eq}}{8\sum\limits_{j=1}^{N}s_j} \tag{4.30}$$

The effective period of the flexible-stiff mixed structure is obtained similarly to Eq.(4.12) as follows.

$$T_e = \frac{2\pi}{\sqrt{3}}\sqrt{\frac{\delta_{m1}\kappa_1}{g(f+1.56)\alpha_{s1}}} \tag{4.31}$$

Referring to Fig.2.1, the energy spectrum is expressed as

$$\left.\begin{array}{ll} \text{for} \quad T_e \le T_G, & V_D = \dfrac{T_e}{T_G}V_{Dm} \\[3mm] \text{for} \quad T_e > T_G, & V_D = V_{Dm} \end{array}\right\} \tag{4.32}$$

Applying Eqs.(4.31) and (4.32), $_s\alpha_1$ is obtained from Eq.(4.29) as follows.

$$\text{for} \quad T_e \le T_G, \quad \alpha_{s1} = \frac{\pi}{\sqrt{6}\,T_G}\sqrt{\frac{\dfrac{\kappa_1}{\sum\limits_{j=1}^{N}s_j}h_{eqm}}{(f+1.56)g}} \tag{4.33}$$

$$\text{for} \quad T_e > T_G, \quad \alpha_{s1} = \frac{h_{eqm}}{8\sum\limits_{j=1}^{N}s_j\delta_{m1}} \tag{4.34}$$

$$\text{where} \quad h_{eqm} = \frac{V_{Dm}^2}{2g}$$

κ_1 is nearly equal to $\sum\limits_{j=1}^{N}s_j$, and to apply the value of $f = 1.0$ to Eq.(4.33) for the system $\alpha_{fi} > \alpha_{si}$ $(f \ge 1.0)$ is a conservative measure to estimate α_{s1}.
Herein, α_{fi} and α_{s1} are defined as

$$\alpha_{fi} = \frac{_fQ_{mi}}{\sum\limits_{i=j}^{N}m_jg}, \quad \alpha_{si} = \frac{_sQ_{Yi}}{\sum\limits_{i=j}^{N}m_jg} \tag{4.35}$$

where $_fQ_{mi}$: maximum shear force in the flexible element in ith story

$_sQ_{Yi}$: maximum yield shear force in the stiff element in ith story

4.5 Illustrative example for the damage-dispersing type

$\alpha_{s1} - N$ relationships under the following conditions are shown in Fig.4.7:

$$\left.\begin{array}{l} T_G = 1.0 \\[3mm] \dfrac{\sum\limits_{j=1}^{N}s_j}{\kappa_1} = 1.0 \\[3mm] f = 1.0 \end{array}\right\} \tag{4.36}$$

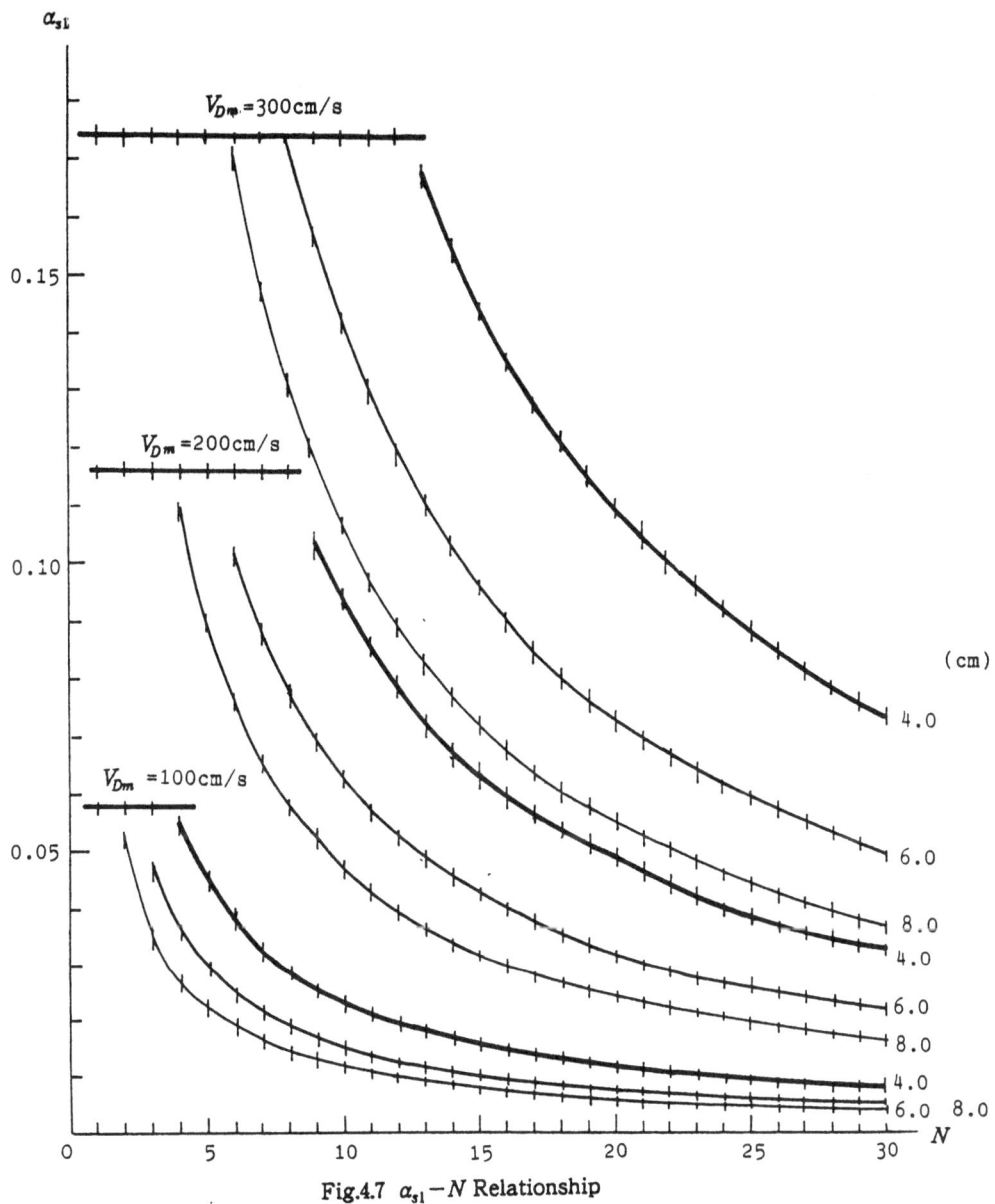

Fig.4.7 $\alpha_{s1} - N$ Relationship

Ten storied frames are taken as illustrative examples.

The frames are specified to be:

- weak-column type ($n = 12$)
- $_f\delta_{Yi} = 3.0\,\text{cm}$

$_f\delta_{Yi}$ is the elastic limit deformation of the flexible element (the structural skeleton) in each story.

The flexible element is not necessarily requested to be rigorously elastic. As a quasi-elastic condition, it is reasonably permitted to apply the following condition (see Fig.5.19).

$$\delta_{m1} = 2{,}0 \, \delta_{Yi} \tag{4.37}$$

Two cases are considered. The first case corresponds a rather stronger frame with sufficient inelastic deformation capacity of $\eta_k = 10.0$. The second case corresponds to a weak frame with a poor inelastic deformation capacity of $\eta_k = 2.0$. Both frames are of weak-column type.

Applying Eq.(3.11), the seismic resistance of the original frame is estimated as shown in Table 4.3. In ordinary frame, the maximum story displacement is related to η_k as follows.

$$\delta_{mi} = \left(1 + \frac{\eta_k}{4}\right)\delta_{Yi} \tag{4.38}$$

In the table, δ_{mk} means the maximum deformation developed in the story where the inelastic deformation capacity is exhausted under the minimum seismic input of $_kV_D$. Under the given condition of Eq.(4.37), the required strength $_s\alpha_1$, is obtained from Fig.4.7. Multiplying $\bar{a}_i$ to $_s\alpha_1$, the required strength of the stiff element in each story is obtained. In Table 4.3, α_i in the original frame corresponds to the maximum strength level developed in the flexible elements. Therefore, the value of f in each story f_i is expressed as

$$f_i = \frac{\alpha_i}{\alpha_{si}} \tag{4.39}$$

The aimed level of seismic input is:

for the first case, $V_{Dm} = 200 \, \text{cm/s}$

for the second case, $V_{Dm} = 100 \, \text{cm/s}$

In both cases, the condition of $f_i > 1.0$ is dully satisfied. Thus, it is seen that the strengthening by adding the stiff elements (dampers) is extremely effective.

Table 4.3 Required Strength in Strengthening

	case 1			$\delta_{mk} = 10.5$ cm $V_D = 200$ cm/s			case 2			$\delta_{mk} = 4.5$ cm $V_D = 100$ cm/s	
	original $\eta_k = 10.0$				strengthened		original $\eta_k = 2.0$				strengthened	
i	α_i	$\bar{a}_i$	p_{io}	$\frac{_iV_D}{\text{cm/s}}$	α_{si}	f_i	α_i	$\bar{a}_i$	p_{io}	$\frac{_iV_D}{\text{cm/s}}$	α_{si}	f_i
10	1.02	2.66	1.20	678	0.168	6.1	0.128	2.66	1.20	110	0.040	3.2
9	0.71	2.01	1.10	412	0.127	5.6	0.089	2.01	1.10	64	0.030	2.9
8	0.46	1.70	0.85	(138)	0.107	4.3	0.058	1.70	0.85	(32)	0.026	2.3
7	0.44	1.52	0.90	165	0.096	4.6	0.055	1.52	0.90	35	0.023	2.4
6	0.45	1.38	1.02	280	0.087	5.2	0.056	1.38	1.02	50	0.021	2.7
5	0.44	1.25	1.10	412	0.079	5.6	0.055	1.25	1.10	70	0.020	2.8
4	0.39	1.15	1.05	327	0.073	5.3	0.049	1.15	1.05	57	0.017	2.9
3	0.43	1.10	1.21	718	0.069	6.2	0.054	1.10	1.21	116	0.017	3.3
2	0.37	1.05	1.11	440	0.066	5.6	0.046	1.05	1.11	74	0.016	2.9
1	0.32	1.00	1.00	262	(0.063)	5.1	0.040	1.00	1.00	48	0.015	2.7

5. Practical Application

5.1 Ordinary method of strengthening

a) A historical building (Marunouchi building)

The building which is shown in Fig.5.1 is a historical building which was designed by an American company and was completed in 1923. Outlined features of the building are shown in Table 5.1. At that time, just before the Great Kanto Earthquake (1923), there was no definite seismic design guideline in Japan. Main skeletons were steel frames as shown in Fig.5.2 and beam-to-column connections were semi-rigid joints.

Table 5.1 Outlined Feature of Marunouchi Building

Site	2-4-1 Marunouchi,Ciyoda-ku,Tokyo
owner	Mitsubishi Estate Co.Ltd
Main purpose	Office, Retails
Total floor space	Approx. 650357 square feet
Floors	1 floors below ground, 8 floors above ground , 1 floors penthouse
Height	Approx. 102 feet
Structure	Steel frame reinforced cocrete
	Steel frame brick built (original)
Completion	February 1923

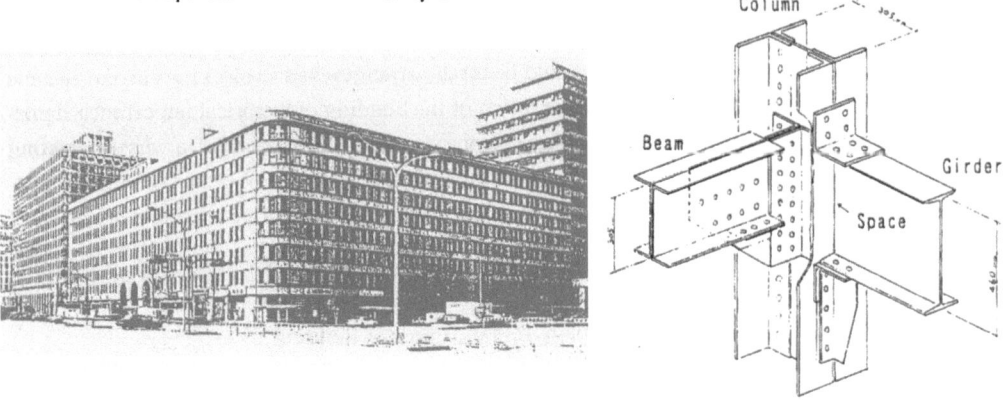

Fig.5.1 Marunouchi Building   Fig.5.2 Typical Column and Girder Connections

During construction, the building met the Uraga-oki Earthquake (1922), and suffered a considerable damage.

After the invaluable experience, various manners of strengthening were introduced such as

1) encasement of beam-to-column connections by concrete to make them rigid (see Fig.5.3)
2) installation of steel diagonal bracings
3) installation of knee-bracings at corners

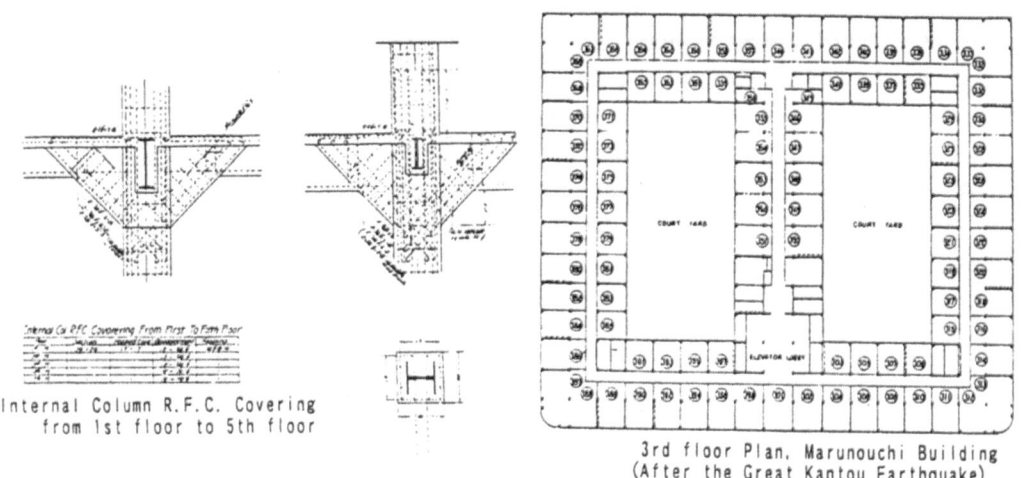

Internal Column R.F.C. Covering
from 1st floor to 5th floor

3rd floor Plan, Marunouchi Building
(After the Great Kantou Earthquake)

Fig.5.3 Covering Skeletons by Concrete

Fig.5.4 Strengthening with Shear Walls

However, immediately after the completion, the building was attacked by the Great Kanto Earthquake and received a heavy damage. After the lessons of the earthquake, the building was thoroughly strengthened with reinforced concrete shear walls which reached a number of four hundreds on the sacrifice of efficiency for office spaces as is seen in Fig.5.4. After the Great Kanto Earthquake, no significant earthquakes happened, and the building survived.

However, it was judged that the building should be totally strengthened to meet the current seismic design practice. After the discussion on preservation of the building of historical importance, demolition of the building was decided. The building disappeared in 1998. In Table 5.2, a very interesting record on the measurement of the natural period of the building is shown. We can see the effectiveness of strengthening and the influence of damage and aging on the shortening and elongation of the natural period.

Table 5.2 Natural Period of Marunouchi Building

observation year	time	east and west direction	north and south direction
December 1921	under construction	1.11 sec	1.14 sec
February 1922	just before completion	0.89	0.91
May 1922	just after Uraga-oki earthquake	1.09	1.01
November 1922	after reinforcement	0.67	0.71
September 1923	just after Great Kanto earthquake	1.11	1.18
May ～ July 1926	after reinforcement (average)	0.50	0.48
June 1997	just before pulling down	0.57	0.60

b) A building for office use

The building shown in Fig.5.5 and Table 5.3 was constructed in 1960s. At that time, the national building code prescribed the seismic design method based on the allowable stress design. In the light of the improved design method after the revision of the building code in 1981, the drawbacks of formerly built buildings were summarized as follows.

1) The yield shear force coefficient distribution was a monotonous one as described by

$$\bar{\alpha}_i = 1.0 \tag{5.1}$$

Thus, in order to prevent the damage concentration, into the upper stories, the upper stories must be strengthened in strength.

2) Since the energy absorption capacity was not checked, the true resistance against earthquakes was uncertain.

In order to strengthen the frames in strength, reinforced concrete shear walls were used; some initially installed shear walls were stiffened by adding thickness and new shear walls were added.
In order to increase energy absorption capacity, steel-panel dampers as shown in Fig.5.6 were introduced. The steel panel damper consists of a rigid surrounding steel frame and a central shear panel

Table 5.3 Outlined Feature of Nihon Building

Site	2-8 Otemachi, Chiyoda-ku, Tokyo
Owner	Mitsubishi Estate Co.Ltd
Main purpose	Office
Total floor spase	Approx. 1,859,889 square feet
Floors	4 floors below ground, 14 floors above ground,
height	165.3feet
Structure	Steel framed reinforced concrete structure
Completion	1962 (PART1), 1965 (PART2), 1968 (PART3)

Fig.5.5 Nihon Building

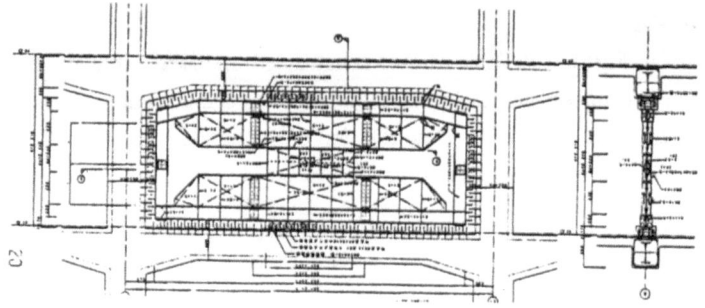

Fig.5.6 Steel Panel Dampers

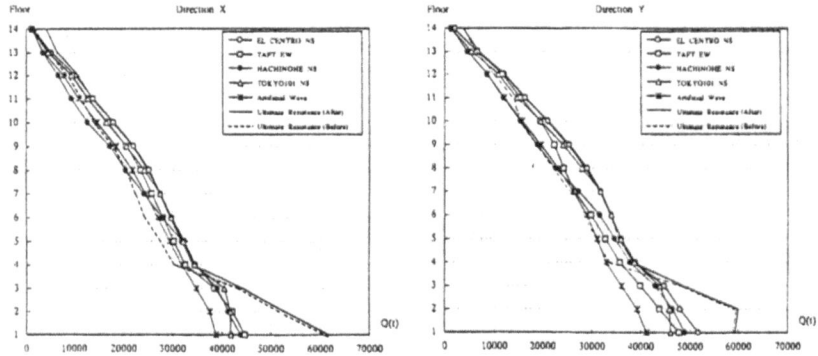

Fig.5.7 Response Maximum Shear Forces

which is made of ultra-mild steel. The central shear panel behaves elastically under a very small story drift, say 1mm and can absorb energy effectively in a form of inelastic strain energy. Thus, the steel panel dampers act as stiff elements in the flexible-stiff mixed structure.

The level of seismic energy input expressed by V_D in Eq.(2.6) is assumed to be

$$V_D > 100.0 \, cm/s \tag{5.2}$$

Numbers of installed shear walls and steel panel dampers are;

   reinforced concrete shear wall : 178
   steel panel damper   : 141

Shear force responses to the aimed level of seismic inputs are shown in Fig.5.7. As is seen in the figure, the increase of the ultimate resistance in the shear strength is about 25% of the original frame. About 5% of them is the contribution of the steel panel dampers, the effectiveness of steel panel dampers in energy absorption will be mentioned in 5.2 (b).

5.2 Advanced method

a) Energy concentrating type

The building shown in Fig.5.8 and Table 5.4 was also constructed in 1960s (Ogura et al, 1997). The structure is of steel framed reinforced concrete type. Strengthening of the upper stories was undertaken by applying the energy absorbing story at the intermediate height (at the eighth floor).

Table 5.4 Outlined Features of Yugawara Training Center

Outline of the building
- Location   ;   Atami city, Shizuoka, Japan
- Year built   ;   1964
- Story   ;   16 (above the ground), 2 (under the ground)
- Total area (m^2) ;   15,658
- Eaves height (m)   ;   44.53

- Design and construction   ;   Taisei Corporation
- Renewal period   ;   Nov. 1996 ~ Apr. 1997

Fig.5.8 Yugawara Training Center

The aimed level of seismic input V_D and the mass ratio M_2 / M_1 are;

$$V_D > 100 \text{ cm/sec} \tag{5.3}$$

$$M_2/M = 0.335 \tag{5.4}$$

The natural periods of the original system are;

$$T_1 = 0.2 \text{ sec}, \ T_2 = 0.45 \text{ sec}, \ T = 0.46 \text{ sec} \tag{5.5}$$

The yield shear force coefficient of the structure at the base is;

$$\alpha_1 = 0.245 \tag{5.6}$$

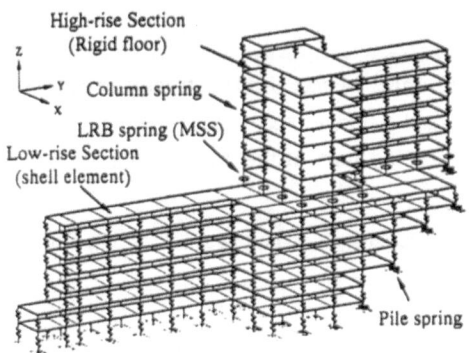

Fig.5.9 Analytical Model

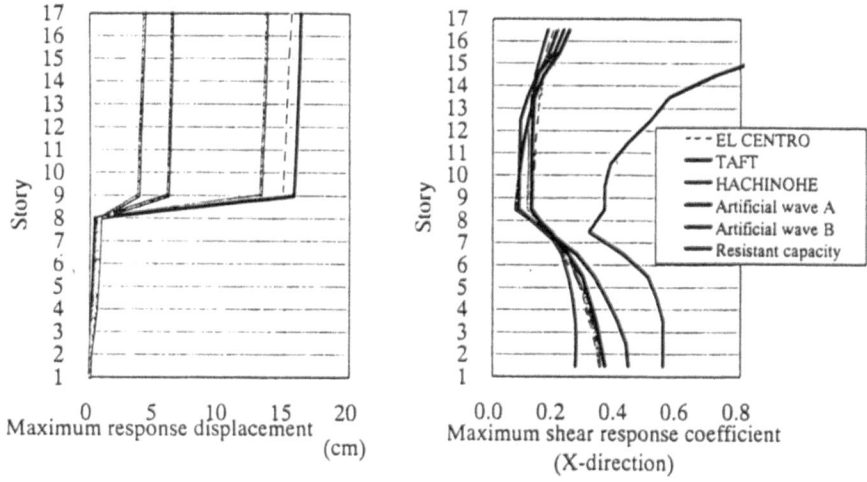

Fig.5.10 Results of Response Analysis

The analytical model for the building is shown in Fig.5.9 and the responses to the aimed level of ground motions are shown in Fig.5.10. At the energy concentrating story, the super structure is supported by 22 sets of laminated rubber bearings with lead plug. Diameters of rubber bearings are 70cm and 80cm.

The lower part of the structure below the eighth floor is installed with reinforced shear walls as major resisting elements to earthquakes.

The very short natural period of the lower part T_l coincides with this fact. To reduce the burden of energy absorption in the lower part is the second objective of introducing the energy concentrating story.

As a reference, the responses of 10 storied   flexible-stiff mixed structure will be shown. First, the original structure is assumed to be characterized by the following condition.

$N = 10$, $m_i / m_1 = 1.0$, $_s\alpha_i = \bar{a}_i \, _s\alpha_1$

$_s\delta_{Yi} = 2.0$ cm, $_s\alpha_1 = 0.4576$

$_f\alpha_{Yi} = 0$, $h = 0$

The aimed level of the seismic input V_E is assumed to be

$$V_E = 150 \text{ cm/sec} \tag{5.7}$$

The natural period of the original system is given by Eq.(4.24) and is obtained as

$$T = 1.0 \text{ sec}$$

Then α_{e1} in Eq.(4.1) is determined as

$$\alpha_{e1} = 0.962$$

Next, the original system is changed into a energy concentrating type of flexible-stiff mixed system by replacing the rigid system of kth story with a flexible-stiff mixed system.
The yield shear force coefficient and the elastic limit deformation in kth story are reduced to one-fourth of the original system, i. e.

$$\left. \begin{array}{l} {_s\alpha_k = \dfrac{_s\alpha_1 \, \bar{a}_k}{4}} \\[3mm] {_s\delta_{Yk} = \dfrac{_s\delta_Y}{4}} \end{array} \right\} \tag{5.8}$$

The elastic rigidity of the flexible element in kth story is determined so as to satisfy the following condition.

$$_f\alpha_k(10) = {_s\alpha_k} \tag{5.9}$$

where $_f\alpha_k(10)$: shear force coefficient of the flexible element at the story drift of 10cm

Under the assumption that the total energy input is finally absorbed by the rigid element in kth story, $_s\delta_{pmk}$ is obtained from the next equation similar to Eq.(4.7)

$$_s\alpha_k \, _s\delta_{pmk} = \frac{h_{eq}}{8}\left(\frac{M}{M_2}\right) \tag{5.10}$$

where $_s\delta_{pmk} = \delta_{mk} - {_s\delta_{Yk}}$

$$h_{eq} = \frac{V_E^2}{2g} = 11.48 \text{ cm} \quad \text{for} \quad V_E = 150 \text{ cm/sec}$$

Three records of ground motion are used:

- Kobe Marine Observatory record in the Hyogoken-nanbu Earthquake
- Hachinohe record in the Tokachi-oki Earthquake
- El Centro record in the Imperial Valley Earthquake

These records are scaled so as to produce the aimed energy input of $V_E = 150$ cm/sec.
The number of story in which the energy concentrates is selected to be 1, 5, 7, 8, 9 and 10.
In Fig.5.11, to what extent the damage is concentrated in one story (kth story) is shown. The limit

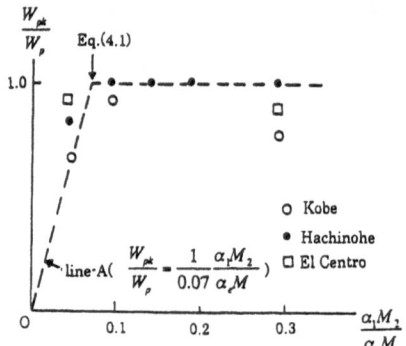

Fig.5.11 Condition for Perfect Damage Concentration

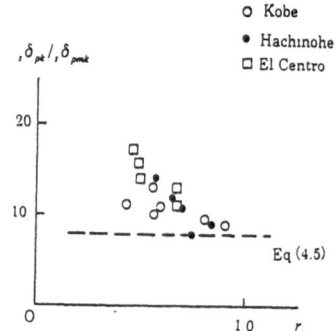

Fig.5.12 $_s\delta_{pk}/_s\delta_{pmk}-r_q$ Relationship

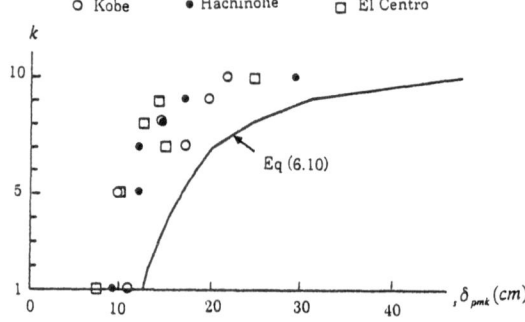

Fig.5.13 Maximum Deformation

condition to bring the perfect concentration given by Eq.(4.1) is verified to be adequate. The imperfection in energy concentration at the right-hand side which corresponds to the case of $k = 5$ is caused by the damage near the top of the system. In these cases, the perfect concentration can be easily attained by strengthening the higher part of the system.

In Fig.5.12, the ratio of $_s\delta_p$ to $_s\delta_{pm}$ in kth story is indicated in relation to $r_q(= {}_fa/_sa)$. The lower bound of the scatter is adequately expressed by Eq.(4.5).

In Fig.5.13, the predicted values of $_s\delta_{pm}$ in kth story by Eq.(5.6) are shown by the solid line.

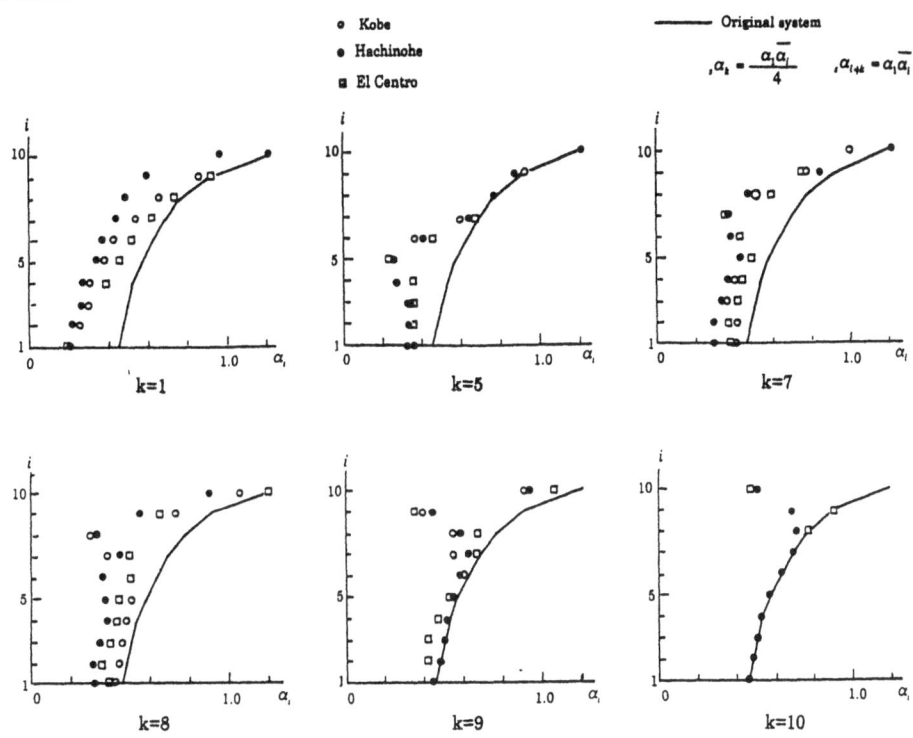

Fig.5.14 Shear Force Response

Table 5.5 Energy Responses

X Direction

	the total energy input (t·cm)	V_E (cm/s)	the energy absorbed isolators (t·cm)	the energy absorption ratio in isolators
EL CENTRO	79372	98	44309	0.56
TAFT	114357	117	56873	0.50
HACHINOHE	71905	93	60350	0.84
Artificial wave A	39488	69	4274	0.11
Artificial wave B	103732	112	16009	0.15

It is clear that the estimated value takes a position of the upper bound of actual responses. The result well coincides with the fact that the lower bound value expressed by Eq.(4.5) was used in estimating the maximum drift.

In Fig.5.14, the maximum shear force coefficients are shown in comparison with the yield shear force level of the original system.

In order to raise the efficiency in concentrating damage into kth story and to keep the other stories elastic, to secure an appropriate strength gap between kth story and the other stories. The condition applied by Eq.(5.8) seems to be adequate. As is seen in Fig.5.14, the maximum shear forces in stories lower than kth story tend to approach to the yield shear forces as the number of k increases. Reversely, the maximum shear forces in stories higher than kth story approaches to the yield shear forces as the number of k decreases.

The energy responses of the practical example shown in Fig.5.8 are listed in Table 5.5.Three natural records and two synthetic waves which are characterized by a very high peak in a shorter period range in the energy spectrum were applied. In this table, the value of V_E and the energy concentrating ratio W_{p8}/W_p are shown. The damage concentration into the eighth story is not complete. Taking the value of $V_E = 120$ cm/s, the applicability of the line-A in Fig.5.11 is checked. Using the values in Eqs. (5.4), (5.5) and (5.6) the following value is obtained.

$$\frac{W_{p8}}{W_p} = \left| 14.3 \, \frac{\alpha_1}{\alpha_{e1}} \cdot \frac{M_2}{M} = 0.70 \right. \tag{5.11}$$

The extent of the damage concentration made by the natural ground motions are well predicted by Eq.(5.11), However, it is shown by the case of synthetic waves that the damage concentration can be smaller than the predicted by Eq.(5.11) under singular spectral characteristics.

The maximum deformation in the damage concentrating story almost agrees with the maximum deformation of the computational model of $k = 8$ in Fig.(5.13) which corresponds to $M_2/M = 0.3$. Another very innovative technique is found in this building in the method of replacing columns in the energy concentrating story with laminated rubber hearings as shown in Fig.5.15.

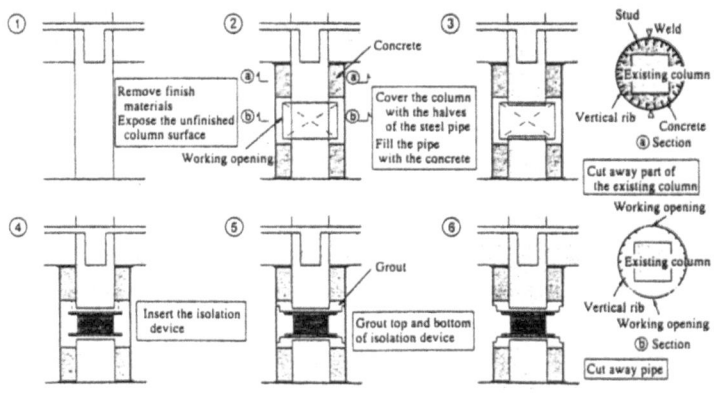

Fig.5.15 Procedure for Inserting Isolation Device

b) Energy dispersing type

The building shown in Fig.5.16 and Table 5.6 was constructed in 1970 (Asano et al, 2000). The major objective of retrofit of this building is the improvement of the structural performance. Namely, the reduction of damage of structural skeletons was intended. For this purpose, the energy absorbing diagonal bracing system consisting of non-buckling inelastic rods (EAB) was applied as shown in Fig.5.17.

The low yield point steel rod covered with an outer tube shown in Fig.5.18 can develop stable hysteretic loops under the constraint of buckling due to the outer tube.

Structural skeletons of the original system were intended to behave almost elastic and thus act as

Table 5.6 Outlined Feature of Original Building

Structure:	Steel framed reinforced concrete
Height:	64 m Structure
Completion:	1970
Target of Design:	Structural safety for earthquakes with ground velocity of 70 cm: Story drift angle less than 1/100~1/150
Designer:	Nikken Sekkei Ltd

Fig.5.16 Original Steel Framed Reinforce Concrete Building

Fig.5.17 Building Stiffened with Energy Absorbing Braces

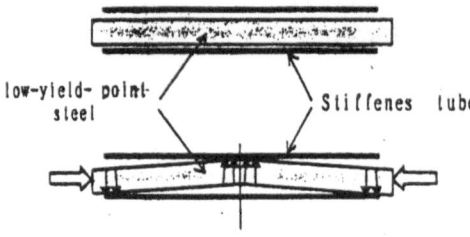

low-yield- point-
steel   Stiffenes tube

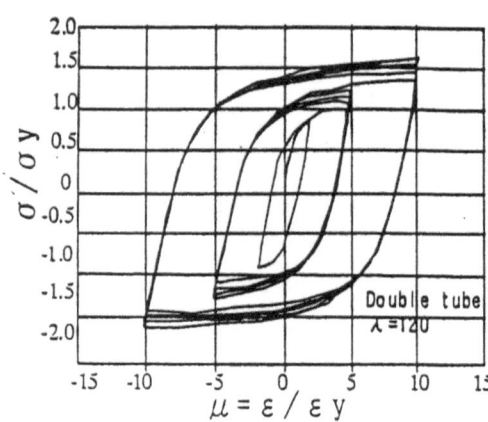

mechanical properties
of low-yield-point steel.

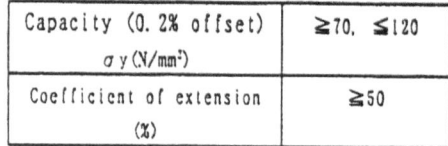

Capacity (0. 2% offset) σ y (N/mm²)	≧70, ≦120
Coefficient of extension (%)	≧50

Fig.5.18 Non-buckling Inelastic Rod

flexible elements within the following range of the seismic input.

$$V_D \doteqdot 150 \text{ cm} / \text{sec} \tag{5.12}$$

The bracing systems were distributed proportionally to the strength distribution of the original system. The level of additional lateral strength due to the bracing system is about 30% of the original.

As a reference, the responses of 15 storied flexible-stiff mixed structure are shown in Fig.5.19. The original structure is a rigid type of structure conditioned below. By adding extremely rigid and inelastic elements, the original system is turned to be flexible elements relatively. The original system which behaves as a flexible element is characterized as:

$$N = 15 \quad m_i / m_1 = 1.0 \quad {}_f\alpha_i = \bar{a}_i \, {}_f\alpha_1$$
$$_f\delta_{Yi} = 2.5 \text{cm}, \, {}_f\alpha_1 = 0.3, \, h = 0 \tag{5.13}$$

The stiff elements are installed in the original system proportionally to the flexible elements, i. e.

$$_s\delta_Y = 0.3 \text{ cm}, \, {}_s\alpha_i = \bar{a}_{i} \, {}_s\alpha_1 \tag{5.14}$$

The level of the yield strength of the stiff element ${}_s\alpha_1$ was selected to be

$$_s\alpha_1 = 0.03, \, 0.06 \quad ({}_f\alpha_1/10, \, {}_f\alpha_1/5)$$

The total energy input is expressed by h_{eq} defined by Eq.(4.6). Similarly, the absorbed energy by the stiff elements is expressed by ${}_s h_{eq}$.

Namely,

$$_s h_{eq} = \frac{_s W_p}{Mg} \qquad (5.15)$$

where $_s W_p$: energy absorbed by stiff elements

$_s\delta_{pm}$ is the maximum value of the maximum apparent plastic deformation found in all of stories of the system $\left(Max \; \{\delta_{mi} - {}_s\delta_Y\} \right)$.

In addition to the normal condition given by Eqs.(5.13) and (5.14), extreme conditions are applied. Under the normal condition in which the strength distribution is proportional to the optimum distribution. The maximum deformation develops in kth story.

To intensify the damage concentration, the strength gap is introduced in the same manner as is found in Eq.(3.10).

When the strength gap is introduced in the stiff elements, the strength distribution is specified as follows.

$$\left. \begin{array}{l} _s\alpha_{i \neq k} = p_d \, \bar{\alpha}_i \, {}_s\alpha_1 \\[2mm] _s\alpha_k = \bar{\alpha}_i \, {}_s\alpha_1 \end{array} \right\} \qquad (5.16)$$

where $p_d = 1.185 - 0.0014N = 1.165$ (for $N = 15$)

p_d is a recommended value to be applied to the practical design purpose (Akiyama, 1985).

Similarly, when the strength gap is introduced in the flexible elements, the strength distribution is specified as follows.

$$\left. \begin{array}{l} _f\alpha_{i \neq k} = p_d \, \bar{\alpha}_i \, {}_f\alpha_1 \\[2mm] _f\alpha_k = \bar{\alpha}_{if} \alpha_1 \end{array} \right\} \qquad (5.17)$$

The responses of the systems in which the strength gap is given in stiff elements are shown by rectangles in Fig.5.19 and the responses of the systems in which the strength gap is given in flexible elements are shown by triangles.

When the optimum damage distribution is attained, the total energy absorbed by the stiff elements are obtained as follows, referring to Eq.(4.29).

$$_s W_p = \Sigma \, s_i \, {}_s W_{p1} = 8 \Sigma \, s_i \cdot Mg \, {}_s\alpha_1 \, {}_s\delta_{pm} \qquad (5.18)$$

where $\Sigma \, s_i$ is given by Eq.(2.36)

Thus, $_s h_{eq}$ is obtained as

$$_s h_{eq} = 8 \Sigma \, s_i \, {}_s\alpha_1 \, {}_s\delta_{pm} \qquad (5.19)$$

As far as the flexible elements remain elastic, it is natural to consider that Eq.(5.15) can give a good estimate of $_s h_{eq}$.

$_s\delta_{pm}$ which corresponds to the elastic limit of the flexible element is given by

$$\delta_{p0} = {}_f\delta_Y - {}_s\delta_Y = 2.2 \text{ cm} \qquad (5.20)$$

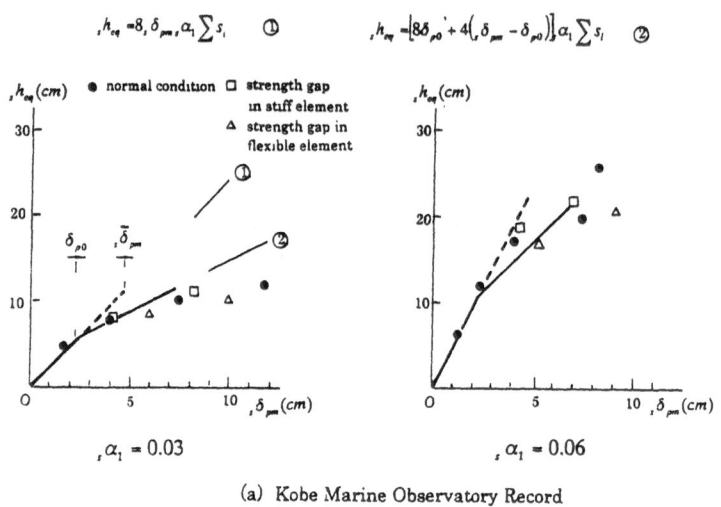

$$_{s}h_{eq} = 8_{s}\delta_{pm\,s}\alpha_{1}\sum s_{i} \quad ① \qquad _{s}h_{eq} = \left[8\delta_{p0}' + 4\left(_{s}\delta_{pm} - \delta_{p0}\right)\right]_{s}\alpha_{1}\sum s_{i} \quad ②$$

(a) Kobe Marine Observatory Record

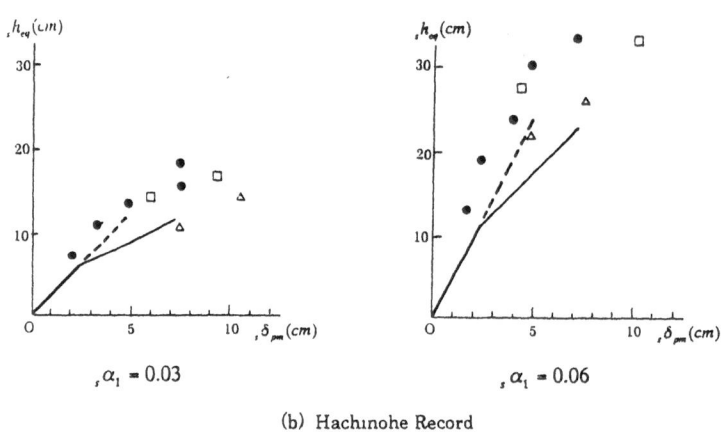

(b) Hachinohe Record

Fig.5.19 Deformation Responses of Flexible-Stiff Mixed Structure

As is seen in Fig.5.19, in the range of $_{s}\delta_{pm}$ up to δ_{p0}, Eq.(5.19) well agrees with the actual responses. In the range of $_{s}\delta_{pm}$ exceeding δ_{p0}, Eq.(5.19) still gives a good estimate of $_{s}\delta_{pm}$ within a certain range of $_{s}\delta_{pm}$.

In Fig.5.19, the solid line is extended up to the following level of $_{s}\delta_{pm}$.

$$_{s}\bar{\delta}_{pm} = 2_{f}\delta_{Y} - _{s}\delta_{Y} = 4.7 \text{ cm} \tag{5.21}$$

More precisely, the fine   line given by the following equation gives a lower bound of $_{s}h_{eq}$ within the range of $_{s}\delta_{pm}$ less than $3_{f}\delta_{Y} - _{s}\delta_{Y} = 7.2$ cm

$$_{s}h_{eq} = \Sigma\, s_{i}\left[8\delta_{p0} + 4\left(_{s}\delta_{pm} - \delta_{p0}\right)\right] \tag{5.22}$$

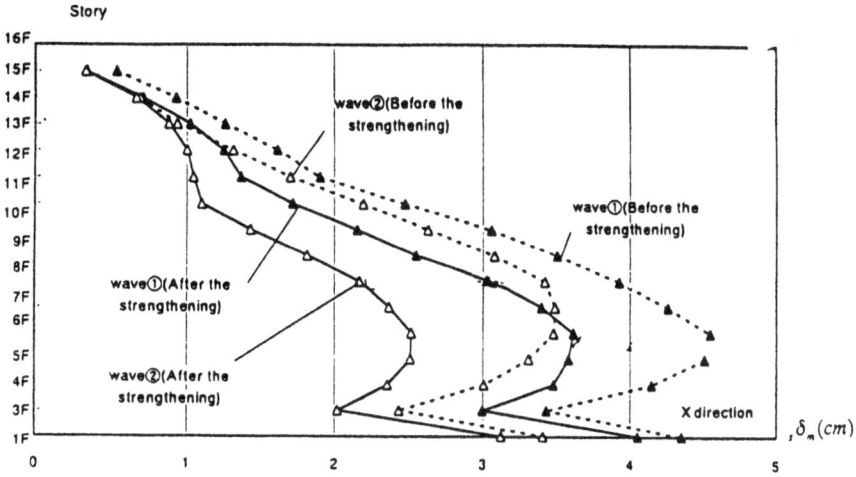

Fig.5.20 Deformation Responses of Braced Building

Thus, the flexible elements behaves as the flexible elements in the range of deformation of $2_f\delta_Y$ to $3_f\delta_Y$ and the energy absorption made by the rigid element reaches the following level of $_sh_{eq}$.

$$_sh_{eq} = 8\sum s_i \cdot {}_s\alpha_1(2_f\delta_Y - {}_s\delta_Y) \tag{5.23}$$

In Fig.5.20, the deformation responses of the braced frame shown in Fig.5.17 are shown. Two types of the ground mention were applied. The energy input attributable to the structural damage was almost equal to the level shown by Eq.(5.12), and the level of $_s\alpha_1$ of the stiffened frame was almost equal to 0.06.

The level of the maximum deformation lies between 2~3cm. These results well agree with the general responses of the flexible-stiff mixed structure shown in Fig.5.19.

6. Conclusion

The ultimate resistance of the structure to earthquake can be well understood in terms of the energy balance between the seismic energy input and the energy absorbed by the structure. The aim of strengthening structures is to improve the structural performances. The structural performances are described by following items.

1) reduction of structural damage
2) reduction of damage of non-structural elements such as claddings and walls
3) reduction of acceleration responses

To attain these aims, it is essential for structures to be equipped with the sufficient energy absorption capacity to absorb the total seismic energy input. In addition, it is very important for the structure to be equipped with marginal elastic elements to stabilize the structural responses.

The structure in which the flexible elastic elements and stiff energy absorbing elements coexist is categorized to be the flexible-stiff mixed structure. The final goal of the seismic design and seismic stiffening must be achieved by the construction of flexible-stiff mixed structures.

Acknowledgement

The author is indebted to Dr. T. Inada, Mitsubishi Real Estate Co., Mr. M. Takayama, Taisei Construction Co., and Mr. T. Kohno, Nikken Sekkei Ltd. for their presenting practical design materials.

REFERENCES

Akiyama, H. (1985) Earthquake-Resistant Limit-State Design for Buildings, University of Tokyo Press.

Housner, G. W. (1956) Limit Design of Structures to Resist Earthquakes, Proc. of 1st WCEE.

Housner, G. W. (1959) Behavior of structures during Earthquake, ASCE, EM4, Oct.

Akiyama, H. (1986) Ds-Values for Multi-Story Frames on the Top Story of Which Damage Concentrates, Transaction of Architectural Institute of Japan, No. 362.

Asano, M., Kohno, T. and Nitayama, N. (2000.6) Seismic Strengthening Using Energy Absorption Brace. US-Japan Symposium on Seismic Rehabilitation of Concrete Structures.

Ogura, K., Maezawa, S., Tsujita, O., Kobayashi, J. and Nakata, Y. (1997) Seismic Isolation Retrofit at Mid-story and Base, AIJ Journal of Architecture and Building Science

Appendix

1 Estimate of structural strength

(a) Axial forces in columns and story shear force

To obtain the yield-shear force of each story, the axial forces in the columns must be known. However, axial forces, because they depend on story shear forces, are not determined uniquely under seismic forces in contrast to the case of vertical loadings. So, an approximated estimation of axial forces should be made. According to response analyses of multi-story frames with the optimum shear force distribution, the maximum overturning moment in each story is found to be approximately equal to the value calculated on the basis of the simultaneous yielding of all stories. Therefore, to analyze axial forces in columns, it is adequate to assume that horizontal forces correspond to the optimum shear force distribution. An elastic analysis is needed to obtain the ratio of the axial force of jth column in the ith story, $_jP_i$, to the story shear force of the ith story, Q_i termed $_ja_i(=_jP_i/Q_i)$.

The change of axial forces is considered to occur only in the exterior columns in moment resistant planar frames or isolated walled structures, and the following approximation may be allowed:

$$_ja_i = \frac{M_{Ovi}}{H_0 Q_i} \tag{A.1}$$

where M_{Ovi} : overturning movement in ith story caused by horizontal forces with the optimum distribution,

Q_i : story shear force under the same horizontal forces,

H_0 : distance between exterior columns.

(b) Decomposition of the frame into story frames

To obtain the yield-shear strength, the spring constant, and the cumulative inelastic deformation ratio of each story, it is convenient if the frame is decomposed into story frames.

When the ith story of a frame, as is shown in Fig.A.1(a), is on the brim of yielding under an earthquake, the upper $(i+1)$ story and the lower $(i-1)$ story rest in almost the same stress sate. Beams and beam-to-column connections on the ith floor carry stresses induced by shear forces in the ith and $(i-1)$ stories. In such a state, it can be considered that approximately half of the stresses in the beam and beam-to-column connections are stresses induced by the story shear force in the ith story. In this context, beam and beam-to-column connections may be decomposed to form story frames as shown in Fig.A.1(b). The story frames are assumed to be connected to each other with

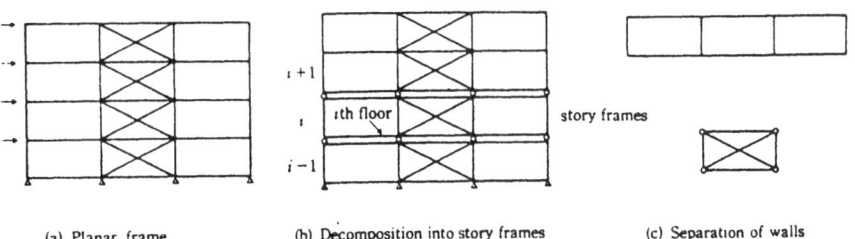

(a) Planar frame   (b) Decomposition into story frames   (c) Separation of walls

Fig.A.1 Decomposition of a Multi-Story Frame into Story Frames

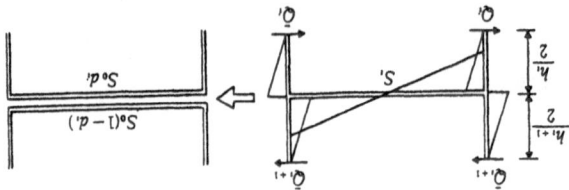

Fig.A.2 Decomposition of Beams

pinned supports, as shown in the figure. Moreover, the story frame is decomposed into the pure moment-resistant frame and the pure shear wall. The pure moment frame is a portion eliminated from the walls. The pure shear wall is formed by wall elements enclosed by a pin-connected rectangular frame consisting of rigid members. The enclosing frame, which is unable to resist horizontal forces by itself, is installed for the convenience of calculating deformation.

The distribution of story shear forces is assumed to be proportional to the yield-shear force distribution prescribed by the optimum yield-shear force coefficient distribution, $\bar{\alpha}_i$. Denoting this standard story shear force distribution associated with $\bar{\alpha}_i$ by $\bar{Q}_i$, $\bar{Q}_i$ can be described by

$$\bar{Q}_i = \frac{\left(\sum\limits_{j=i}^{N} m_j\right)\bar{\alpha}_i}{M} \tag{A.2}$$

Since the upper beam in the ith story resists the moment, $_bM_i$, which is produced by the ith story shear force and the moment, $_bM_{i+1}$, which is produced by the $(i+1)$th story shear force, it is natural to assume that the upper beam in the ith story should decompose so as to be proportional to $_bM_i$, and $_bM_{i+1}$. Assuming the shear span of a column to be one-half of the column length, as shown in Fig.A.2 the dividing ratio for the observed beam, d_i, should be obtained as follows:

$$d_i = \frac{h_i\bar{Q}_i}{h_i\bar{Q}_i + h_{i+1}\bar{Q}_{i+1}} \tag{A.3}$$

where h_i, h_{i+1} : the story heights of the ith and $(i+1)$th stories.

Denoting structural properties of the observed beam, including strength and stiffness, by s_0, structural properties thus decomposed become:

$$\left.\begin{array}{l} S_i = S_0 d_i \\ S_{i+1} = S_0(1-d_i) \end{array}\right\} \tag{A.4}$$

where S_i, S_{i+1} : structural properties of decomposed beam in the ith and $(i+1)$th stories.

When the column bases in the first story are pin-ended, the shear span for the columns in the first becomes h_1. Therefore, d_1, for this case, should be replaced by

$$d_1 = \frac{2h_1\bar{Q}_1}{2h_1\bar{Q}_1 + h_2\bar{Q}_2} \tag{A.5}$$

The above discussion can be applied in exactly the same manner to the decomposition of panel zones in beam-to-column connections.

(c) **Yield-shear force of the story**

Assuming that the effect of distributed loads over beams and the $P-\delta$ effect are negligible, a method of calculating the yield strength of story frames is described in the following section. Moment-resistant frames are dealt with as a typical example.

The sum of the yield strength of each story frame, q_Y, parallel to the seismic force comprises the yield-shear force of a story. On the assumption that the effect of the distributed loads on the beams is negligible, the maximum bending moment is attained at the ends of members. Accordingly, when the joints of members are sufficiently strong, the positions in which plastic hinges are formed are restricted to:

(1) Column ends,
(2) Beam ends, adjacent to the column observed,
(3) Panel zone of beam-to-column connection.

The plastic hinge moment, M_k, which represents the kth joint of a story frame, is the smallest value of the fully plastic moment of the column, M_c, the sum of the fully plastic moment of beams on both sides of the joint, M_b, or the fully plastic moment of the panel zone of the beam-to-column connection, M_p. Knowing the hinge moments in all joints, q_Y is obtained as

$$q_Y = \frac{\Sigma M_k}{h} \tag{A.6}$$

where Σ : summation over all the joints of a story frame,
h : the height of the story frame.

The fully plastic moment of each member depends on the axial force. The axial force due to vertical loads is known prior to the calculation of story yield-shear strength, but the axial force due to horizontal forces depends on the magnitude of shear forces. Nevertheless, the ratio of the axial force of each column to horizontal seismic forces, $_j\alpha_i$, is already known. Therefore, at first, q_Y is determined with the neglect of axial forces. Next, the fully plastic moment is revised, applying axial forces given by Eq.(A.6). Repeating the revision of hinge moments several times, q_Y is converged to an accurate value.

2 **Estimate of structural inelastic deformation capacity**

A moment frame comprises columns, beams, and beam-to-column connections. So it is of primary importance to clarify the correspondence between the deformation characteristics of structural members and the total story deformation. The load-deformation relation of the story frame can be obtained by synthesizing these members' behaviors and watching the compatibility of deformations between members. Nevertheless, an exact solution requires a procedure that is too complicated. So, to further simplify, the story frame shown in Fig.A.2 is reduced to a frame with a single span (termed the reduced unit frame) as shown in Fig.A.3.

(a) **Reduced unit frames**

The behavior of the collected body of unit frames transformed from the story frame is similar to that of the story frame. The transformation is made by dividing columns and panel zones of beam-

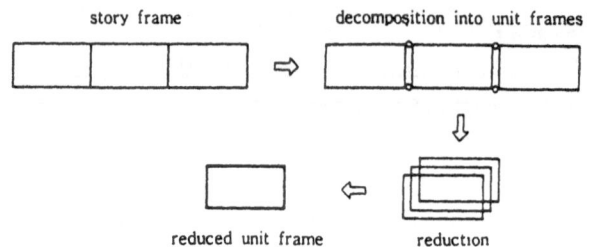

Fig.A.3 Reduction of Story Frames

to-column connections into equal parts, as done when transforming a multi-story frame into individual story frames.

When constituent members of a unit frame are identical for every kind of member, the collected
body of unit frames can be further reduced to single unit frames. The fully plastic moment, $_kM_Y$,
and the flexural rigidity, $_kK$, of each member of this reduced unit frame are given by the following
equations, in which the subscript k designates kind of member (where k = c, b or p for column,
beam and panel zone of beam-to-column connections respectively):

$$
\left.
\begin{aligned}
\text{for column,} \qquad\qquad & _cM_Y = \frac{\Sigma M_{cY}}{2}, \quad _cK = \frac{\Sigma _ck}{2} \\[2mm]
\text{for beams,} \qquad\qquad & _bM_Y = \frac{\Sigma M_{bY}}{2}, \quad _bK = \frac{\Sigma _bk}{2} \\[2mm]
\begin{aligned}\text{for panel zones of beam–}\\[-1mm]\text{to–column connections,}\end{aligned} \quad & _pM_Y = \frac{\Sigma M_{pY}}{2}, \quad _pK = \frac{\Sigma _pk}{2}
\end{aligned}
\right\}
\qquad (\text{A.7})
$$

where Σ : the sum over all numbers of a story frame that are the same kind,

M_{cY}, M_{bY}, M_{pY} : fully plastic moments of the column, the beam, and the panel zone of
  a story frame,

$_ck$, $_bk$, $_pk$: rigidities of the column, the beam and the panel zone of a story frame,
  respectively.

Flexural rigidities are defined as:

$$
\left.
\begin{aligned}
_ck &= \frac{M_{cY}}{h\phi_{cY}} \\[3mm]
_bk &= \frac{M_{bY}}{l\phi_{bY}} \\[3mm]
_pk &= \frac{M_{pY}}{\gamma_Y}
\end{aligned}
\right\}
\qquad (\text{A.8})
$$

where h : height of story,

l : length of member,

γ_Y : shear yield strain,

ϕ_{cY}, ϕ_{bY} : curvatures for M_{cY} and M_{bY}.

For steel structures, $_ck$, $_bk$ and $_pk$ are expressed as

$$_ck = \frac{EI_c}{h}$$

$$_bk = \frac{EI_b}{l} \qquad \qquad (A.9)$$

$$_pk = GV_p$$

where I_c, I_b : second moment of inertia,

V_p : the effective volume of panel zone,

G : shear modulus

For reinforced concrete structures and steel reinforced concrete structures, flexural rigidities $_bk$, $_ck$ are obtained from the curvature at the yield strength state. ϕ_Y is given approximately as

$$\phi_Y = \frac{\varepsilon_Y}{D} \qquad \qquad (A.10)$$

where D : height of section,

ε_Y : elastic strain at the yield point of reinforcing bar.

In these structures, the deformation of panel zones of beam-to-column connections, being comparably small, may be neglected. Then, $_pk$ need not be specified.

In general frames, a story frame cannot be transformed into equal unit frames. On the other hand, Eq.(A.7) implies the sum of the averaged value of strength and stiffness of constituent members of a story frame, and some error in averaging may usually be permitted in the structural analysis. Conversely it can be claimed that the unit frame equipped with the strength and rigidity given by Eq.(A.7) is the reduced unit frame.

When the column bases in the lowest story are pin-connected or clamped by rigid foundation beams, the lower beams of the story frame need not be taken into account. Therefore, the following revision to Eq.(A.7) should be made:

$$_bM_Y = \Sigma M_{bY}, \quad K_b = \Sigma_bk$$

$$_pM_Y = \frac{\Sigma M_{pY}}{2}, \quad K_p = \frac{\Sigma_pk}{2} \qquad \qquad (A.11)$$

To relate the deformation capacity of individual members to that of the frame, it is necessary to know the spring constant o the unit frame under the horizontal force.

The reduced unit frames are classified into three types as shown in Fig. A.4. The first is a general frame with lower and upper beams. The second and third correspond to the lowest story of the multi-story frame. In the second case, where the column bases are clamped, a hinge is introduced at the midpoint of the column for the sake of simplifying calculations.

For this case, another simplification can be made by replacing the unit frame with fixed column bases by the ordinary one shown in the left hand side of Fig.A.4. The beam and the beam-to-column connections which are inserted at the base of the unit frame should be the same as those at the upper part of the unit frame. Although this simplification may yield a slight softening in the rigidity of the frame, it has been proved that the simplification produces no significant error in the damage estimations of the multi-story frame.

Considering flexural deformations of columns and beams and shear deformations of panel zones of

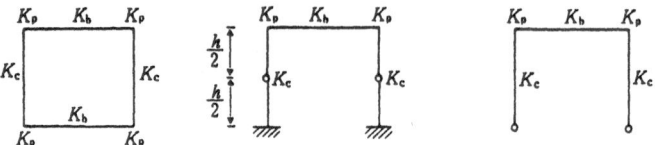

Fig. A.4 Reduced Unit Frames

Table A.1   Deformation Characteristics of Reduced Unit Frames

	C	a_c	a_b	a_p
General story	$\dfrac{4}{h^2}$	$\dfrac{1}{6}$	$\dfrac{1}{6}$	1
Lowest story				
Fixed base	$\dfrac{4}{h^2}$	$\dfrac{1}{6}$	$\dfrac{1}{12}$	$\dfrac{1}{2}$
Pinned base	$\dfrac{4}{h^2}$	$\dfrac{1}{3}$	$\dfrac{1}{6}$	1

beam-to-column connection, the spring constant for each case is obtained as

$$_R k = \frac{C}{\dfrac{a_c}{K_c} + \dfrac{a_b}{K_b} + \dfrac{a_p}{K_p}} \tag{A.12}$$

where a_c, a_b, a_p and C are as listed in Table A.1

(b) Cumulative inelastic deformation ratio of reduced unit frame

Actually, all members of the frame can deform elastically. The elastic limit deformation of the unit frame, in which only a single type of member can deform elastically or inelastically, is denoted by $_k\delta_Y$ the elastic limit deformation of the frame in which all members can deform elastically is written δ_Y. The ratio between them is derived from Eq.(A.12) as

$$\frac{_k\delta_Y}{\delta_Y} = \frac{\dfrac{a_k}{K_k}}{\dfrac{a_c}{K_c} + \dfrac{a_b}{K_b} + \dfrac{a_p}{K_p}} = {}_k a_m \tag{A.13}$$

The cumulative inelastic deformation of the individual member in the unit frame can be given by $_k\bar{\eta}_u {}_k\delta_Y$. Taking account of all members' elastic deformation, the corrected values of $_k\bar{\eta}_u$ and $_k\bar{\eta}'_u$ are given by dividing the cumulative inelastic deformation by the revised elastic limit deformation, δ_Y:

$$_k\bar{\eta}'_u = \frac{_k\bar{\eta}_u {}_k\delta_Y}{\delta_Y} = {}_k a_m {}_k\bar{\eta}_u \tag{A.14}$$

The ultimate average cumulative deformation ratio of the reduced unit frame is given by

$$_R\bar{\eta}_u = \text{Min}\{_k\bar{\eta}_u'\} \tag{A.15}$$

In the following paragraphs, the value of $_k a_m$ for ordinary frames will be discussed.

The stiffness of members can be obtained by the following equation, similar to Eq.(A8), wherein the yield strength of each member in the reduced unit frame is assumed to take a constant value, that is, $_cM_Y = {_b}M_Y = {_p}M_Y = M_0$:

$$
\left.\begin{array}{l}
_cK = \dfrac{M_0}{h\phi_{cY}} \\[3mm]
_bK = \dfrac{M_0}{l\phi_{bY}} \\[3mm]
_pK = \dfrac{M_0}{\gamma_Y}
\end{array}\right\} \tag{A.16}
$$

(a) **Reinforced concrete structures and steel reinforced concrete structures**

In evaluating stiffness, Eq.(A1) is applied using sectional depth of the column and beam of the reduced unit frame, $_cD$ and $_bD$ respectively, for the value of D. In this type of structure, the deformation of the beam-to-column connection can be neglected, and because of an abundant deformation capacity of beams, the cumulative inelastic deformation is limited exclusively by that of columns, $_c\bar{\eta}_u'$. The value of $_ca_m$ becomes

$$_ca_m = \frac{a_c}{a_c + a_b\left(\dfrac{l}{h}\dfrac{_cD}{_bD}\right)} \tag{A.17}$$

Assuming that $_cDl/_bDh = 1.0$, the values of $_ca_m$ are calculated as is shown in Table A.2.

(b) **Steel structure**

γ_Y in Eq.(A.16) can be written as

$$\gamma_Y = \frac{\tau_Y}{G} = \sigma_Y/\left\{\frac{\sqrt{3}E}{2(1+\nu)}\right\} = 1.5\varepsilon_Y \tag{A.18}$$

where ν : Poisson's ratio $= 0.3$

ϕ_{cY} and ϕ_{bY} are approximated as

$$
\left.\begin{array}{l}
\phi_{cY} = \dfrac{1.5\varepsilon_Y}{_cD} \\[3mm]
\phi_{bY} = \dfrac{2\varepsilon_Y}{_bD}
\end{array}\right\} \tag{A.19}
$$

Assuming that $h/_cD = l/_bD = 8.0$, the values of $_ka_m$ are obtained by averaging, as shown in Table A.2

When inelastic deformations take place only in beams or panel zones of beam-to-column connections, the excessive development of inelastic deformation in single story is prevented by a rather uniform development of inelastic deformation over all stories. Therefore, in this case, it will lead to a too-conservative estimate of the energy absorption capacity of the frame, to take account of an

Table A.2 Values of a_m

	RC Structures* SRC Structures**	Steel structures		
General story	$_c a_m$	$_c a_m$	$_b a_m$	$_p a_m$
Lowest story				
Fixed base	0.5	0.32	0.43	0.24
Pinned base	0.67	0.49	0.33	0.184

* RC : reinforced concrete.
**SRC : steel reinforced concrete.

Table A.3 Design Values of $_k a_m$

	RC Structures SRC Structure	Steel structures		
General story	$_c a_m$	$_c a_m$	$_b a_m$	$_p a_m$
Lowest story				
Fixed base	0.5	0.35	0.9(0.45)	0.5(0.25)
Pinned base	0.7	0.5	0.7(0.35)	0.4(0.2)

evaluation of the damage concentration inherent to the weak-column type of structure. In this situation, the values of $_k a_m$ for these weak-beam types of structures can be increased by a reasonable amount. In multi-story frames of this type, the damage occurs uniformly over at least two or three stories. Eventually, the values of $_b a_m$ and $_p a_m$ listed in Table A.2 may be doubled. Using this inference, rounded values for practical design purposes are obtained as shown in Table A.3. The values in parenthesis indicate the values used for the first story of the frame.